FORENSIC CHEMISTRY

FORENSIC CHEMISTRY

Editor

Sieghild Walther

scitus academics

Forensic Chemistry

Edited by **Sieghild Walther**

This Book is Printed on Acid-Free Paper. ♾

ISBN: 9781681178929
Library of Congress Control Number: 2017948920

www.scitusacademics.com

Notice

Printed in United States of America.

Preface

One of the most important aspects of criminal justice is forensic science, or the practice of scientifically examining physical evidence collected from the scene of a crime or a person of interest in a crime. Who hasn't heard of the tales of Sherlock Holmes? The crime resolving talents of Sir Arthur Conan Doyle's hero inspires many till today to take up a profession in forensics and assist in outsmarting criminals. But around a decade ago, the activities of Holmes had inspired Dr. Edmond Locard to set up the world's first forensic laboratory in France in 1910 equipped with just a microscope and spectroscope. Since then, over the last century, powered by tremendous advancements in analytical chemical techniques, forensic chemistry has progressed by leaps and bounds and is now revolutionizing the criminal justice system. Chemical analysis has an important role in law enforcement and forensics. The purity of a material could be detected using spectroscopy techniques and this could be major analysis in approving results. To detect evidences and in the field of narcotics, forensic chemistry play an important role. Illegal drugs threat to society and global market, where in forensic science; the forensic chemistry is widely used to explore these crimes.

This volume 'Forensic Chemistry' reports forensic analytical technique, where forensic chemistry is used to estimate age of an unknown human body will be estimated. It also include the application and or development of any molecular and atomic spectrochemical technique, electrochemical techniques, sensors, surface characterization techniques, mass spectrometry, nuclear magnetic resonance, chemometrics and statistics, and separation sciences (e.g. chromatography) that provide insight into the forensic analysis of materials. This book will be useful to practitioners of forensic medicine, experts, pathologists, law makers, investigating authorities, undergraduate and postgraduate medical school, graduates of medicine.

Table of Contents

CHAPTER 1

Forensic Analysis of Three Pets: Victims of Carbamate Intoxication

Tália Missen Tremori*, Laila Massad Ribas, Mara Rita Rodrigues Massad, Luiz Maurício Montoya Flórez, Elan Cardozo Paes de Almeida and Noeme Sousa Rocha

Department of Veterinary Clinic, Campus Botucatu, UNESP - Univ Estadual Paulista, Brazil

ABSTRACT

Cases of intoxication are common in the history of the veterinary medicine, mainly in dogs and cats. Appropriate recognition of the clinical signs and pathological findings could characterize cases of intoxication, however laboratorial methods in forensic toxicology is fundamental to establish a definitive diagnosis of the toxic agent. In the present work was made analysis of three cases of pet animals criminally intoxicated by carbamate. Necropsy helps to uncover causes mortis, were verified cardiac respiratory insufficiency and hypovolemic shock in all cases. Histopathology exam of liver, kidney and brain described the main lesions like congestion, degeneration and haemorrhage. Organs show signs of autolysis and putrefaction. Therefore, diagnostic techniques helps veterinarian to make reports for litigation in animal criminal intoxication.

Keywords: Veterinary forensic medicine; Expertise; Toxicology; Forensic histopathology

INTRODUCTION

Over the years, forensic veterinary medicine has experienced a significant increase of performance and effectiveness also because a raise of social sensitivity to animal cruelty. Non-accidental intoxication is frequent, mainly exogenous. However, animal crimes often remain unreported [1,2].

Criminal intoxication falls among prohibited conducts indicted by the Environmental Crimes Law, No 9.605/98. Specifically, Article 32 typifies as a crime to perform acts of abuse, to injure or mutilate wildlife, domestic or domesticated, native or exotic, providing imprisonment from three months to one year, and additionally, a respective fine [3,4].

According to previous studies, fatal intoxication occurs mainly due to pesticides, being 50.4% of cases caused by carbamates, 18.9% by anticoagulant rodenticides, 5.1% by organophosphate and 3.4% by non-anticoagulant rodenticides [5-7].

Non-violent deaths in animals should always be investigated to verify possible intoxication, because veterinarians routinely facing these cases [5,8-12].

On these terms, toxicology allows determining chemical action mechanisms in the body, clinical signs, post mortemcharacteristics (microscopic and macroscopic), necropsy findings, and toxic agent detection in order to get a reliable poisoning diagnosis. Evidence identification is extremely important to facilitate forensic investigation and necropsy report [1,13,14].

Toxicants are substances able to damage health, or even result in death. Depending on gravity of signs and symptoms, poisoning syndrome is classified as mild, moderate or severe. Some factors, alone or in combination, contribute to its manifestations such as dose quantity, individual strength, exposure time, and the route of administration [15,16].

If the animal dies, forensic necropsy and histopathologycal analysis, associated with a drug testing, are needed to reach an accurate diagnosis. Thus, the confirmation of suspected cases increases [7,9,13,17,18].

Finally, necropsy report plays a key role by clarifying facts, and defining death cause. Likewise, forensic histopathology through a comprehensive tissue analysis is quite helpful for interpretation [10,19,20].

MATERIAL AND METHODS

Three cases of pet intoxication were selected from records of the archives at Veterinary Pathology Service-FMVZ from July 2014 to January 2015. Although more cases have been found furthermore this study was requested only that were comply with all requirements: forensic necropsy reports, microscopy analysis, judicial requests, and toxicological examination.

The reports were documented by photographs (JPEG Image DSCH9) before undergoing necropsy. The autopsy technique was the modified Ghon, which consists of the opening of the skull and thoracoabdominal cavity to extract organs in anatomical and functionally related blocks, totaling four. Picked organs were removed from each block and examined in detail. Some of them were photographed (JPEG Image DSC-H9) for registry purposes [20].

Concurrently to necropsy, samples were collected for physical and chemical analysis, and histopathologycal conventional exam, regardless time of death. For the physical-chemical examination, duplicate samples of liver and kidney (30 gm) were harvested from each animal, then identified, packed and, sometimes, refrigerated.

One of the duplicate specimens was sent to the toxicology laboratory with the request form and death certificate, which included possible poisoning agents, in order to identify and quantify their chemical compounds. The other duplicate was kept in custody until obtaining toxicological official results, and later discarded, according biosecurity protocols.

For histological analysis, samples of kidney, liver and brain were placed into individual vials containing 10% buffered formalin, at minimum proportion of five times specimen size. It remained so until processing, according to conventional methods. Thickness of cuts ranged from 4.0 to 5.0. Staining was made by using Hematoxylin and Eosin (H&E).

For morphological expression of lesions in brain, kidney and liver, from both macroscopic and microscopic perspectives, the criteria contained in referential textbooks prevailed. In this case, characterizations were focused on lesions caused by the action of toxic compounds, especially on those causing degeneration and other types of non-reversible damage to cells or tissues comprising such organs [19,21,22].

RESULTS

The first animal was a four-year male dog, non-defined breed (NDB), positive for carbamate by a qualitative drug testing. Macroscopic study showed a whitish and frothy fluid at the final third of the trachea, extending to larger bronchi. Lung displayed extensive reddish and crackling areas. Stomach had a moderate foamy yellowish liquid along with a substance of granular aspect and blackened shade. A discrete hepatomegaly with an intense red color was found at liver. Pancreas had a softened consistency. Sulfhemoglobin stains were on the outer surface of spleen, which characterize a decomposition process. Moderate blood vessels with a dark red color were also detected in brain and skeletal muscles of the skull.

On microscopic examination, liver revealed a marked microvacuolation and degenerate hepatocytes near lobular vein center with the presence of "ghost" cells. A significant trabecular congestion was verified by the presence of large quantities of erythrocytes in blood vessels. Kidneys had congestion and microcalcifications.

The cortical and medullary vessels were bleeding with an intense and diffuse blockage. Tubular cells showed a severe vacuolation and a moderate quantity of hyaline material in the glomerular space. In some renal regions, degenerate cells detached from the basement membrane and massive pyknotic nuclei were present, all related to autolysis of parenchyma.

Brain evaluation unveiled a moderate disseminated intravascular coagulation (DIC), absence of inflammation, rare satellite cells nearby to neurons. In the gray matter, a region with bone tissue, head trauma indications and bruises appeared. This coincides with lesions verified by microscopy of erythema on temporal muscles covering skullcap. Besides, moderate bleeding was located in meninges and cerebral cortex (**Figure 1**).

A histopathologycal examination of lungs disclosed edema, pulmonary congestion and a severe CID, mainly due the presence of adhered red blood cells at the vessel wall, beyond a critical bleeding to the alveolar sacks by eritrocytes extravasation. The second case is related to an 8-month female feline, also NDB, positive for carbamate by a qualitative toxicology exam. At necropsy, an off-white and foamy liquid in the final third of the trachea was observed, extending to bronchial area with a deposition of blood with hypostasis at the left side of lung. Likewise, there was a moderate amount of unidentified food in the stomach, mixed with blackened granules. Encephalon presented evident lesions in blood vessels (**Figure 2**).

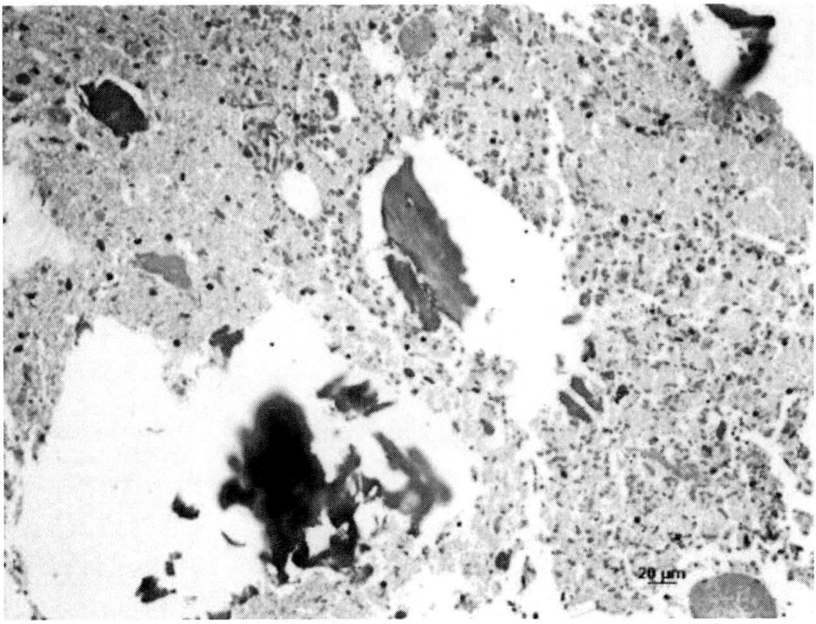

Figure 1: Dog brain: Bone tissue in the cerebral cortex hemorrhage (H&E 20X).

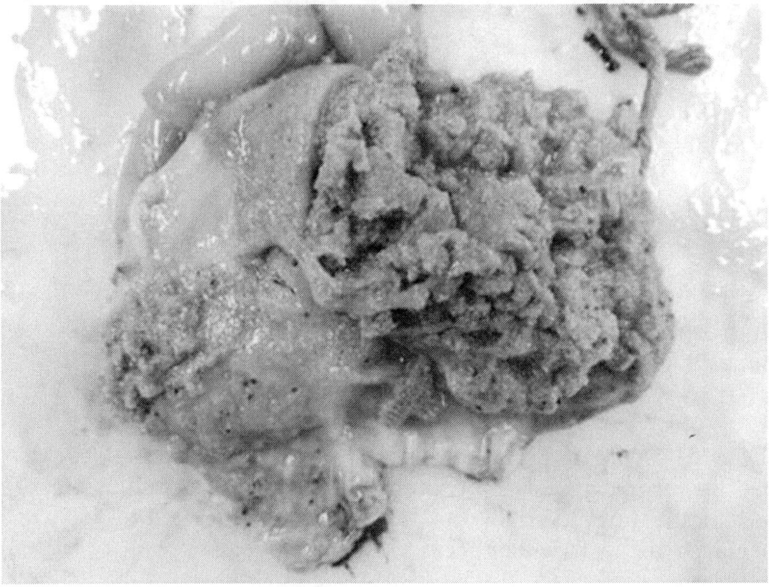

Figure 2: Cat stomach: Blackened particles intermingled with food content (carbamate).

As for microscopic examination of liver, vacuoles close to portal space featured a pronounced fatty degeneration, and macroglandular steatosis. Brownish spots indicated biliary stasis (cholestasis), and a mild vascular congestion.

Kidneys demonstrated vascular congestion within medullary and profuse hemorrhage areas. Vacuoles at renal tubule cells of the cortex pointed to a minor fatty degeneration, and cell separation from basal membrane (**Figure 3**). In the brain, there was an incipient congestion and a moderately diffuse lung edema, followed by intra-alveolar bleeding.

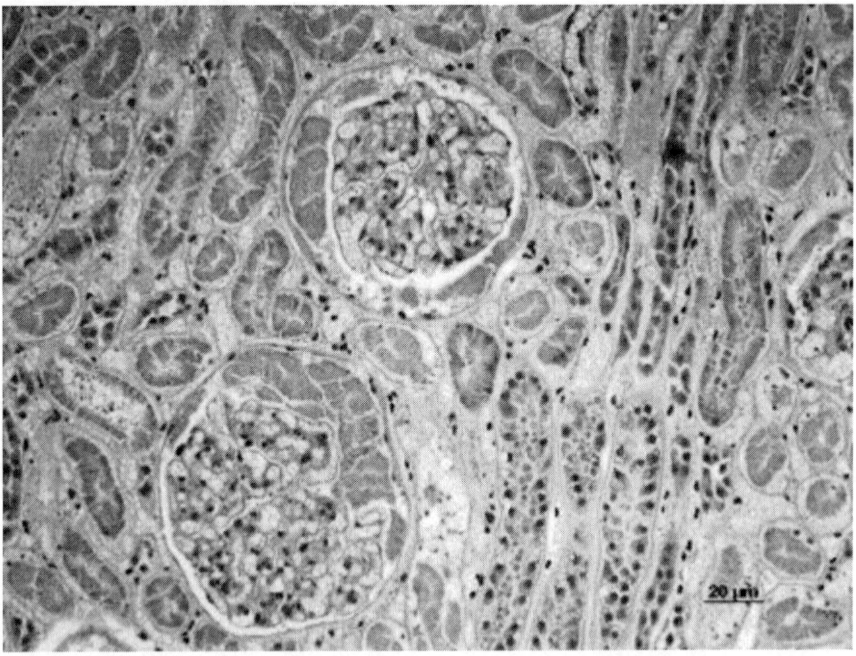

Figure 3: Cat rim cortical region: Detachment from the basement membrane (H&E 20X).

The third considered sample was a 12-year female canine, SRD, also tested positive for carbamate. After forensic exam, a whitish-reddish frothy liquid was in the final third of the trachea, extending to the bronchi. Lungs displayed yellow irregular areas to multifocal coalescing.

Spleen had multiple nodules with whitish appearance, firm consistency, and irregular capsular surface, of approximately 1.5 cm diameter in average. Liver exhibited bulging lips, distended gallbladder, and filled of bile with lumpy appearance. Stomach lumen was filled by blackish liquid, greenish paste content in the bowel, presence of edema in the gastric and intestinal wall. Renal examination revealed capsular adherence in both kidneys, reddish subcapsular region, and expanded pelvis. Urinary bladder contained a whitish thick fluid.

Under microscope, liver tissue showed diffuse multifocal areas, large quantities of degenerate hepatocytes, mainly near the centrilobular region. Vascular congestion and diffuse areas with moderate bleeding also were watched. Kidneys presented a moderate infiltrate multifocal lymphoplasmacytic at the cortical region, extending to spinal area, vascular congestion with a slight hemorrhage. In fact, both organs suffered from acute tubular necrosis (ATN).

Brain was slightly congested with some red blood cells adhered to the capillary internal layer, and a light bleeding. A sharp pulmonary edema was characterized by eosinophilic amorphous material, prominent hemorrhage at interstitial region and vascular congestion.

DISCUSSION

Histological knowledge is essential for a proper understanding of the structure and functioning of cells, tissues and organs. After removing fragments for histopathological evaluation, samples must be preserved by fixation in order to prevent enzymatic tissue digestion, or autolysis, by the own cells. Commonly, forensic necropsies and specimens sent for examination are under decomposition. However, once received by the forensic staff, all material must be processed, and supported, in any case, by a technical report. If the sample quality affects the outcome, this information should be amended [23].

Liver acts as an interface between the digestive tract and blood. In this organ, nutrients are absorbed through the portal circulation, via the portal vein and hepatic artery, and later processed and stored.

In addition, the liver is the main organ performing the biotransformation. This chemical modification comprises a set of enzymatic reactions that alter activity and physical properties of various endogenous and exogenous compounds. Hepatic biotransformation develops two phases involving primarily the cytochrome P450 enzymes, mainly located in the microsomes of the endoplasmic reticulum of hepatocytes, especially those closer to the centrolobular vein (Rappaport zone III). The first phase accomplishes substance oxidation by cytochrome P450. Then, the resulting product is combined for excretion at the second phase. As a result, toxic or non-toxic metabolite formation occurs [24].

If, during the poisoning, the dose ingested by the animal is high, degeneration is possible to be seen in areas other than the region near the centrolobular vein. In acute poisoning, as the studied cases, zone I of the liver is the most affected by an intense deterioration of hepatic stem cells. This decomposition process is characterized by voids in the cytoplasm known as vacuoles.

Acute intoxication is not sufficient to trigger edematous responses. Consequently, inflammatory cells resulting from other processes will further exacerbate liver degeneration.

After death, liver autolysis begins immediately, due to the large amount of enzymes. Thus, the cadaveric phenomena change, rapidly and easily, organ structure, which can compromise histological evaluation.

The coagulative hepatocellular necrosis also occurs commonly in toxic injury, which may be focal or multifocal. In this case, the tissue architecture is normal. Hepatocytes with karyolytic, karyorrhexistic, and piknotic cores, and acidophilic cytoplasm are present [22].

Cholestasis (biliary steatosis) occurs by hepatocyte deterioration. In this way, a common finding is established in the hepatic histology of animals dying by poisoning, since chemical compounds highly promote functional cell degradation in this organ. Likewise, the deposition of hemosiderin is usual because an increased destruction of red blood cells in the tissue. In the case of a fatal acute poisoning, observing hepatic fibrosis is unlikely. Nevertheless, if the animal survives, this lesion can be one of those irreversible since Ito cells, located between hepatocytes, gather vitamin A. When damaged, these cells lose fat granules and synthesize collagen, which is important to the fibrogenesis process.

Among post-mortem changes, liver coloration is very sensitive. It can be modified by the following facts: cadaveric lividity; print marks caused by pressure of ribs, leaving some areas with paler coloring; hemoglobin impregnation, becoming dark red; false melanotic in contact with the bowel, by eliminating hydrogen sulfide combined with the iron released by erythrocytes lysates causes greenish color; and gall soaking leaving yellowish spots [22].

Kidneys perform glomerular filtration, reabsorption, and tubular secretion. Chemicals typically induce renal injury for different reasons: glomerular epithelial cells can undergo a direct lesion; changes occurred in renal blood flow; and immunocomplex formation, especially by acute exposure, such as the case of toxic agents.

As noted, the tubular epithelium cells, primarily in the proximal tubules, are more susceptible to ischemia and nephrotoxins because its high metabolic activity and the amount of glomerular filtrate reabsorbed during urine formation process [25].

In toxic damages, hyperemia is manifested with blood stasis. Macroscopically, cuts reveal increased blood flowing on the surface, actually redder. Microscopically, all blood vessels, especially capillaries, are congested, matching the literature [26].

Acute tubular necrosis (ATN) is the leading cause of kidney failure. Generally, ATN results from degenerative nephrotoxic lesions to kidney epithelial tubules. The microscopic appearance of kidney varies depending on damage grade, exposure time to the toxic agent, and the interval between lesion and death. Histological examination disclosed an extensive necrosis of proximal tubules with, or without, basal membrane preservation. Lesions with greater intensity were in the cortex, and into a lesser extent, in the renal marrow [27,28].

An important aspect is to realize the existence of deadly diseases of neurological origin with, or without, macroscopic or microscopic changes on nervous tissue, such as intoxication by strychnine, carbamate, and organophosphate. As a result, pathologist must proceed through toxicology techniques [29].

Finally, correlation among findings of illegal poisoning cases here described matches what literature points out. Therefore, strengthening forensic veterinary pathology becomes a very important matter for the suitable completion of necropsy report [30].

CONCLUSION

Based on macroscopic and microscopic interpretation of anatomopathological findings related to illegal poisoning of dogs and cats, it can be concluded that the use of histopathological and toxicology techniques were complementary. The analysis of cell and tissue alterations in some organs, along with toxicant agent tests, facilitated the understanding of criminal intoxication in selected samples. Thus, applying proper techniques of veterinary legal medicine can contribute to produce material evidence for the benefit of judicial processes involving animals.

ACKNOWLEDGEMENT

CAPES (Coordination and Improvement of Higher Education Personnel Development) - Brazil.

REFERENCES

1. Gwaltney-Brant SM (2007) Epidemiology of animal poisoning. Veterinary Toxicology, Elsevier, pp: 67-73.
2. Marlet EF, Maiorka PC (2010) Retrospective analyzes of cruelty toward dogs and cats in the city of São Paulo. Braz J Vet Res Anim Sci 47: 385-394.
3. Brazil Law no. 9605 (1998) Law of Environmental Crimes. Brasília: Official Federal Gazette, Brazil.
4. Brazil Law No. 5517 (1968) Regulates the exercise of Veterinary Medicine in Brazil: Official Gazette of the Union, Brazil.
5. Wang Y, Kruzik P, Helsberg A, Helsberg I, Raush WD (2007) Pesticide poisoning in domestic animals and livestock in Austria: A 6 years retrospective study. Forensic Sci Int 169: 157-160.
6. Xavier FG, Righi DA, Spinosa HS (2007) Fatal poisoning in dogs and cats: A 6 year report in a veterinary pathology service. Braz J Vet Res Anim Sci 44: 304-309.
7. Medeiros RJ, Monteiro FO, Silva GC, Nascimento AJr (2009) Poisoning dogs and cats attended in the Veterinarian College in Federal Fluminense University during the period of 2002 to 2008. Cienc Rural 39: 2105-2110.
8. Bulcão RP, Tonello R, Piva RJ, Schmitt GC, Emanuelli T, et al. (2010) Intoxication in dogs and cats: toxicological diagnosis using thin layer chromatography and high pressure liquid chromatography with ultraviolet detection in stomach samples. Cienc Rural 40: 1109-1113.
9. Byard RW, Boardman W (2011) The potential role of forensic pathologists in veterinary forensic medicine. Forensic Sci Med Pathol 7: 231-232.
10. Cooper JE, Cooper ME (2008) Forensic veterinary medicine: a rapidly evolving discipline. Forensic Sci Med Pathol 4: 75-82.
11. Berny P, Caloni F, Croubele S, Sachana M, Vandenbroucke V, et al. (2010) Animal poisoning in Europe. Part 2: Companion animals. The Vet J 183: 255-259.
12. Mcdonough SP, Gerdin J, Wuenschmann A, Mcewen BJ, Brooks JW (2015) Illuminating Dark Cases Veterinary Forensic Pathology Emerges. Vet Pathol 52: 5-6.
13. Cooper JE, Cooper ME, Future trends in forensic veterinary medicine. Semin Avian Exot Pet Med. 7: 210-217.
14. Marlet EF, Yoshida AS, Gorniak SL, Maiorka PC (1998) Elaboration of expertise report in veterinary medicine. CFMV Journal 55: 12-19.
15. Munro R (1998) Forensic necropsy. Semin Avian Exot Pet 7: 201-209.
16. Below E, Lignitz E (2003) Cases of fatal poisoning in post-mortem examinations at the Institute of Forensic Medicine in Greifswald–Analysis of five decades of post-mortems. Forensic Sci Int 133: 125-131.
17. Klaassen CD, Watkins JB (2012) Fundaments in toxicology. McGraw-Hill Education, New York, USA.
18. Klaassen CD, Casarett LJ, Doull J, Klaasen CD, Casarett et al. (2013) Casarett and Doull's toxicology : the basic science of poisons. (8th edtn), McGraw-Hill Education, New York, USA.
19. Cummings PM, Trelka DP, Springer KM (2011) Poisoning. In: Atlas of forensic histopathology. Cambridge University Press, New York, USA.
20. Prestes Junior LCL, Ancillotti R (2009) Manual of medical-legal necropsy techniques. Blonde publisher, Brazil.
21. Jubb KVVF, Kennedy PC, Palmer NC (2007) Pathology of Domestic Animals, Saunders Elsevier, Edinburgh, USA.
22. Volmer PA, Meerdink GL (2002) Diagnostic Toxicology for the Small Animal practitioner. Vet Clin North Am 32: 357-365.
23. Junqueira LCU, Junqueira LMMS (1983) Basic techniques of cytology and histology. Santos publisher, Portugal.

24. Carlton WW, Mcgavin MD (1998) Special Veterinary Pathology of Thompson. (2nd Edn), Artmed Publishers, Porto Alegre, Brazil.
25. Serakide R, Sistema Urinário (2010) In: Santos RL, Alessi AC Veterinary Pathology. Roca Publishers, Sao Paulo, Brazil pp: 291-336.
26. Kumar V, Abbas AK, Fausto N, Aster JC (2009) Robbins and Cotran Pathologic Basis of Disease (9th Edn), Saunders Company, Philadelphia, USA.
27. Rocha NS, Bacchi CE, Carvalho M, Schmitt D, Schmitt F et al.(1994) Immunohistochemical characterization of probable intravascular haematopoesis in the vasa recta of the renal marrow in acute tubular necrosis. Pathol Res Prac 190: 1066-1070.
28. Mcgavin MD, Zachary JF (2013) Pathologic basis of Veterinary Pathology. (5th edn), Elsevier publishers, São Paulo, Brazil.
29. Graça DL, Alessi AC, Ecco R, Viott AM, Patologia Do Sistema Nervoso (2010) In: Santos RL, Alessi AC Veterinary Pathology. Roca Publishers, Sao Paulo, Brazil pp: 525-610.
30. Tremori TM, Rocha NS (2013) The examination of the corpus delicti in veterinary expertise. Journal of Continuing Education in Animal Science of CRMV-SP. Legal Veterinary Medicine 11: 30-35.

CHAPTER 2

Comprehensive Review of the Detection Methods for Synthetic Cannabinoids and Cathinones

Akira Namera[1] • Maho Kawamura[2] • Akihiro Nakamoto[2] • Takeshi Saito[3] • Masataka Nagao[1]

[1]*Department of Forensic Medicine, Institute of Biomedical and Health Sciences, Hiroshima University, 1-2-3 Kasumi, Minamiku, Hiroshima 734-8553, Japan*

[2]*Forensic Science Laboratory, Hiroshima Prefectural Police Headquarters, Hiroshima, Japan*

[3]*Department of Emergency and Critical Care MedicineTokai University School of MedicineIseharaJapan*

ABSTRACT

A number of *N*-alkyl indole or indazole-3-carbonyl analogs, with modified chemical structures, are distributed throughout the world as synthetic cannabinoids. Like synthetic cannabinoids, cathinone analogs are also abused and cause serious problems worldwide. Acute deaths caused by overdoses of these drugs have been reported. Various analytical methods that can cope with the rapid changes in chemical structures are required for routine analysis and screening of these drugs in seized and biological materials for forensic and clinical purposes. Although many chromatographic methods to analyze each drug have been published, there are only a few articles summarizing these analytical methods. This review presents the various colorimetric detections, immunochemical assays, gas chromatographic–mass spectrometric methods, and liquid chromatographic–mass spectrometric methods proposed for the analysis of synthetic cannabinoids and cathinones.

Keywords: Synthetic cannabinoids Cannabimimetics Cathinones GC–MS-MS LC–MS-MS Analytical methods

INTRODUCTION

Currently, many illegal drugs are abused worldwide, with serious social problems arising as a consequence. Although various stimulants and narcotics have been in use to date, new drugs targeting cannabinoid receptors have been abused since their existence in herbal mixtures was disclosed in 2008 [1]. HU-210, a synthetic classical cannabinoid, and cyclohexylphenols were commonly used as recreational drugs, but mainstream use has since changed to *N*-alkyl indole-3-carbonyl derivatives, such as drugs of the JWH and AM series (Figure 1), because their activities are stronger than those of the conventional cannabinoids. These compounds are called cannabimimetics or synthetic cannabinoids and can be purchased as "spice" or "K2" in the drug market or via the Internet. Cathinones, also known as "bath salts" or "plant food," are psychoactive drugs and are also abused as recreational drugs. The parent compound, cathinone, is a well-known stimulant, and can be isolated from the khat plant or produced by synthetic means. Cathinone analogs with high selectivity and strong activity for serotonin receptors and monoamine transporters have been distributed in the drug market (Figure 2). The prevalence of cannabinoid and cathinone abuse in many countries has been reviewed elsewhere [2, 3, 4, 5, 6, 7].

Although the same substances are distributed throughout the world, the times at which they are abused tend to vary depending on whether the substances are controlled by local laws. As shown in the reviews [2, 3, 4, 5, 6, 7], new analogs appear in the drug market just after the preceding drug comes under regulation. Although many such substances are controlled in countries throughout the world, the regulations are usually limited by the structures of the drugs. Therefore, when the structure of a side chain or substitution is slightly different from that of the regulated drug, the analog is regarded as being beyond the scope of the regulation. These emerging drugs always show psychoactive actions because their chemical structures are similar to those of the drugs being controlled. However, the detailed pharmacological activities of these analogs are not known, which makes access easy and use of these drugs very dangerous to human health.

Although many researchers have focused on the development of detection methods, only a few analytical reviews that summarize the systematic identification and quantification techniques for these drugs have appeared [8, 9, 10]. In this review, we summarize the various techniques for the detection of synthetic cannabinoids and cathinones that have been published up to 2014, including colorimetric, immunochemical, and chromatographic methods.

Indole 3-carbonyl derivatives

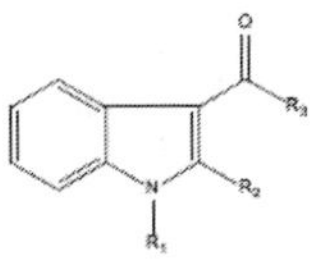

	R_1	R_2	R_3
JWH-015	*n*-Propyl	Methyl	1-Naphthyl
JWH-018	*n*-Pentyl	H	1-Naphthyl
AM-1220	(1-Methyl-2-piperidinyl)methyl	H	1-Naphthyl
JWH-200	2-(4-Morpholinyl)ethyl	H	1-Naphthyl
JWH-203	*n*-Pentyl	H	2-Chloro-benzyl
JWH-250	*n*-Pentyl	H	2-Methoxy-benzyl
JWH-251	*n*-Pentyl	H	2-Methoxy-benzyl
RCS-4	*n*-Pentyl	H	4-Methoxy-phenyl
AB-001	*n*-Pentyl	H	Adamantyl
AM-1248	(1-Methyl-2-piperidinyl)methyl	H	Adamantyl
AM-679	*n*-Pentyl	H	2-Iodo-phenyl
AM-694	5-Fluoro-*n*-pentyl	H	2-Iodo-phenyl
AM-2233	(1-Methyl-2-piperidinyl)methyl	H	2-Iodo-phenyl
AM-1241	(1-Methyl-2-piperidinyl)methyl	H	2-Iodo-5-nitro-phenyl
-[a]	5-Fluoro-*n*-pentyl	H	3-Pyridinoyl

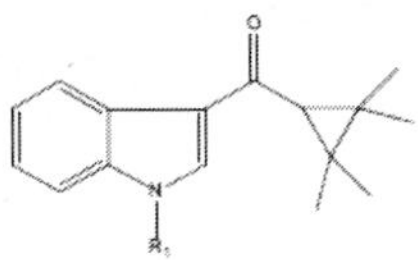

	R_1
UR-144	*n*-Pentyl
XLR-11	5-Fluoro-*n*-pentyl
XLR-12	4,4,4-Trifluoro-butyl
AB-005	(1-Methyl-2-piperidinyl)methyl
A-834735	(Tetrahydro-2*H*-pyran-4-yl)methyl
A-796260	2-(4-Morpholinyl)ethyl

[a] No abbreviated name available

Pyrrole 3-carbonyl derivatives

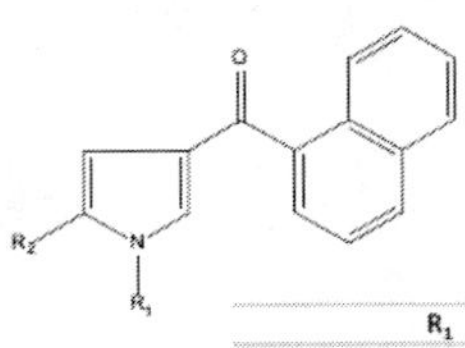

	R_1	R_2
JWH-030	*n*-Pentyl	H
JWH-307	*n*-Pentyl	2-Fluoro-phenyl

Indazole 3-carbonyl derivative

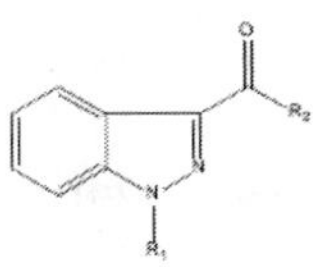

	R_1	R_2
THJ-2201	5-Fluoro-*n*-pentyl	1-Naphthyl

Indole 3-carbonylamide derivatives

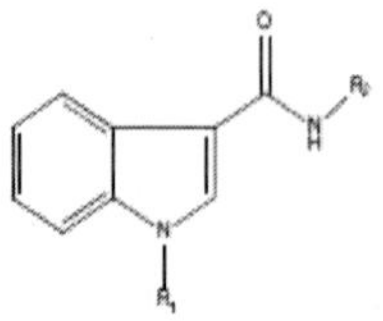

	R_1	R_2
APICA	*n*-Pentyl	Adamantyl
5F-APICA	5-Fluoro-*n*-pentyl	Adamantyl
NNEI(MN-24)	*n*-Pentyl	1-Naphthyl
5F-NNEI	5-Fluoro-*n*-pentyl	1-Naphthyl

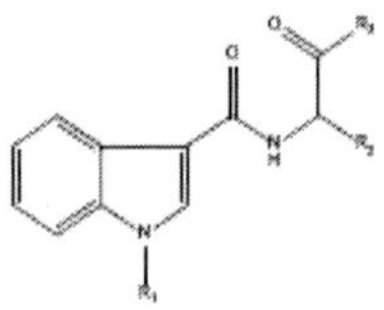

	R_1	R_2	R_3
ADBICA	*n*-Pentyl	*t*-Butyl	NH_2
PX1	5-Fluoro-*n*-pentyl	Benzyl	NH_2

Indole 3-carbonylester derivatives

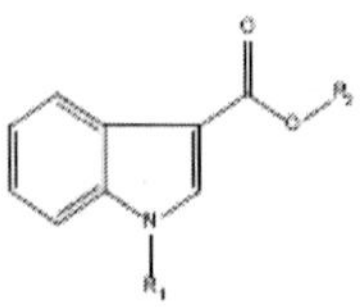

	R_1	R_2
NM-2201	5-Fluoro-*n*-pentyl	1-Naphthyl
FDU-PB-22	4-Fluoro-benzyl	1-Naphthyl
QUPIC(PB-22)	*n*-Pentyl	8-Quinolinyl
5F-PB-22	5-Fluoro-*n*-pentyl	8-Quinolinyl
QUCHIC(BB-22)	Cyclohexylmethyl	8-Quinolinyl
FUB-PB-22	4-Fluoro-benzyl	8-Quinolinyl

Indazole 3-carbonylamide derivatives

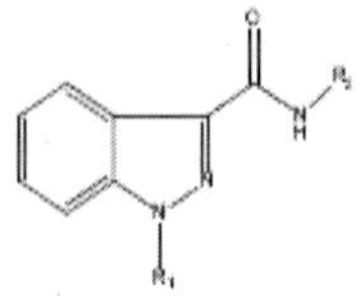

	R_1	R_2
APINACA	*n*-Pentyl	Adamantyl
5F-APINACA	5-Fluoro-*n*-pentyl	Adamantyl
MN-18	*n*-Pentyl	1-Naphthyl
5F-MN-18	5-Fluoro-*n*-pentyl	1-Naphthyl

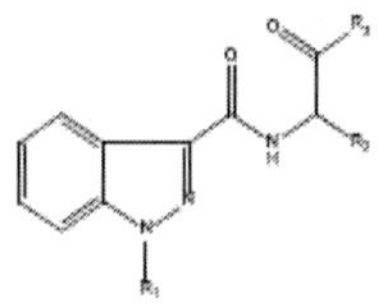

	R_1	R_2	R_3
AB-PINACA	*n*-Pentyl	*i*-Propyl	NH_2
5F-AB-PINACA	5-Fluoro-*n*-pentyl	*i*-Propyl	NH_2
ADB-PINACA	*n*-Pentyl	*t*-Butyl	NH_2
AB-FUBINACA	4-Fluoro-benzyl	*i*-Propyl	NH_2
ADB-FUBINACA	4-Fluoro-benzyl	*t*-Butyl	NH_2
AB-CHMINACA	Cyclohexylmethyl	*i*-Propyl	NH_2
AMB	*n*-Pentyl	*i*-Propyl	Methoxy
5F-AMB	5-Fluoro-*n*-pentyl	*i*-Propyl	Methoxy

Indazole 3-carbonylester derivatives

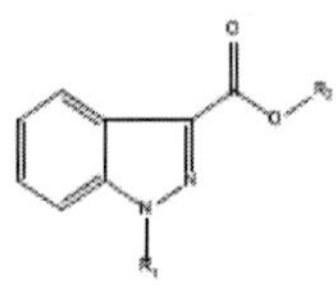

	R_1	R_2
5F-SDB-005	5-Fluoro-*n*-pentyl	1-Naphthyl
5F-NPB-22	5-Fluoro-*n*-pentyl	8-Quinolinyl

Figure 1. Structures of synthetic cannabinoids

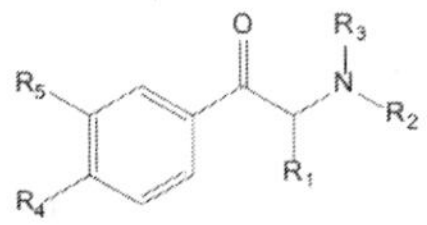

Name	R_1	R_2	R_3	R_4	R_5
Cathinone	Methyl	H	H	H	H
Methcathinone	Methyl	Methyl	H	H	H
N,N-Dimethylcathinone	Methyl	Methyl	Methyl	H	H
Mephedrone	Methyl	Methyl	H	Methyl	H
3,4-Dimethylmethcathinone	Methyl	Methyl	H	Methyl	Methyl
4-Ethylmethcathinone	Methyl	Methyl	H	Ethyl	H
Methylone	Methyl	Methyl	H	Methylenedioxy	
Ethcathinone	Methyl	Ethyl	H	H	H
Ethylone	Methyl	Ethyl	H	Methylenedioxy	
Amfepramone	Methyl	Ethyl	Ethyl	H	H
Buphedrone	Ethyl	Methyl	H	H	H
4-Methylbuphedrone	Ethyl	Methyl	H	Methyl	H
Butylone	Ethyl	Methyl	H	Methylenedioxy	
N-Ethylbuphedrone	Ethyl	Ethyl	H	H	H
Eutylone	Ethyl	Ethyl	H	Methylenedioxy	
Pentedrone	Propyl	Methyl	H	H	H
α-Ethylaminopentiophenone	Propyl	Ethyl	H	H	H

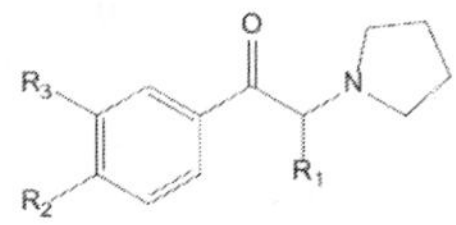

Name	R_1	R_2	R_3
α-PPP	Methyl	H	H
α-PBP	Ethyl	H	H
α-PVP	Propyl	H	H
α-PHP	Butyl	H	H
PV8	Pentyl	H	H
PV9	Hexyl	H	H
MPPP	Methyl	Methyl	H
MOPPP	Methyl	Methoxy	H
MPBP	Ethyl	Methyl	H
Pyrovalerone	Propyl	Methyl	H
Naphyrone	Propyl	Phenyl	
MPHP	Butyl	Methyl	H
MDPPP	Methyl	Methylenedioxy	
MDPBP	Ethyl	Methylenedioxy	
MDPV	Propyl	Methylenedioxy	

Figure 2. Structures of cathinones

SYNTHETIC CANNABINOIDS

Colorimetric detection

The Duquenois–Levine color test, which is used to identify classical cannabinoids such as Δ^9-tetrahydrocannabinol, is negative for the synthetic cannabimimetics. The van Urk color test, which is used to identify indole-containing drugs of abuse, is also negative for these compounds. The use of 2,4-dinitrophenylhydrazine, which reacts with a keto moiety, is capable of reacting with synthetic cannabimimetics, such as the naphthoylindole, phenylacetylindole, benzoylindole, and cyclopropylindole classes, either in powder form or adsorbed onto plant material, and a positive test solution turns from yellow to orange. Although the LOD concentration was not detailed in the article, the solution tested contained at least 10 mg of cannabimimetic powder suspended in methanol (1 ml) [11]. The Marquis reagent, which reacts with all nitrogen-containing drugs, is positive for cyclohexylphenols and the JWH series. Although Dragendorff reagent is also positive for the JWH series, its LOD concentration is higher than that of Marquis reagent. Fast blue BB reacts with cyclohexylphenols, and the LOD concentration is not lower than that of Marquis reagent [12]. Iodoplatinate is also used as a detection reagent after TLC [13]. Although it is possible to detect synthetic cannabinoids with each reagent in these screening tests, it is difficult to detect small amounts or mixtures of synthetic cannabinoids.

Immunochemical detection

ELISAs developed in-house could be calibrated at 5 ng/ml with the 5-OH and 4-OH metabolites of JWH-018 and JWH-250, respectively, and evaluated for the detection of synthetic cannabinoids in urine [14]. Recently, some commercially available immunoassay kits, such as DrugCheck K2/Spice Test, DrugSmart Cassette, and RapiCard InstaTest, have been developed for the detection of these drugs in urine. These devices are more useful than the colorimetric methods, because they do not require special reagents or tools, and the results are obtained easily and quickly. The devices also can detect older types of synthetic cannabinoids, such as JWH-018 or JWH-073, but, unfortunately, new designer drugs such as QUPIC and AB-CHMINACA cannot be detected.

GC–MS detection

Typical mass spectra of synthetic cannabinoids are shown in Figure 3. Molecular (M^+) and/or fragment ions observed by full scan data acquisition of GC–MS reflect the structures of the synthetic cannabinoids [13, 15, 16]. As shown in Figure 4, the fragmentation pathways of naphthoylindoles have been well studied for the identification of synthetic cannabinoids by GC–MS [12, 15]. Therefore, the identification of synthetic cannabinoids is facilitated by comparison of the spectra with commercial and open databases.

In naphthoylindoles, the carbonyl group fragment ions, which are caused by α-cleavage of the alkylamino group of the indole, are typically observed. In addition, $[M-17]^+$ is certainly observed in naphthoylindoles. For example, fragment ions at *m/z* 284 and 214 are observed in JWH-018, corresponding to those of the indole moiety caused by α-cleavage of the *N*-pentyl of indole and naphthoyl. Fragment ions at *m/z* 127 and 155 are observed in JWH-018, corresponding to the naphthalene group caused by the α-cleavage of the carbonyl group. Moreover, ions at *m/z* 324 are observed as $[M-17]^+$ (Figure 3a). Like naphthoylindoles, fragment ions caused by α-cleavage of the alkylamino group of the indole and carbonyl groups are shown, although $[M-17]^+$ is not observed for benzoylindoles. For example, fragment ions at *m/z* 264 and 214 are observed for RCS-4, caused by α-cleavage of *N*-pentyl of the indole and 4-methoxybenzoyl. The ions at m/z 127 and 155, which are caused by naphthyl and naphthoyl moieties of naphthoylindoles (Figure 3a), and the ion at m/z 135 caused by the 4-methoxybenzoyl moiety (Figure 3b) are useful as precursor ions for identification of these drugs by GC–MS-MS. On the other hand, the methylpiperidine moiety is bound to the nitrogen of the indole, and the ion at m/z 98 is observed as the base peak (Figure 3d). Unlike naphthoyl and benzoyl indoles, the base peak of the fragment ion caused by the *N*-alkylindole 3-carbonyl moiety for phenylacetyl (Figure 3c), cyclopropyl, or adamantyl (Figure 3f) indoles, is only shown in each full scan spectrum. Analogs, in which the indole skeleton is changed to an indazole, such as THJ-018, have also appeared on the market. In these analogs, molecular and *N*-dealkylated ions are typically observed in the spectrum (Figure 3e).

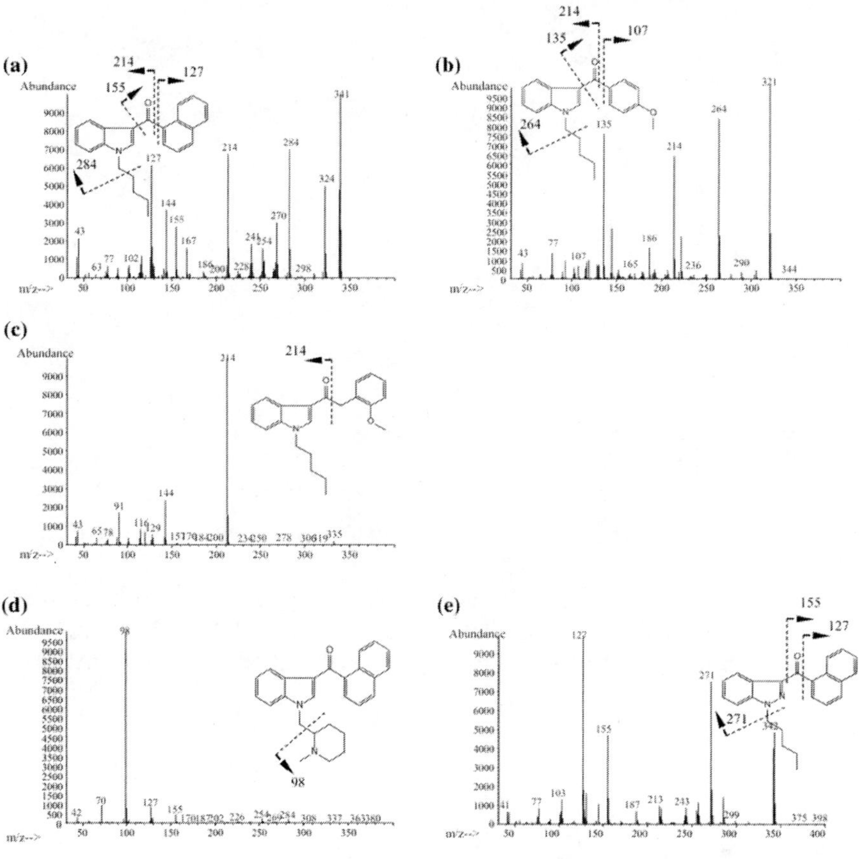

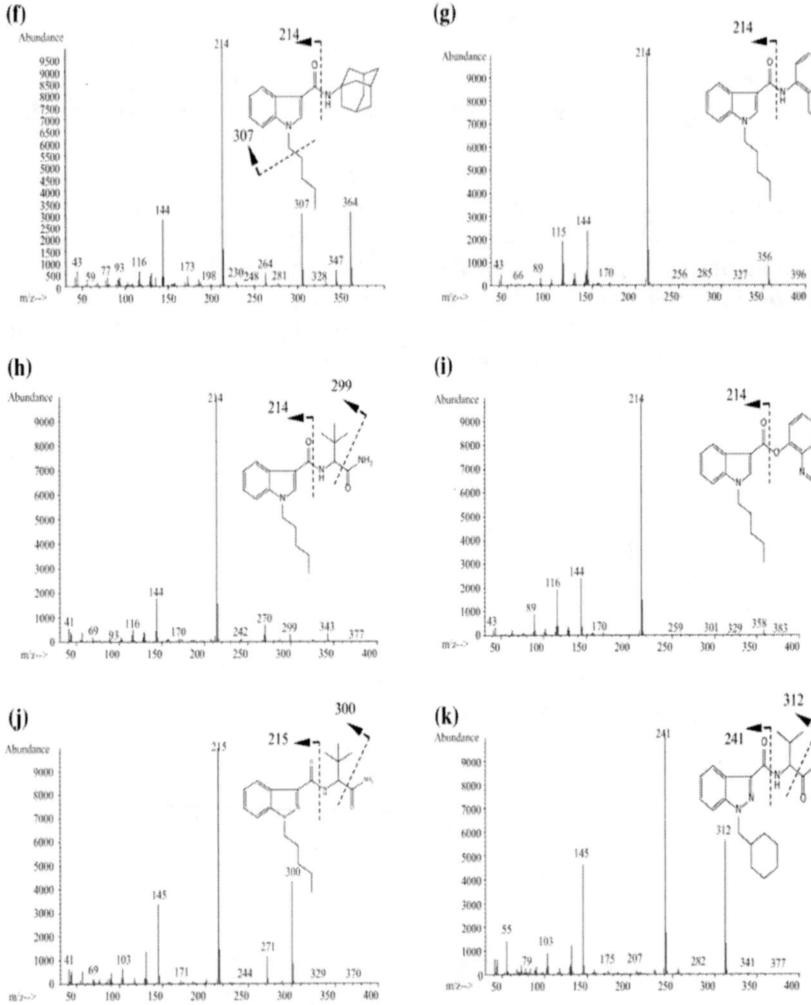

Figure 3. Typical mass spectra of synthetic cannabinoids obtained by GC–MS. **a** JWH-018, **b** RCS-4, **c** JWH-250, **d** AM-1220, **e** THJ-018, **f** APICA, **g** NNEI, **h** ADBICA, **i** QUPIC (PB-22), **j** ADB-PINACA, **k** AB-CHMINACA

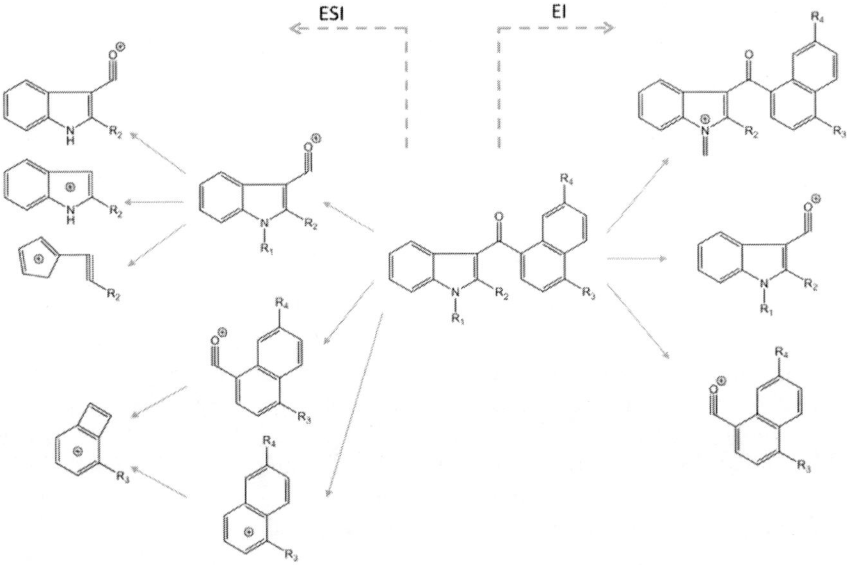

Figure 4. Probable fragmentation pathways of synthetic cannabinoids by electrospray ionization and electron ionization (modified from references [12, 31])

Recently, amide- or ester-type analogs bonded with an *N*-alkylindole or *N*-alkylindazole 3-carbonyl moiety have appeared on the market [17, 18]. In these analogs, the abundance of the molecular ion is low, and the fragment ion caused by the indoyl (or indazoyl) moiety is observed as a base peak (Figure 3f–k). Although the fragment ion caused by elimination of the terminal CO–NH_2 is lower than that of the cleavage of the amide moiety in indole analogs, such as ADBICA (Figure 3h) [19], the fragment ion caused by elimination of terminal CO–NH_2 is as intense as that of the cleavage of the amide moiety in indazole analogs, such as ADB-PINACA and AB-CHMINACA (Figure 3j, k) [20]. The substitution of the indole skeleton with the indazole moiety, such as in THJ-018 and THJ-2201 [21], has also been observed in these analogs. In these analogs, molecular and *N*-dealkylated ions are typically observed in the spectrum.

For example, in the simultaneous analysis of synthetic cannabinoid species, 10 mg of ground powder of the dried leaves was extracted with 10 ml of methanol under ultrasonication for 10 min. The extracts were centrifuged for 5 min at 3,000 rpm, and the supernatants were filtered and used for GC–MS analysis. The LODs were 0.5–1.0 mg/l, and linearity was obtained at concentrations up to 100 mg/l [16]. In another article [22], herbal samples (approximately 50 mg) were put into 10-ml headspace vials, and the vials were capped with 20-mm magnetic crimp seal caps with PTFE/silicone septa. The samples were incubated at 200 °C with pulse-agitation at 250 rpm. A StableFlex carboxen/ polydimethy lsiloxane fiber was inserted into the headspace for 5 min for extraction. The fiber was then injected into the GC inlet for 15 min to desorb the analytes. The LOD of synthetic cannabinoid in the samples was at least 20 μg.

The tentative identification of synthetic cannabinoids appears easy, but similar mass spectra are sometimes obtained by GC–MS because regio- and ring-substituted analogs are still distributed on the market. The misidentification of these analogs arises when using only the information from the mass spectra. When tandem and high-resolution MS are used to identify the conformational isomers or regioisomers, such misidentification does not occur [23, 24, 25, 26, 27, 28]. Moreover, identification of cyclopropyl or ester analogs, such as UR-144 or QUPIC, is usually not possible because cyclopropyl analogs are heat-unstable and are easily degraded in the injection port of the GC instrument [29, 30].

LC–MS-MS detection

Many research groups have used LC–MS-MS for determination of synthetic cannabinoids in herbs and biological samples, and some have studied the fragmentation of synthetic cannabinoids in detail [15, 31]. The probable fragmentation pathways are shown in Figure 4. Because the protonated molecular ion is only observed by LC–MS, and the information acquired by LC–MS is lesser than that for GC–MS, it is necessary to obtain other data that reflect the chemical structures by LC–MS-MS or TOFMS. Fragment ions are observed by product ion scanning when the protonated molecular ion is used as the precursor ion. In naphthoylindole, ions at *m/z* 127 and 155 are generated by naphthyl and naphthoyl moieties. However, information about the indole moiety tends to be not revealed by LC–MS-MS. On the other hand, the *N*-alkyl moiety of a synthetic cannabinoid is mainly modified for excretion into urine as a metabolite. Therefore, LC–MS-MS is a useful methodology to search for metabolites of synthetic cannabinoids in urine.

Recently, packages containing mixtures of multiple synthetic cannabinoids have been sold commercially, even though the package ingredients have been largely unknown to both sellers and buyers. In this aspect, the LC–MS-MS screening method is helpful in some estimation of the ingredients. Kneisel and Auwärter [32] demonstrated the simultaneous detection of 30 synthetic cannabinoids in serum; the LODs and LOQs were 0.01–2.0 and 0.1–2.0 ng/ml, respectively. There are many applications for analysis of synthetic cannabinoids in urine, hair, and oral fluids [33, 34, 35, 36, 37]. The typical published methods for analysis of synthetic cannabinoids in biological materials are summarized in Table 1 [38, 39, 40, 41, 42, 43, 44, 45, 46, 47, 48, 49, 50, 51, 52, 53, 54]. Simple LLE is usually used for the extraction of synthetic cannabinoids from biological materials because of the high hydrophobicity of the drugs. The chromatographic conditions are generally simple and do not require a special technique; octadecyl-type columns were used as analytical columns and analyses were performed in gradient mode.

Table 1. LC–MS or LC–MS-MS conditions for synthetic cannabinoids in biological materials

Target(s)	Sample(s)	Purification(s)	Column(s)	Mobile phase	LOD (ng/ml)	Linear range (ng/ml)	Reference(s)
JWH-018	Serum	LLE	Luna C18 (2) (150 mm, 2 mm ID, 5 μm) (Phenomenex)	10 mM ammonium acetate (0.1 % acetic acid, pH 3.2), methanol	0.07	0.21–20	[38]
JWH-018, JWH-073, JWH-019, JWH-250	Blood	LLE	Acquity UPLC HSS T3 (100 mm, 2.1 mm ID, 1.8 μm) (Waters)	1 % formic acid, methanol (1 % formic acid)	0.006–0.016	0.1–20	[39, 40]
Aminoalkylindoles, methanandamide	Serum	LLE	Luna phenyl hexyl (50 mm, 2 mm ID, 5 μm) (Phenomenex)	2 mM ammonium formate (0.2 % formic acid), methanol	0.1	0.1–2, 0.3–2 (methanandamide)	[41]
JWH-018, JWH-073, metabolites	Urine	Dilution (hydrolysis)	Zorbax Eclipse XDB-C18 (150 mm, 4.6 mm ID, 5 μm) (Agilent)	0.1 % formic acid, acetonitrile (0.1 % formic acid)	<2.0	2–100	[42]
Metabolites of JWH-018 and JWH-073	Urine	SPE (hydrolysis)	Zorbax Eclipse XDB-C18 (150 mm, 4.6 mm ID, 5 μm) (Agilent)	0.1 % formic acid/acetonitrile (0.1 % formic acid) (45:55), isocratic	<0.1	0.1–100	[43]
Metabolites of 8 synthetic cannabinoids	Urine	LLE (hydrolysis)	AQUASIL C18 (100 mm, 2.1 mm ID, 5 μm) (Thermo Scientific)	5 mM ammonium acetate, methanol/acetonitrile (1:1, 5 mM ammonium acetate)		0.1–10	[44]
Metabolites of JWH-018 and JWH-073	Urine	LLE (hydrolysis)	Acquity UPLC HSS T3 (100 mm, 2.1 mm ID, 1.8 μm) (Waters)	0.1 % formic acid (0.1 %), acetonitrile (0.1 % formic acid)		4–400	[45]
30 Synthetic cannabinoids	Serum	LLE	Luna phenyl hexyl (50 mm, 2 mm ID, 5 μm) (Phenomenex)	0.2 % formic acid (2 mM ammonium formate), methanol	0.01–2.0	0.1–2.0 (2–40 : JWH-387)	[32, 46–48]
Metabolites of 7 synthetic cannabinoids	Urine	LLE (hydrolysis)	Luna C18 (150 mm, 2 mm ID, 5 μm) (Phenomenex)	0.2 % formic acid (2 mM ammonium formate), methanol			[49]
22 Synthetic cannabinoids	Hair	Ethanol ext	Luna phenyl hexyl (50 mm, 2 mm ID, 5 μm) (Phenomenex)	0.2 % formic acid (2 mM ammonium formate), methanol	0.5 pg/mg		[37]
JWH-018, JWH-073	Blood	LLE	Acquity UPLC BEH C18 (50 mm, 2.1 mm ID, 1.8 μm) (Waters)	0.1 % formic acid, acetonitrile (0.1 % formic acid)	0.01	0.05–50	[50]
UR-144, metabolites, pyrolysis product	Urine	LLE (hydrolysis)	Zorbax Eclipse XDB-C18 (150 mm, 2.1 mm ID, 3.5 μm) (Agilent)	20 mM ammonium formate buffer (pH5), acetonitrile			[29]
UR-144, metabolites	Blood, urine	PP	Kinetex C18 (100 mm, 4.6 mm ID, 2.6 μm) (Phenomenex), Ascentis express C18 (7.5 cm, 2.1 mm ID, 2.7 μm) (Supelco)	0.1 % formic acid, acetonitrile (0.1 % formic acid)	0.15 (blood)	0.5–100 (blood)	[51]
9 Synthetic cannabinoids, 20 metabolites	Urine	PP (hydrolysis)	XB-C18 (50 mm, 3.0 mm ID, 2.6 μm) (Kinetex)	0.1 % formic acid, acetonitrile (0.1 % formic acid)	0.5–10		[34]
MAM-2201	Blood	SPE	InertSustain C18 HP (100 mm, 3 mm ID, 3 μm) (GL Sciences)	0.1 % acetic acid, acetonitrile	1	2.5–100	[52]
5F-PB-22	Blood, serum	LLE	Acquity UPLC BEH C18 (100 mm, 2.1 mm ID, 1.7 μm) (Waters)	0.1 % formic acid, acetonitrile (0.1 % formic acid)	0.1	0.5–10	[53]
Indole derivative synthetic cannabinoids	Saliva, urine, blood	PP (blood) LLE (urine, saliva) (hydrolysis)	Ascentis C18 (150 mm, 2.1 mm, 5 μm) (Supelco)	0.1 % formic acid (5 mM ammonium formate), acetonitrile (0.1 % formic acid)	0.1–0.5		[54]

The identification of an unknown drug in a biological material without information is almost always difficult, even if analysis is carried out with LC–MS-MS. To overcome this situation, high-resolution MS or TOFMS become helpful tools for tentative estimation of parent drugs and their metabolites of synthetic cannabinoids.

To clarify the chemical structure of an unknown drug in a herbal blend product that contains more than several milligrams of the drug, GC–MS, LC–MS-MS, and high-resolution MS (or TOFMS) can be used to estimate the structure. The target compound is then purified by preparative LC or preparative TLC to obtain more than several milligrams of the compound of high purity, which is then analyzed by NMR spectroscopy [17, 18, 19, 20, 21]. The detailed chemical structure can be elucidated by the above laborious instrumental analyses.

CATHINONES

Colorimetric detection

The Marquis reagent, which reacts with all nitrogen-containing drugs, is negative for cathinones, such as cathinone and mephedrone, but is positive for cathinone analogs that have a methylenedioxy moiety in each molecule. The cathinone analogs with a methylenedioxy moiety also react with the Chen reagent, which changes to orange in positive tests. Although the LOD concentration is not reported, the described test solution contained cathinone powder (at least 10 mg) suspended in methanol (1 ml) [55, 56, 57]. The combination of Marquis, Ehrlich, Simon, Lieberman–Burehand, and Mandelin reagents is useful for the detection of cathinones in samples. Like synthetic cannabinoids, the identification of these compounds, of course, cannot be performed using these methods; moreover, the detection of small amounts or mixtures of cathinones is difficult.

Immunochemical detection

Some researchers have tried to detect cathinones in urine using immunoassay technology [58, 59]. Some articles revealed false-positive results by immunoassays; for example, MDPV was cross-reactive with phencyclidine [60]. Therefore, specific detection of cathinones by a commercial immunoassay is not yet possible.

GC–MS-MS detection

Mass spectral profiles of cathinones are very simple in the positive mode of GC–MS, because only the base peak originating from the immonium ion in each molecule is observed. The probable fragmentation pathways of cathinones are described in previous articles [61, 62, 63, 64] and are shown in Figure 5. However, this phenomenon makes the identification of cathinones difficult. To help identify cathinones, other information, such as tandem mass spectrometric data, are usually used because more structural information about the molecule is obtained.

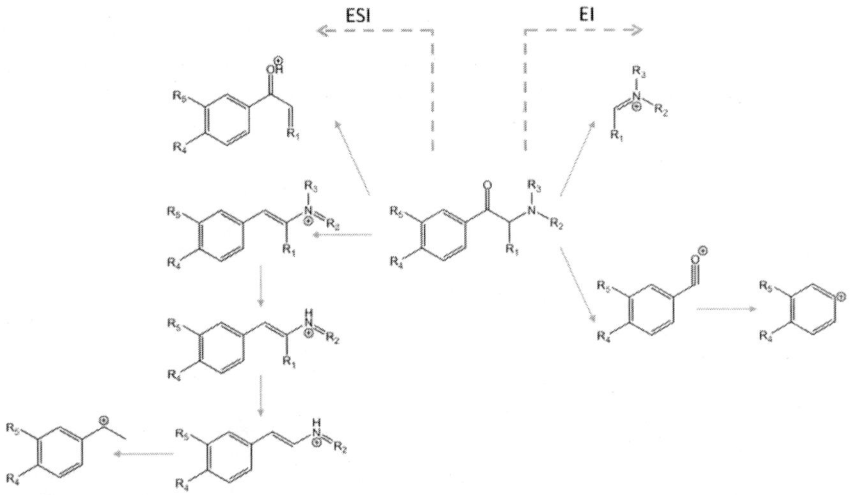

Figure 5. Probable fragmentation pathways of cathinones by electrospray ionization and electron ionization (modified from references [61, 62, 65, 66])

Zuba [61] introduced the systematic identification of cathinones using the mass spectra obtained. First, it should be checked whether the molecular ion is observed. The immonium ion (m/z = 16 + 14 n, n = 1, 2, 3,...) is then checked in the EI spectrum. If the immonium ion is found in the spectrum, the substance could be a straight-chained cathinone. If not, it is checked whether the ion for a pyrrolidine ring is observed (m/z = 70 + 14 n, n = 1, 2, 3,...). If this ion is found in the spectrum, the substance could be a cathinone with a pyrrolidine ring in the molecule [61]. There are various regioisomers in cathinones. To identify the cathinones, it is necessary to assign both the location and length of the bonded alkyl chain. Moreover, the ring-substituted moiety is also needed to be assigned. Zuba [61] demonstrated the following rules: the fragment ions reflecting the ring-substituted moiety are observed at m/z 77 and 105 for a nonsubstituted phenyl ring, at m/z 91 and 119 for a methylphenyl ring, and at m/z 121 and 149 for a methylenedioxyphenyl ring. Matsuta et al. [62] demonstrated the detailed analysis of MS data obtained by GC–EI-MS for identification of cathinones and specified indexing information. However, the ionization rate of the fragments in ring-substituted cathinones is remarkably weaker than that of the immonium ion. Other information obtained by TOFMS or CI-MS is helpful to delineate the molecular structure [63, 64, 65, 66]. The identification of the regioisomer of the fluorinated cathinones was demonstrated using CI-MS [67]. However, this phenomenon was suggested to be limited to these analogs.

The published methods for analysis of cathinones in biological materials are summarized in Table 2 [68, 69, 70, 71, 72, 73, 74, 75, 76, 77, 78, 79]. Simple LLE is usually used for extraction of cathinones from biological materials. The chromatographic conditions are also simple and do not usually require a special technique.

Table 2. GC–MS conditions for cathinones in biological materials

Target(s)	Sample(s)	Purification	Derivatization(s)	Column(s)	LOD(s) (ng/ml)	Linear range (ng/ml)	Reference
MDPV, metabolites	Urine	SPE (hydrolysis)	Methyl, acetyl, trimethylsilyl	HP-1 (12 m, 0.2 mm ID, 0.33 μm) (Agilent)	–	–	[68]
MDPV, metabolites	Cellular fraction, urine	LLE (hydrolysis)	Trimethylsilyl	5 % Phenyl-methylsilicone (17 m, 0.2 mm ID, 0.33 μm) (J and W)	2	10–2,000	[69]
Mephedrone, MDPV	Blood, urine	LLE		DB-1 (30 m, 0.32 mm ID, 0.25 μm) (Agilent)	–	–	[70]
MDPV	Urine	LLE	Heptafluorobutyryl	HP-5MS (12 m, 0.2 mm ID, 0.33 μm) (Agilent), ZB-5MS (12 m, 0.2 mm ID, 0.33 μm) (Phenomenex)	10	20–2,000	[71]
Methylone	Blood	LLE	Heptafluorobutyryl	RTx-5 MS (30 m, 0.25 mm ID, 0.25 μm) (Restek)	50	100–2,000	[72]
α-PVP, pyrovalerone (PV), MDPV	Blood	SPME		InertCap 5 (30 m, 0.25 mm ID, 0.25 μm) (GL Sciences)	0.5 (PV, PVP), 1.0 (MDPV)	1–200	[73]
MDPV, α-PVP, α-PBP	Blood	LLE (Extrelut)		InertCap 5MS/NP (30 m, 0.25 mm ID, 0.25 μm) (GL Sciences)	1	2–2,000	[74]
MDPV	Blood, tissue, urine	SPE		Zebron Guardian ZB-50 (10 m, 0.18 mm ID, 0.18 μm) (Phenomenex)	–	10–2,000	[75]
MDPV	Blood, urine	LLE		Rtx-5 ms (30 m, 0.25 mm ID, 0.25 μm) (Restek)	–	–	[76]
3,4-Dimethylmethcathinone, metabolites	Urine	LLE	Trifluoroacetyl	DB-5MS (30 m, 0.25 mm ID, 0.25 μm) (Agilent)	–	–	[77]
16 Synthetic cathinones	Urine	LLE	Trifluoroacetyl	CP7684 (10 m, 0.15 mm ID, 0.12 μm) (Agilent)	–	–	[78]
α-PVP, metabolites	Urine	LLE	Trimethylsilyl	DB-5MS (30 m, 0.25 mm ID, 0.25 μm) (Agilent)	–	–	[79]

LC–MS-MS detection

The strategy for the detection of cathinones by LC–MS-MS is almost same as that for synthetic cannabinoids; almost all methods use MRM or SRM mode for sensitive determination. The probable fragmentation pathways are shown in Figure 5. The $[M+H]^+$ ion is selected as a precursor ion, and three product ions that reflect the chemical structures of the cathinones are selected. Using this method, 30–50 drugs are monitored simultaneously in samples [80, 81, 82]. The published methods for analysis of cathinones in biological materials are summarized in Table 3 [68, 69, 74, 75, 77, 79, 82, 83, 84, 85, 86, 87, 88, 89, 90, 91, 92, 93,

94, 95, 96, 97, 98]. Simple LLE is usually used for the extraction of cathinones from biological materials. The chromatographic conditions are also simple and do not require a special technique.

Table 3. LC–MS or LC–MS-MS conditions for cathinones in biological materials

Target(s)	Sample(s)	Purification(s)	Column(s)	Mobile phase(s)	LOD(s) (ng/ml or g)	Linear range (ng/ml or g)	Reference
MDPV, metabolites	Urine	SPE (hydrolysis)	Hypersil Gold column (10 mm, 2.1 mm ID, 1.9 μm) (Thermo Scientific)	10 mM ammonium formate (0.1 % formic acid), acetonitrile (0.1 % formic acid)			[68]
MDPV, metabolites	Cellular fraction, urine	LLE (hydrolysis)	Zorbax Eclipse Plus C18 (100 mm, 2.1 mm ID, 1.8 μm) (Agilent)	0.1 % formic acid, acetonitrile (0.1 % formic acid)			[69]
MDPV	Serum	SPE	Phenyl-hexyl (50 mm, 3.0 mm ID, 3 μm) (Phenomenex)	10 mM ammonium acetate (0.1 % formic acid), methanol	3	10–500	[83]
Mephedrone	Plasma	LLE	Synergi Fusion (150 mm, 4.6 mm ID)(Phenomenex); Spherisorb (150 mm, 4.6 mm ID) (Waters)	10 % acetonitrile (25 mM triethylammonium phosphate buffer), isocratic	39	78–10,000	[84]
9 Cathinones	Blood	PP	Prodigy Phenyl-3 (150 mm, 2.0 mm ID, 5 μm) (Phenomenex)	0.1 % formic acid, methanol	0.5–3	10–400	[82]
7 Cathinones	Hair	LLE	Kinetex PFP (50 mm, 2 mm ID, 2.6 μm) (Phenomenex)	5 mM ammonium formate (pH 3.5), methanol (5 mM ammonium formate)	10–50 pg/mg		[85]
Butylone	Blood, liver	SPE	Allure PFP (50 mm, 2.1 mm ID, 5 μm) (Restek)	0.02 % formic acid (2 mM of ammonium formate), acetonitrile	25 (blood)	50–2,000 (blood)	[86]
4-Methylethcathinone	Blood, urine	LLE	Zorbax SB-C18 (50 mm, 2.1 mm ID, 1.8 μm) (Agilent)	0.1 % formic acid, acetonitrile (0.1 % formic acid)	0.96 (blood), 0.68 (urine)	10–1,000	[87]
MDPV, α-PVP, α-PBP	Hair	LLE (Extrelut)	Phenyl-hexyl (150 mm, 2.1 mm ID, 3 μm) (Agilent)	10 mM ammonium formate (0.1 % formic acid, pH 3.3)/methanol (65:35), isocratic	0.02 ng/10-mm	0.05–50 ng/10-mm	[74]
Mephedrone	Blood	PP	Zorbax SB-C18 (50 mm, 2.1 mm ID, 1.8 μm) (Agilent)	0.1 % formic acid, acetonitrile (0.1 % formic acid)	0.08	1–100	[88]
MDPV, mephedrone	Blood, plasma, urine	SPE	Kinetex PFP (50 mm, 2.1 mm ID, 1.8 μm) (Phenomenex)	2 mM ammonium formate (2 % formic acid), acetonitrile (0.1 % formic acid)	2	5–2,000	[89]
Mephedrone	Blood, urine	LLE	Zorbax SB-C18 (150 mm, 2.1 mm ID, 3.5 μm) (Agilent)	0.1 % formic acid, methanol (0.1 % formic acid)	1 (blood), 2 (urine)	20–2,000	[90]
10 Cathinones	Blood, other specimens	LLE	Zorbax XDB-C18 (150 mm, 4.6 mm ID, 5 μm) (Agilent)	5 mM ammonium acetate, methanol/acetonitrile	–	5–200	[91]
MDPV	Hair	SPE	Zorbax Eclipse Plus C18 (50 mm, 2.1 mm, 1.8 μm) (Agilent)	0.1 % formic acid, acetonitrile	2.0 pg/mg	2–3,000 pg/mg	[75]
MDPV	Blood	PP	Zorbax SB-C18 (50 mm, 2.1 mm ID, 1.8 μm) (Agilent)	0.1 % formic acid, acetonitrile (0.1 % formic acid)	0.5	5–500	[92]
Buphedrone	Blood	PP	Zorbax SB-C18 (50 mm, 2.1 mm ID, 1.8 μm) (Agilent)	0.1 % formic acid, acetonitrile (0.1 % formic acid)	0.3	1–1,000	[93]
3,4-Dimethylmethcathinone, metabolites	Urine	PP (hydrolysis)	L-column2 ODS (150 mm, 1.5 mm ID, 5 μm) (Chemicals Evaluation and Research Institute)	10 mM ammonium formate buffer (pH 5), methanol	–	10–5,000	[77]
MDPV, metabolites	Plasma	PP (hydrolysis)	Synergy polar-RP (100 mm, 2 mm ID, 2.5 μm) (Phenomenex)	0.1 % formic acid, acetonitrile (0.1 % formic acid)	0.1	0.25–1,000	[94]
α-PBP	Blood, urine, tissues	QuEChERS	Zorbax Eclipse Plus C18 (100 mm, 2.1 mm ID, 1.8 μm) (Agilent)	10 mM ammonium formate (0.1 % formic acid), acetonitrile	0.05 (blood, urine), 0.1 (tissues)	8.6–4,280	[95]
3,4-Dimethylmethcathinone, metabolites	Blood, urine	QuEChERS	Shim-pack XR-ODS III (50 mm, 2.0 mm ID, 1.6 μm) (Shimadzu); L-column2 ODS (150 mm, 1.5 mm ID, 5 μm) (Chemical Evaluation and Reaearch Institute)	10 mM ammonium formate, methanol; 10 mM ammonium formate (pH 5.0), methanol	1.03 (blood), 1.37 (urine)	5–400	[96]
MDPV metabolattes	Urine	PP, LLE, SPE (hydrolysis)	Atlantis T3 (150 mm, 2.1 mm) (Waters)	10 mM ammonium formate buffer (0.1 % formic acid), acetonitrile (0.1 % formic acid)			[97]
α-PVP, metabolites	Urine	PP (hydrolysis)	L-column2 ODS (150 mm, 1.5 mm ID, 5 μm) (Chemicals Evaluation and Research Institute)	10 mM ammonium formate (pH 5), methanol	–	10–10,000	[79]
PV9	Blood, urine	QuEChERS	Zorbax Eclipse Plus C18 (100 mm, 2.1 mm ID, 1.8 μm) (Agilent)	10 mM ammonium formate (0.1 % formic acid), acetonitrile	0.05	10–1,000	[98]

In the same way as identification by GC–MS, other information obtained by TOFMS or tandem MS is needed to clarify the molecular structure. An authentic drug or library database is needed to identify the drugs. Moreover, the probable fragmentation pathways of cathinones are described in the previous articles [61, 65, 66, 99]. These data are helpful in identifying the drugs.

As described in the section on synthetic cannabinoids, we occasionally encounter a dubious product that contains more than several milligrams of an unknown cathinone-like compound. In such a case, GC–MS, LC–MS, high-resolution MS (or TOFMS), and finally NMR spectroscopy are used to clarify the detailed chemical structure of the compound.

CONCENTRATIONS IN THE CASES OF ABUSE

Synthetic cannabinoids

The common method of consumption of synthetic cannabinoids is smoking, which is the same as for conventional cannabis. The maximum concentrations of synthetic cannabinoids in serum are reached in less than 10 min after smoking [38]. The drugs absorbed in the body are metabolized smoothly, and the concentrations decrease rapidly. Moreover, there is also a report that cannabinoids accumulate in the adipose tissue because of their high lipophilicity [52]. Therefore, detection of the drug from serum is usually difficult. Synthetic cannabinoids absorbed in the human body are metabolized to hydroxyl or carboxyl derivatives of the aromatic ring or *N*-alkyl side chain [100]. It is difficult to identify the parent drug and its metabolites in blood by GC–MS alone because the fragmentation of the metabolites is similar to that of the parent drug and analogs. Moreover, the concentration of the unchanged synthetic cannabinoids in blood is very low, and the number of metabolites that are commercially available is small. Low sensitivity is a limitation for the determination of synthetic cannabinoids in blood by GC–MS.

Although the concentration is influenced by the sampling time after drug intake and by the intake amount, concentrations of these drugs in serum were reported in the range of 0.1–190 ng/ml in poisoning cases [46]. In fatal cases, the concentrations of the drugs in blood were 0.1–199 ng/ml for JWH-018 and 0.1–68.3 ng/ml for JWH-073 [50], 12 ng/ml for AM-2201 [100], 1.1–1.5 ng/ml for 5F-PB-22 [53], and 12.4 ng/ml for MAM-2201 [52].

Cathinones

Unlike synthetic cannabinoids, the most common method of consumption of cathinones are insufflation (snorting) or ingestion. Inhalation, sublingual and rectal administration, and intramuscular or intravenous injection have also been reported. Unlike synthetic cannabinoids, the concentration of cathinones in blood is thought to vary because of the many modes of administration used by abusers. Only the blood concentration at one point and at several points have been quantified, and there is no report on continuous monitoring of the profile of the drug concentration in blood. The fatal concentration of the drug in blood was reported to be around 400 ng/ml [75]. The stability of cathinones in blood samples is clearly influenced by pH, as well as in the final extracts. In blood samples preserved with NaF/potassium oxalate, the measured concentrations of cathinone, methcathinone, ethcathinone, mephedrone, and flephedrone declined by ca. 30 % after 2 days of storage at 20 °C [82].

Some groups have studied the metabolic pathways of cathinones [68, 94, 101, 102, 103]. Unlike synthetic cannabinoids, the parent cathinones are detected easily in biological materials and are selected as the target because the unchanged parent drugs are rapidly excreted in urine. Cathinones are ionized in the body, and the reabsorption rate is low in the kidney because of low hydrophobicity. The excretion profile of α-PBP and α-PVP in human urine was determined after an intravenous injection, and the elimination half-life in urine was approximately 12 h. Moreover, the excreted amount in urine was influenced by urinary pH, like a psycho-stimulant [104]. To analyze these drugs in biological materials, it is necessary to remove endogenous substances from each sample and enrich the content of the drug. As shown in Table 3, LLE is usually used for extraction of the drugs from biological materials. The quantification of the metabolites is important to predict the hazardous properties of the metabolites. However, because there are few metabolites marketed, no detailed study about their pharmacological activity or toxicity has been conducted.

In fatal cases, the concentrations of the drugs in blood were: 560–3,300 [72], 272 [105], and 60–1,120 ng/ml [106] for methylone; 1.2–22 [107], 5.1 [108], and 5.5 μg/ml [88] for mephedrone; 55.2 ng/ml for α-PBP [95]; 486 [73] and 654 ng/ml [109] for α-PVP; 180 ng/ml [98] for PV9; 170 [70], 82 [110], 1,200 [74], 440 [75], 17–38 [92], and 700 ng/ml [111] for MDPV.

CONCLUSIONS

The number of abusers of synthetic cannabinoids and cathinones has increased remarkably worldwide. The chemical structures of the distributed drugs are skillfully changed so that the drugs may pass through screenings for detection. Simple screening methods are required for detection of these drugs in seized and biological materials. There are currently no commercial kits or devices for the routine screening of these drugs. Colorimetric, immunochemical, and chromatographic methods have been introduced in this review; a suitable method must be chosen for each laboratory. Although various human sample matrices are available for testing, urine and blood are of the first choices. However, many of these drugs, especially unchanged synthetic cannabinoids, exist in urine and blood for only a short period. Therefore, other matrices that can prove the consumption of these drugs, such as hair and saliva, are likely to receive more attention in the future.

ACKNOWLEDGMENTS

The authors are thankful to Professors Akira Ishii, Chief Editor, and Osamu Suzuki, Emeritus Chief Editor, Forensic Toxicology, for providing us an opportunity to write this review.

Conflict of Interest

There are no financial or other relations that could lead to a conflict of interest.

REFERENCES

1. Auwärter V, Dresen S, Weinmann W, Müller M, Pütz M, Ferreirós N (2009) 'Spice' and other herbal blends: harmless incense or cannabinoid designer drugs? J Mass Spectrom 44:832–837
2. Maxwell JC (2014) Psychoactive substances—some new, some old: a scan of the situation in the US. Drug Alcohol Depend 134:71–77
3. Kikura-Hanajiri R, Uchiyama N, Kawamura M, Goda Y (2013) Changes in the prevalence of synthetic cannabinoids and cathinone derivatives in Japan until early 2012. Forensic Toxicol 31:44–53
4. Chung H, Choi H, Heo S, Kim E, Lee J (2014) Synthetic cannabinoids abused in South Korea: drug identifications by the National Forensic Service from 2009 to June 2013. Forensic Toxicol 32:82–88
5. Seely KA, Patton AL, Moran CL, Womack ML, Prather PL, Fantegrossi WE, Radominska-Pandya A, Endres GW, Channell KB, Smith NH, McCain KR, James LP, Moran JH (2013) Forensic investigation of K2, Spice, and "bath salt" commercial

preparations: a 3-year study of new designer drug products containing synthetic cannabinoid, stimulant, and hallucinogenic compounds. Forensic Sci Int 233:416–422

6. Barratt MJ, Cakic V, Lenton S (2013) Patterns of synthetic cannabinoid use in Australia. Drug Alcohol Rev 32:141–146
7. Caudevilla-Gálligo F, Ventura M, Indave Ruiz BI, Fornís I (2013) Presence and composition of cathinone derivatives in drug samples taken from a drug test service in Spain (2010–2012). Hum Psychopharmacol 28:341–344
8. Presley BC, Jansen-Varnum SA, Logan BK (2013) Analysis of synthetic cannabinoids in botanical materials: a review of analytical methods and findings. Forensic Sci Rev 25:27–46
9. Elsohly MA, Gul W, Wanas AS, Radwan MM (2014) Synthetic cannabinoids: analysis and metabolites. Life Sci 97:78–90
10. Favretto D, Pascali JP, Tagliaro F (2013) New challenges and innovation in forensic toxicology: focus on the "New Psychoactive Substances". J Chromatogr A 1287:84–95
11. Isaacs RCA (2014) A structure–reactivity relationship driven approach to the identification of a color test protocol for the presumptive indication of synthetic cannabimimetic drugs of abuse. Forensic Sci Int 242:135–141
12. Zaitsu K, Katagi M, Nakanishi K, Shima N, Kamata H, Kamata T, Nishioka H, Miki A, Tatsuno M, Iwamura T, Sato T, Tsuchihashi H, Suzuki K (2011) Comprehensive analytical methods of the synthetic cannabinoids appearing in the illicit drug market (in Japanese with English abstract). Jpn J Forensic Sci Tech 16:73–90
13. Logan BK, Reinhold LE, Xu A, Diamond FX (2012) Identification of synthetic cannabinoids in herbal incense blends in the US. J Forensic Sci 57:1168–1180
14. Arntson A, Ofsa B, Lancaster D, Simon JR, McMullin M, Logan B (2013) Validation of a novel immunoassay for the detection of synthetic cannabinoids and metabolites in urine specimens. J Anal Toxicol 37:284–290
15. Hudson S, Ramsey J (2011) The emergence and analysis of synthetic cannabinoids. Drug Test Anal 3:466–478
16. Choi H, Heo S, Choe S, Yang W, Park Y, Kim E, Chung H, Lee J (2013) Simultaneous analysis of synthetic cannabinoids in the materials seized during drug trafficking using GC–MS. Anal Bioanal Chem 405:3937–3944
17. Uchiyama N, Kawamura M, Kikura-Hanajiri R, Goda Y (2012) Identification of two new-type synthetic cannabinoids, *N*-(1-adamantyl)-1-pentyl-1*H*-indole-3-carboxamide (APICA) and *N*-(1-adamantyl)-1-pentyl-1*H*-indazole-3-carboxamide (APINACA), and detection of five synthetic cannabinoids, AM-1220, AM-2233, AM-1241, CB-13 (CRA-13), and AM-1248, as designer drugs in illegal products. Forensic Toxicol 30:114–125
18. Uchiyama N, Matsuda S, Wakana D, Kikura-Hanajiri R, Goda Y (2013) New cannabimimetic indazole derivatives, *N*-(1-amino-3-methyl-1-oxobutan-2-yl)-1-pentyl-1*H*-indazole-3-carboxamide (AB-PINACA) and *N*-(1-amino-3-methyl-1-oxobutan-2-yl)-1-(4-fluorobenzyl)-1*H*-indazole-3-carboxamide (AB-FUBINACA) identified as designer drugs in illegal products. Forensic Toxicol 31:93–100
19. Uchiyama N, Matsuda S, Kawamura M, Kikura-Hanajiri R, Goda Y (2013) Two new-type cannabimimetic quinolinyl carboxylates, QUPIC and QUCHIC, two new cannabimimetic carboxamide derivatives, ADB-FUBINACA and ADBICA, and five synthetic cannabinoids detected with a thiophene derivative α-PVT and an opioid receptor agonist AH-7921 identified in illegal products. Forensic Toxicol 31:223–240
20. Uchiyama N, Shimokawa Y, Matsuda S, Kawamura M, Kikura-Hanajiri R, Goda Y (2014) Two new synthetic cannabinoids, AM-2201 benzimidazole analog (FUBIMINA) and (4-methylpiperazin-1-yl)(1-pentyl-1*H*-indol-3-yl)methanone (MEPIRAPIM), and three phenethylamine derivatives, 25H-NBOMe 3,4,5-trimethoxybenzyl analog, 25B-NBOMe, and 2C-N-NBOMe, identified in illegal products. Forensic Toxicol 32:105–115

21. Uchiyama N, Shimokawa Y, Kawamura M, Kikura-Hanajiri R, Hakamatsuka T (2014) Chemical analysis of a benzofuran derivative, 2-(2-ethylaminopropyl)benzofuran (2-EAPB), eight synthetic cannabinoids, five cathinone derivatives, and five other designer drugs newly detected in illegal products. Forensic Toxicol 32:266–281
22. Cox AO, Daw RC, Mason MD, Grabenauer M, Pande PG, Davis KH, Wiley JL, Stout PR, Thomas BF, Huffman JW (2012) Use of SPME-HS-GC–MS for the analysis of herbal products containing synthetic cannabinoids. J Anal Toxicol 36:293–302
23. Ernst L, Krüger K, Lindigkeit R, Schiebel HM, Beuerle T (2012) Synthetic cannabinoids in "spice-like" herbal blends: first appearance of JWH-307 and recurrence of JWH-018 on the German market. Forensic Sci Int 222:216–222
24. Kneisel S, Westphal F, Bisel P, Brecht V, Broecker S, Auwärter V (2012) Identification and structural characterization of the synthetic cannabinoid 3-(1-adamantoyl)-1-pentylindole as an additive in 'herbal incense'. J Mass Spectrom 47:195–200
25. Harris DN, Hokanson S, Miller V, Jackson GP (2014) Fragmentation differences in the EI spectra of three synthetic cannabinoid positional isomers: JWH-250, JWH-302, and JWH-201. Int J Mass Spectrom 368:23–29
26. DeRuiter J, Smith FT, Abdel-Hay K, Clark CR (2014) Analytical differentiation of 1-alkyl-3-acylindoles and 1-acyl-3-alkylindoles: isomeric synthetic cannabinoids. Anal Chem 86:3801–3808
27. Shevyrin V, Melkozerov V, Nevero A, Eltsov O, Morzherin Y, Shafran Y (2014) 3-Naphthoylindazoles and 2-naphthoylbenzoimidazoles as novel chemical groups of synthetic cannabinoids: chemical structure elucidation, analytical characteristics and identification of the first representatives in smoke mixtures. Forensic Sci Int 242:72–80
28. Ibáñez M, Bijlsma L, van Nuijs ALN, Sancho JV, Haro G, Covaci A, Hernández F (2013) Quadrupole-time-of-flight mass spectrometry screening for synthetic cannabinoids in herbal blends. J Mass Spectrom 48:685–694
29. Grigoryev A, Kavanagh P, Melnik A, Savchuk S, Simonov A (2013) Gas and liquid chromatography–mass spectrometry detection of the urinary metabolites of UR-144 and its major pyrolysis product. J Anal Toxicol 37:265–276
30. Tsujikawa K, Yamamuro T, Kuwayama K, Kanamori T, Iwata YT, Inoue H (2014) Thermal degradation of a new synthetic cannabinoid QUPIC during analysis by gas chromatography–mass spectrometry. Forensic Toxicol 32:201–207
31. Sekuła K, Zuba D, Stanaszek R (2012) Identification of naphthoylindoles acting on cannabinoid receptors based on their fragmentation patterns under ESI–QTOFMS. J Mass Spectrom 47:632–643
32. Kneisel S, Auwärter V (2012) Analysis of 30 synthetic cannabinoids in serum by liquid chromatography-electrospray ionization tandem mass spectrometry after liquid–liquid extraction. J Mass Spectrom 47:825–835
33. Scheidweiler KB, Huestis MA (2014) Simultaneous quantification of 20 synthetic cannabinoids and 21 metabolites, and semi-quantification of 12 alkyl hydroxy metabolites in human urine by liquid chromatography–tandem mass spectrometry. J Chromatogr A 1327:105–117
34. Wohlfarth A, Scheidweiler KB, Chen X, Liu H, Huestis MA (2013) Qualitative confirmation of 9 synthetic cannabinoids and 20 metabolites in human urine using LC–MS/MS and library search. Anal Chem 85:3730–3738
35. Kneisel S, Auwärter V, Kempf J (2013) Analysis of 30 synthetic cannabinoids in oral fluid using liquid chromatography-electrospray ionization tandem mass spectrometry. Drug Test Anal 5:657–669
36. Kneisel S, Speck M, Moosmann B, Corneillie TM, Butlin NG, Auwärter V (2013) LC/ESI–MS/MS method for quantification of 28 synthetic cannabinoids in neat oral fluid and its application to preliminary studies on their detection windows. Anal Bioanal Chem 405:4691–4706

37. Hutter M, Kneisel S, Auwärter V, Neukamm MA (2012) Determination of 22 synthetic cannabinoids in human hair by liquid chromatography-tandem mass spectrometry. J Chromatogr B 903:95–101
38. Teske J, Weller JP, Fieguth A, Rothämel T, Schulz Y, Tröger HD (2010) Sensitive and rapid quantification of the cannabinoid receptor agonist naphthalen-1-yl-(1-pentylindol-3-yl)methanone (JWH-018) in human serum by liquid chromatography–tandem mass spectrometry. J Chromatogr B 878:2659–2663
39. Kacinko SL, Xu A, Homan JW, McMullin MM, Warrington DM, Logan BK (2011) Development and validation of a liquid chromatography–tandem mass spectrometry method for the identification and quantification of JWH-018, JWH-073, JWH-019, and JWH-250 in human whole blood. J Anal Toxicol 35:386–393
40. Yeakel JK, Logan BK (2013) Blood synthetic cannabinoid concentrations in cases of suspected impaired driving. J Anal Toxicol 37:547–551
41. Dresen S, Kneisel S, Weinmann W, Zimmermann R, Auwärter V (2011) Development and validation of a liquid chromatography–tandem mass spectrometry method for the quantitation of synthetic cannabinoids of the aminoalkylindole type and methanandamide in serum and its application to forensic samples. J Mass Spectrom 46:163–171
42. Moran CL, Le VH, Chimalakonda KC, Smedley AL, Lackey FD, Owen SN, Kennedy PD, Endres GW, Ciske FL, Kramer JB, Kornilov AM, Bratton LD, Dobrowolski PJ, Wessinger WD, Fantegrossi WE, Prather PL, James LP, Radominska-Pandya A, Moran JH (2011) Quantitative measurement of JWH-018 and JWH-073 metabolites excreted in human urine. Anal Chem 83:4228–4236
43. Chimalakonda KC, Moran CL, Kennedy PD, Endres GW, Uzieblo A, Dobrowolski PJ, Fifer EK, Lapoint J, Nelson LS, Hoffman RS, James LP, Radominska-Pandya A, Moran JH (2011) Solid-phase extraction and quantitative measurement of omega and omega-1 metabolites of JWH-018 and JWH-073 in human urine. Anal Chem 83:6381–6388
44. de Jager AD, Warner JV, Henman M, Ferguson W, Hall A (2012) LC–MS/MS method for the quantitation of metabolites of eight commonly used synthetic cannabinoids in human urine—an Australian perspective. J Chromatogr B 897:22–31
45. Yanes EG, Lovett DP (2012) High-throughput bioanalytical method for analysis of synthetic cannabinoid metabolites in urine using salting-out sample preparation and LC–MS/MS. J Chromatogr B 909:42–50
46. Hermanns-Clausen M, Kneisel S, Szabo B, Auwärter V (2013) Acute toxicity due to the confirmed consumption of synthetic cannabinoids: clinical and laboratory findings. Addiction 108:534–544
47. Hermanns-Clausen M, Kneisel S, Hutter M, Szabo B, Auwärter V (2013) Acute intoxication by synthetic cannabinoids—four case reports. Drug Test Anal 5:790–794
48. Musshoff F, Madea B, Kernbach-Wighton G, Bicker W, Kneisel S, Hutter M, Auwärter V (2014) Driving under the influence of synthetic cannabinoids ("Spice"): a case series. Int J Legal Med 128:59–64
49. Hutter M, Broecker S, Kneisel S, Auwärter V (2012) Identification of the major urinary metabolites in man of seven synthetic cannabinoids of the aminoalkylindole type present as adulterants in herbal mixtures using LC–MS/MS techniques. J Mass Spectrom 47:54–65
50. Shanks KG, Dahn T, Terrell AR (2012) Detection of JWH-018 and JWH-073 by UPLC-MS-MS in postmortem whole blood casework. J Anal Toxicol 36:145–152
51. Adamowicz P, Zuba D, Sekuła K (2013) Analysis of UR-144 and its pyrolysis product in blood and their metabolites in urine. Forensic Sci Int 233:320–327
52. Saito T, Namera A, Miura N, Ohta S, Miyazaki S, Osawa M, Inokuchi S (2013) A fatal case of MAM-2201 poisoning. Forensic Toxicol 31:333–337

53. Behonick G, Shanks KG, Firchau DJ, Mathur G, Lynch CF, Nashelsky M, Jaskierny DJ, Meroueh C (2014) Four postmortem case reports with quantitative detection of the synthetic cannabinoid, 5F-PB-22. J Anal Toxicol 38:559–562
54. Mazzarino M, de la Torre X, Botrè F (2014) A liquid chromatography–mass spectrometry method based on class characteristic fragmentation pathways to detect the class of indole-derivative synthetic cannabinoids in biological samples. Anal Chim Acta 837:70–82
55. Wood DM, Dargan PI (2012) Mephedrone (4-methylmethcathinone): what is new in our understanding of its use and toxicity. Prog Neuro-Psychopharmacol Biol Psychiatry 39:227–233
56. Uchiyama N, Kawamura M, Kamakura H, Kikura-Hanajiri R, Goda Y (2008) Analytical data of designated substances (Shitei-yakubutsu) controlled by the pharmaceutical affairs law in Japan, part II: color test and TLC (in Japanese with English abstract). Yakugaku Zasshi 128:981–987
57. Dal Cason TA (1997) The characterization of some 3,4-methylenedioxycathinone (MDCATH) homologs. Forensic Sci Int 87:9–53
58. Swortwood MJ, Hearn WL, DeCaprio AP (2014) Cross-reactivity of designer drugs, including cathinone derivatives, in commercial enzyme-linked immunosorbent assays. Drug Test Anal 6:716–727
59. Ellefsen KN, Anizan S, Castaneto MS, Desrosiers NA, Martin TM, Klette KL, Huestis MA (2014) Validation of the only commercially available immunoassay for synthetic cathinones in urine: randox drugs of abuse V biochip array technology. Drug Test Anal 6:728–738
60. Penders TM, Gestring RE, Vilensky DA (2012) Intoxication delirium following use of synthetic cathinone derivatives. Am J Drug Alcohol Abuse 38:616–617
61. Zuba D (2012) Identification of cathinones and other active components of legal highs by mass spectrometric methods. Trends Anal Chem 32:15–30
62. Matsuta S, Katagi M, Nishioka H, Kamata H, Sasaki K, Shima N, Kamata T, Miki A, Tatsuno M, Zaitsu K, Tsuboi K, Tsuchihashi H, Suzuki K (2014) Structural characterization of cathinone-type designer drugs by EI mass spectrometry (in Japanese with English abstract). Jpn J Forensic Sci Tech 19:77–89
63. Brandt SD, Sumnall HR, Measham F, Cole J (2010) Analyses of second-generation 'legal highs' in the UK: initial findings. Drug Test Anal 2:377–382
64. Brandt SD, Freeman S, Sumnall HR, Measham F, Cole J (2011) Analysis of NRG 'legal highs' in the UK: identification and formation of novel cathinones. Drug Test Anal 3:569–575
65. Fornal E (2013) Identification of substituted cathinones: 3,4-methylenedioxy derivatives by high performance liquid chromatography–quadrupole time of flight mass spectrometry. J Pharm Biomed Anal 81–82:13–19
66. Jankovics P, Váradi A, Tölgyesi L, Lohner S, Németh-Palotás J, Koszegi-Szalai H (2011) Identification and characterization of the new designer drug 4′-methylethcathinone (4-MEC) and elaboration of a novel liquid chromatography–tandem mass spectrometry (LC–MS/MS) screening method for seven different methcathinone analogs. Forensic Sci Int 210:213–220
67. Westphal F, Junge T (2012) Ring positional differentiation of isomeric *N*-alkylated fluorocathinones by gas chromatography/tandem mass spectrometry. Forensic Sci Int 223:97–105
68. Meyer MR, Du P, Schuster F, Maurer HH (2010) Studies on the metabolism of the α-pyrrolidinophenone designer drug methylenedioxy-pyrovalerone (MDPV) in rat and human urine and human liver microsomes using GC–MS and LC–high-resolution MS and its detectability in urine by GC–MS. J Mass Spectrom 45:1426–1442
69. Strano-Rossi S, Cadwallader AB, de la Torre X, Botrè F (2010) Toxicological determination and in vitro metabolism of the designer drug methylenedioxypyrovalerone (MDPV) by gas chromatography/mass

spectrometry and liquid chromatography/quadrupole time-of-flight mass spectrometry. Rapid Commun Mass Spectrom 24:2706–2714
70. Spiller HA, Ryan ML, Weston RG, Jansen J (2011) Clinical experience with and analytical confirmation of "bath salts" and "legal highs" (synthetic cathinones) in the US. Clin Toxicol 49:499–505
71. Ojanperä IA, Heikman PK, Rasanen IJ (2011) Urine analysis of 3,4-methylenedioxypyrovalerone in opioid-dependent patients by gas chromatography–mass spectrometry. Ther Drug Monit 33:257–263
72. Pearson JM, Hargraves TL, Hair LS, Massucci CJ, Frazee CC 3rd, Garg U, Pietak BR (2012) Three fatal intoxications due to methylone. J Anal Toxicol 36:444–451
73. Saito T, Namera A, Osawa M, Aoki H, Inokuchi S (2013) SPME–GC–MS analysis of α-pyrrolidinovaleorophenone in blood in a fatal poisoning case. Forensic Toxicol 31:328–332
74. Namera A, Urabe S, Saito T, Torikoshi-Hatano A, Shiraishi H, Arima Y, Nagao M (2013) A fatal case of 3,4-methylenedioxypyrovalerone poisoning: coexistence of α-pyrrolidinobutiophenone and α-pyrrolidinovalerophenone in blood and/or hair. Forensic Toxicol 31:338–343
75. Wyman JF, Lavins ES, Engelhart D, Armstrong EJ, Snell KD, Boggs PD, Taylor SM, Norris RN, Miller FP (2013) Postmortem tissue distribution of MDPV following lethal intoxication by bath salts. J Anal Toxicol 37:182–185
76. Wright TH, Cline-Parhamovich K, Lajoie D, Parsons L, Dunn M, Ferslew KE (2013) Deaths involving methylenedioxypyrovalerone (MDPV) in Upper East Tennessee. J Forensic Sci 58:1558–1562
77. Shima N, Katagi M, Kamata H, Matsuta S, Nakanishi K, Zaitsu K, Kamata T, Nishioka H, Miki A, Tatsuno M, Sato T, Tsuchihashi H, Suzuki K (2013) Urinary excretion and metabolism of the newly encountered designer drug 3,4-dimethylmethcathinone in humans. Forensic Toxicol 31:101–112
78. Uralets V, Rana S, Morgan S, Ross W (2014) Testing for designer stimulants: metabolic profiles of 16 synthetic cathinones excreted free in human urine. J Anal Toxicol 38:233–241
79. Shima N, Katagi M, Kamata H, Matsuta S, Sasaki K, Kamata T, Nishioka H, Miki A, Tatsuno M, Zaitsu K, Ishii A, Sato T, Tsuchihashi H, Suzuki K (2014) Metabolism of the newly encountered designer drug α-pyrrolidinovalerophenone in humans: identification and quantitation of urinary metabolites. Forensic Toxicol 32:59–67
80. Swortwood MJ, Boland DM, DeCaprio AP (2013) Determination of 32 cathinone derivatives and other designer drugs in serum by comprehensive LC-QQQ-MS/MS analysis. Anal Bioanal Chem 405:1383–1397
81. Concheiro M, Anizan S, Ellefsen K, Huestis MA (2013) Simultaneous quantification of 28 synthetic cathinones and metabolites in urine by liquid chromatography–high resolution mass spectrometry. Anal Bioanal Chem 405:9437–9448
82. Sørensen LK (2011) Determination of cathinones and related ephedrines in forensic whole-blood samples by liquid chromatography-electrospray tandem mass spectrometry. J Chromatogr B 879:727–736
83. Kriikku P, Wilhelm L, Schwarz O, Rintatalo J (2011) New designer drug of abuse: 3,4-methylenedioxypyrovalerone (MDPV). Findings from apprehended drivers in Finland. Forensic Sci Int 210:195–200
84. Maskell PD, De Paoli G, Seneviratne C, Pounder DJ (2011) Mephedrone (4-methylmethcathinone)-related deaths. J Anal Toxicol 35:188–191
85. Rust KY, Baumgartner MR, Dally AM, Kraemer T (2012) Prevalence of new psychoactive substances: a retrospective study in hair. Drug Test Anal 4:402–408
86. Rojek S, Kłys M, Strona M, Maciow M, Kula K (2012) "Legal highs"—toxicity in the clinical and medico-legal aspect as exemplified by suicide with bk-MBDB administration. Forensic Sci Int 222:e1–e6
87. Gil D, Adamowicz P, Skulska A, Tokarczyk B, Stanaszek R (2013) Analysis of 4-MEC in biological and non-biological material-three case reports. Forensic Sci Int 228:e11–e15

88. Adamowicz P, Tokarczyk B, Stanaszek R, Slopianka M (2013) Fatal mephedrone intoxication—a case report. J Anal Toxicol 37:37–42
89. Johnson RD, Botch-Jones SR (2013) The stability of four designer drugs: MDPV, mephedrone, BZP and TFMPP in three biological matrices under various storage conditions. J Anal Toxicol 37:51–55
90. Cosbey SH, Peters KL, Quinn A, Bentley A (2013) Mephedrone (methylmethcathinone) in toxicology casework: a Northern Ireland perspective. J Anal Toxicol 37:74–82
91. Marinetti LJ, Antonides HM (2013) Analysis of synthetic cathinones commonly found in bath salts in human performance and postmortem toxicology: method development, drug distribution and interpretation of results. J Anal Toxicol 37:135–146
92. Adamowicz P, Gil D, Skulska A, Tokarczyk B (2013) Analysis of MDPV in blood—determination and interpretation. J Anal Toxicol 37:308–312
93. Zuba D, Adamowicz P, Byrska B (2013) Detection of buphedrone in biological and non-biological material—two case reports. Forensic Sci Int 227:15–20
94. Anizan S, Ellefsen K, Concheiro M, Suzuki M, Rice KC, Baumann MH, Huestis MA (2014) 3,4-Methylenedioxypyrovalerone (MDPV) and metabolites quantification in human and rat plasma by liquid chromatography–high resolution mass spectrometry. Anal Chim Acta 827:54–63
95. Wurita A, Hasegawa K, Minakata K, Gonmori K, Nozawa H, Yamagishi I, Suzuki O, Watanabe K (2014) Postmortem distribution of α-pyrrolidinobutiophenone in body fluids and solid tissues of a human cadaver. Leg Med 16:241–246
96. Usui K, Aramaki T, Hashiyada M, Hayashizaki Y, Funayama M (2014) Quantitative analysis of 3,4-dimethylmethcathinone in blood and urine by liquid chromatography–tandem mass spectrometry in a fatal case. Leg Med 16:222–226
97. Bertol E, Mari F, Berto RB, Mannaioni G, Vaiano F, Favretto D (2014) A mixed MDPV and benzodiazepine intoxication in a chronic drug abuser: determination of MDPV metabolites by LC–HRMS and discussion of the case. Forensic Sci Int 243:149–155
98. Hasegawa K, Wurita A, Minakata K, Gonmori K, Nozawa H, Yamagishi I, Suzuki O, Watanabe K (2014) Identification and quantitation of a new cathinone designer drug PV9 in an "aroma liquid" product, antemortem whole blood and urine specimens, and a postmortem whole blood specimen in a fatal poisoning case. Forensic Toxicol 32:243–250
99. Bijlsma L, Sancho JV, Hernández F, Niessen WMA (2011) Fragmentation pathways of drugs of abuse and their metabolites based on QTOF MS/MS and MSE^E accurate-mass spectra. J Mass Spectrom 46:865–875
100. Patton AL, Chimalakonda KC, Moran CL, McCain KR, Radominska-Pandya A, James LP, Kokes C, Moran JH (2013) K2 toxicity: fatal case of psychiatric complications following AM2201 exposure. J Forensic Sci 58:1676–1680
101. Springer D, Fritschi G, Maurer HH (2003) Metabolism of the new designer drug alpha-pyrrolidinopropiophenone (PPP) and the toxicological detection of PPP and 4′-methyl-alpha-pyrrolidinopropiophenone (MPPP) studied in rat urine using gas chromatography–mass spectrometry. J Chromatogr B 796:253–266
102. Sauer C, Peters FT, Haas C, Meyer MR, Fritschi G, Maurer HH (2009) New designer drug alpha-pyrrolidinovalerophenone (PVP): studies on its metabolism and toxicological detection in rat urine using gas chromatographic/mass spectrometric techniques. J Mass Spectrom 44:952–964
103. Tyrkkö E, Pelander A, Ketola RA, Ojanperä I (2013) In silico and in vitro metabolism studies support identification of designer drugs in human urine by liquid chromatography/quadrupole-time-of-flight mass spectrometry. Anal Bioanal Chem 405:6697–6709
104. Namera A, Konuma K, Kawamura M, Saito T, Nakamoto A, Yahata M, Ohta S, Miyazaki S, Shiraishi H, Nagao M (2014) Time-course profile of urinary excretion

of intravenously administered α-pyrrolidinovalerophenone and α-pyrrolidinobutiophenone in a human. Forensic Toxicol 32:68–74
105. Kovács K, Tóth AR, Kereszty EM (2012) A new designer drug: methylone related death. Orv Hetil 153:271–276
106. Cawrse BM, Levine B, Jufer RA, Fowler DR, Vorce SP, Dickson AJ, Holler JM (2012) Distribution of methylone in four postmortem cases. J Anal Toxicol 36:434–439
107. Torrance H, Cooper G (2010) The detection of mephedrone (4-methylmethcathinone) in 4 fatalities in Scotland. Forensic Sci Int 202:e62–e63
108. Lusthof KJ, Oosting R, Maes A, Verschraagen M, Dijkhuizen A, Sprong AG (2011) A case of extreme agitation and death after the use of mephedrone in The Netherlands. Forensic Sci Int 206:e93–e95
109. Hasegawa K, Suzuki O, Wurita A, Minakata K, Yamagishi I, Nozawa H, Gonmori K, Watanabe K (2014) Postmortem distribution of α-pyrrolidinovalerophenone and its metabolite in body fluids and solid tissues in a fatal poisoning case measured by LC–MS–MS with the standard addition method. Forensic Toxicol 32:225–234
110. Murray BL, Murphy CM, Beuhler MC (2012) Death following recreational use of designer drug "bath salts" containing 3,4-methylenedioxypyrovalerone (MDPV). J Med Toxicol 8:69–75
111. Kesha K, Boggs CL, Ripple MG, Allan CH, Levine B, Jufer-Phipps R, Doyon S, Chi P, Fowler DR (2013) Methylenedioxypyrovalerone ("bath salts"), related death: case report and review of the literature. J Forensic Sci 58:1654–1659

CHAPTER 3

Pneumomediastinum and Cervical Soft Tissue Emphysema as an Important Vital Sign in Hanging: An Autopsy Based Prospective Study

Pawan Mittal* **and Gaurav Sharma**

Department of Forensic Medicine, B.P.S. Govt. Medical College for Women, Haryana

ABSTRACT

Objective: To determine the frequency and reliability of cervical soft tissue emphysema and pneumomediastinum in autopsy cases of hanging.

Material and Method: We prospectively studied 30 autopsy cases of hanging including 18 males and 12 females with age ranging from 19 to 65 years. The control group consisted of age and sex matched 30 cases that died due to various natural pathologies. Along with this, a small group of five cases consisting of homicidal asphyxial deaths from ligature and manual strangulation was also studied for this purpose. Various other known vital phenomenon in hanging were also recorded in the form of hemorrhages in the subcutaneous tissues, laryngeal hyoid skeleton fractures, conjunctival and facial petechial hemorrhages, hemorrhages over clavicular origin of sternocleidomastoid muscle, carotid intimal tears (Amussat's sign), Simon's bleeding and hemorrhages in the respiratory auxiliary muscles of respiration. The dissection and preparation of the neck was done in layers. None of the cases underwent cardiopulmonary resuscitation. Signs of putrefaction were absent.

Results: The soft tissue emphysema was seen in 22 cases (73.3%) of hanging as fine soapy air bubbles in the mediastinum that could be traced further in the superficial and deep tissues of the neck up to the level of ligature mark beyond which it was missing. Among two other groups, 3 controls depicted this finding while no traces of emphysema were seen in any of the strangulation cases.

Conclusion: Cervical soft tissue emphysema along with pneumomediastinum is a frequent and important vital finding in cases of death due to hanging. Whenever possible, the vitality of the sign should be considered in conjunction with other known vital signs of hanging. However a proper care should be taken during dissection of cervical and thoracic structures to avoid artefactual introduction of air in the soft tissues. The interpretation must be based on sound medical grounds and observations.

Keywords: Hanging; Emphysema; Pneumomediastinum; Sternocleidomastoid; Air; Autopsy

INTRODUCTION

Pneumomediastinum or mediastinal emphysema implies the presence of air in the mediastinum. It is a well-known phenomenon in the clinical practice where it has been described in multiple conditions causing alveolar rupture such as Valsalva maneuver, parturition, postoperative vomiting, acute exacerbation of bronchial asthma [1-4]. The clinical diagnosis of the mediastinal and soft tissue emphysema is primarily based upon radiological findings. There is a paucity of forensic literature highlighting the autopsy demonstration of such emphysema. Limited forensic studies are available that have shown this finding to be a vital determinant in making the conclusion of hanging at autopsy [5,6]. Postmortem imaging has demonstrated the finding of pneumomediastinum and CSTE, with gas collection and air following visceral spaces around trachea, oesophagus, large vessels of the neck and even the cervical spinal canal in cases of hanging using MSCT and MRI protocols [6,7]. The proposed mechanism for this finding includes either direct or indirect trauma to the cervical airways because of ligature pressure and stretching, or rupture of terminal alveoli caused by elevated intra-thoracic pressure with subsequent dissection of air in the mediastinal and cervical soft tissues around visceral structures [8,9] . The latter mechanism is known as Macklin effect [10]. The aim of this study was to determine the frequency of pneumomediastinum and CSTE at autopsy in cases of deaths due to hanging.

MATERIAL AND METHODS

A prospective study was carried out that included 30 cases of suicidal hanging that were brought for autopsy to our forensic medicine department over a period from October 2015 to August 2016. Detailed information of the circumstances was gained from police agency, enquiry from relatives and witnesses, history of previous suicidal attempts and suicidal notes. A visit to the death scene was also made in few cases, wherever deemed necessary. The sample included 18 males and 12 females with ages ranging from 19 to 65 years. The cases consisted of both complete and partial and typical and atypical hangings. In the latter instance there were no cases classified as anterior hanging, i.e., with knot over the submental mandibular region. The knot site was determined from the crime scene photographs and course of ligature mark at autopsy. All the cases were declared dead on arrival in the hospital emergency unit without any cardiopulmonary resuscitation or treatment being carried out. No additional trauma was present in any of the cases.

The post mortem interval was between 12 to 24 hours, determined on the basis of postmortem changes and information regarding when the victim was last seen alive. Toxicology examination of the viscera and body fluids did not reveal ethyl alcohol or any other common poison in any subject. The histopathological examination of organs did not show any macroscopic or microscopic evidence of chronic or significant pathology of the respiratory tract or other organs. The dissection and preparation of the neck was done in layers as described in the forensic literature [11]. The skin, subcutaneous tissues, muscles and deep structures of the neck were dissected and reflected in layer-wise manner keeping traction at the minimum, to avoid artefactual sucking of air in the tissues. The mediastinal and CSTE was recorded and stated to be positive by finding an almost continuous stream of fine soapy air bubbles in the superficial and/or deep soft tissues of the mediastinum and neck. This could be easily observed in most cases over the pulmonary viscera, and around the trachea, oesophagus, neck vessels and dorsal surface of the upper layer strap muscles and ventral aspect of deep strap muscles.

Other vital findings of hanging were recorded as hemorrhages in the subcutaneous tissues, laryngeal hyoid skeleton fractures, conjunctival and facial petechial hemorrhages, hemorrhages over the clavicular origin of sternocleidomastoid muscle, carotid intimal tears (Amussat's sign), Simon's bleeding and hemorrhages in the respiratory auxiliary muscles of respiration, so that along with one or more of these findings, the soft tissue emphysema was considered to be vital in origin. An oblique ligature mark with typical appearance consistent with knot site was present in each case.

A group of five cases of deaths from homicidal asphyxia were also autopsied during this period and evaluated for similar evidence of soft tissue emphysema. The cases consisted of two deaths from ligature strangulation, two cases of manual strangulation and one involving both ligature and manual strangulation with blunt head trauma. All cases were young males in their 30's. There was no evidence of chest trauma in any case. No CPR was performed and putrefaction was absent. Layer-wise dissection of the neck was performed in each case.

A control group of 30 non-hanging age and sex matched cases (age range 19 to 65 years; 18 males and 12 females) was formed who died from a variety of natural cardiac and non-cardiac causes. The causes of death included: 12 cases of complicated coronary atherosclerosis, six cases belonged to acute left ventricular failure, four young adults died of cardiomyopathy and myocarditis, four cases of hypertensive intracranial hemorrhage, three from bleeding peptic ulcer and one from bleeding oesophageal varices. The basic methodology and criteria of selection for these cases were the same as the hanging cases. All were declared dead on arrival in the emergency unit without any cardiopulmonary resuscitation being carried out. The cases lack any signs of initial putrefaction, the postmortem interval was between 12 to 24 hours and toxicology screening did not reveal ethanol or any other common poison. There was no autopsy evidence of trauma in any of the subjects. The dissection and preparation of the neck was done in layers and the same autopsy technique, as for hanging cases, was used for all these cases.

The finding of CSTE and pneumomediastinum was confirmed by two independent observers.

RESULTS

CSTE and pneumomediastinum was seen in 22 out of 30 cases (73.3%) of hanging. The finding was found to be independent of the age and sex of the individual. The presence of CSTE and pneumomediastinum was not dependent upon the type of hanging and suspension point. As we did not get any case belonging to anterior hanging, any contribution of the respiratory asphyxia to the proposed carotid artery occlusion/carotid sinus related vagal death could not be evaluated [12]. In the neck the air bubbles could be well appreciated up to the level of mark (at lower margin) above which they were missing (**Figure 1**).

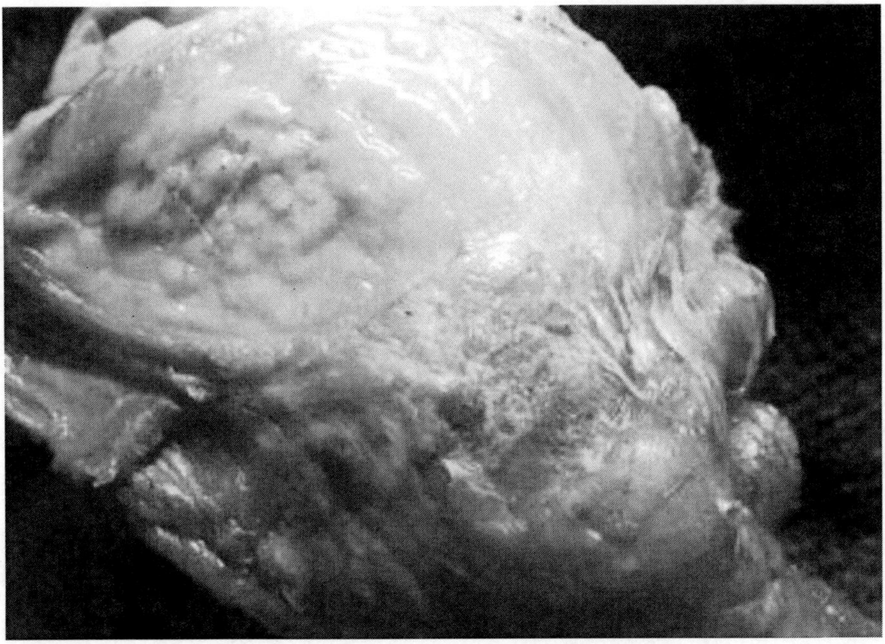

Figure 1: Air bubbles at the level of mark (at lower margin).

A further evaluation for the specific location of emphysema showed that CSTE and pneumomediastinum occur as coexistent findings, as none of the cases revealed these as independent findings which also suggest the presence of a pressure gradient between ruptured alveoli and the upper respiratory tract. A correlation of the phenomenon was made with Simon's bleeding (13/30 positive 43.3%) and clavicular periosteal hemorrhage (18/30 positive-60%) at the sternocleidomastoid muscle origin.

Soft tissue emphysema with Simon's bleeding was found in 9 of 13 cases (69.23%) while the coexistence with clavicular periosteal hemorrhage was seen in 13 of 18 cases (72.22%).

In the neck the air bubbles were found to be prominent over the clavicles near the root of the neck and on the dorsal surface of upper layer strap muscles and ventral surface of deep strap muscles, after reflecting them (**Figure 2**).

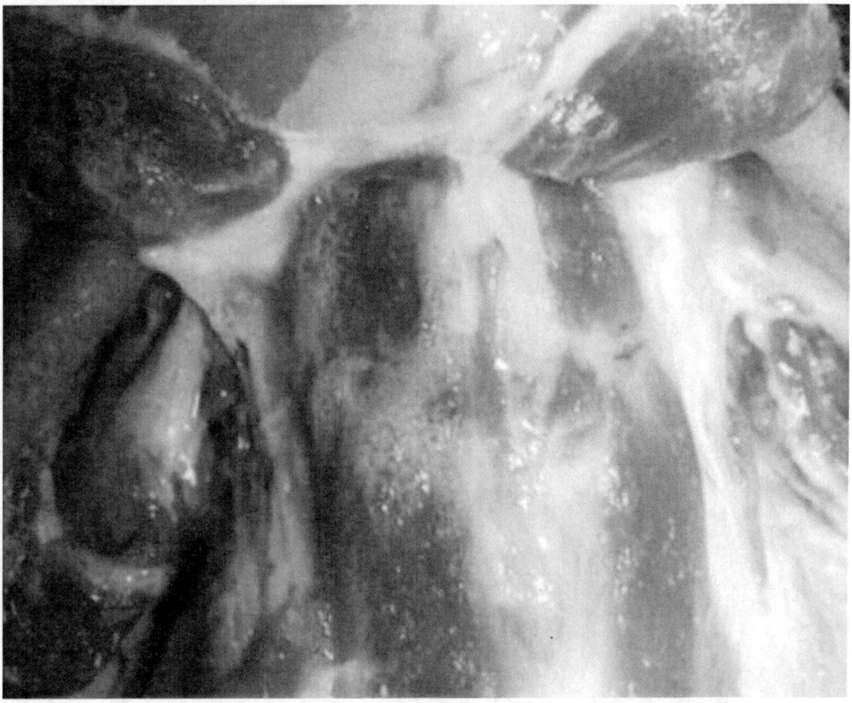

Figure 2: Air bubbles at the clavicles near the root of the neck and on the dorsal surface of upper layer strap muscles and ventral surface of deep strap muscles.

In the neck muscles, the dorsum aspect of sternocleidomastoid muscle depicted this finding most frequently (**Figure 3**). Pneumomediastinum was prominent in midline mediastinal soft tissues (**Figure 4**).

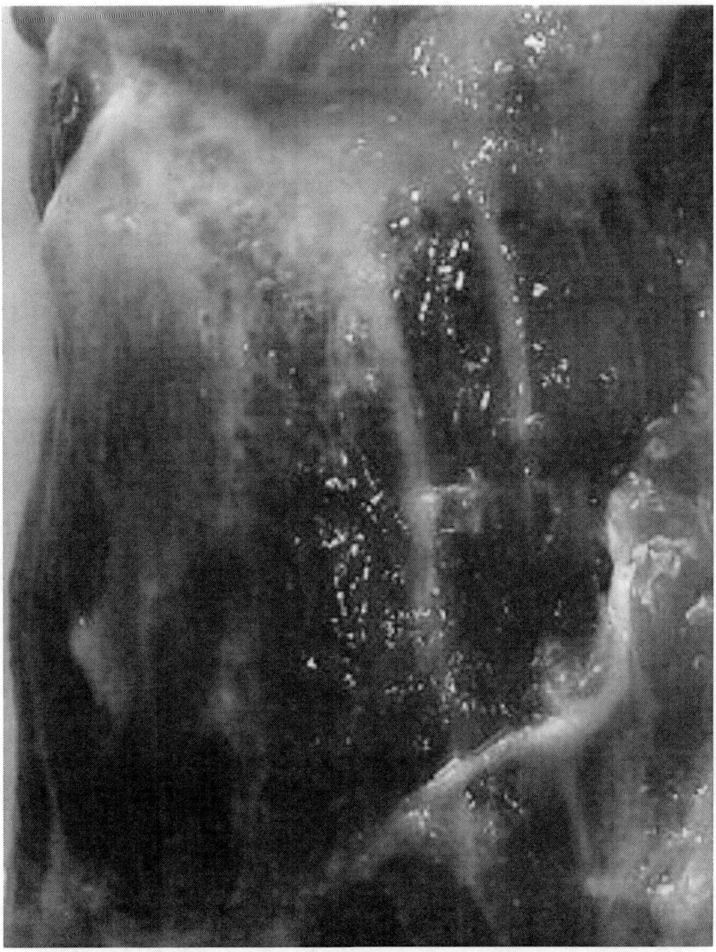

Figure 3: The dorsum aspect of sternocleidomastoid muscle.

Furthermore the evidence of emphysema could be well appreciated during autopsy from the crackling sound under the dissection knife. The autopsy did not reveal any mucosal lesions of either the upper or lower airways. All the findings were noted and photographed in situ as well as after separating them from their anatomical attachments.

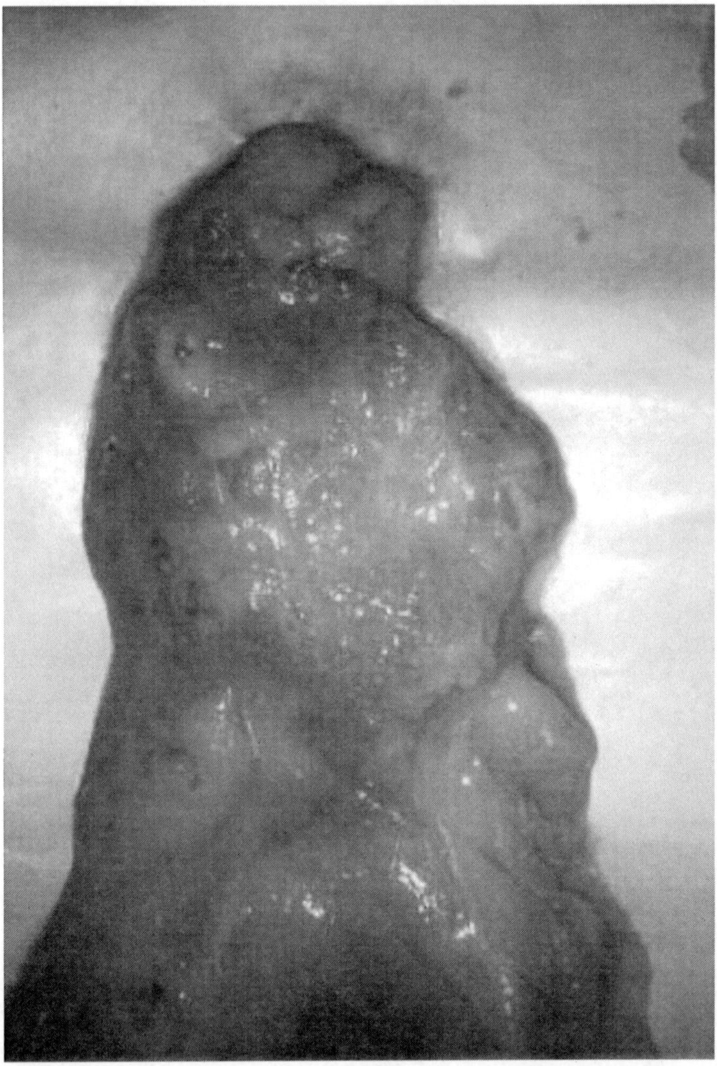

Figure 4: Pneumomediastinum in midline mediastinal soft tissues.

In five strangulation cases, toxicology analysis revealed a blood alcohol concentration of 65 mg% and 110 mg% in one case of manual strangulation and one case of strangulation with head trauma respectively. Typical findings consisted of showers of petechial hemorrhages and congestion syndrome above the level of neck compression, hyposphagma, strap muscles hemorrhages and fractures of the laryngeal framework.

Evidence of any cardiopulmonary pathology was ruled out from gross and microscopic histopatholgical examination. Layer-wise and careful dissection of the neck and chest did not reveal any evidence of pneumomediastinum (**Figure 5**), CSTE (**Figure 6**) or air bubbles around mark of strangulation (**Figure 7**).

CSTE and pneumomediastinum was recorded in three control cases (10%); two cases of acute LVF and one case of hypertensive cerebral hemorrhage. However it was neither too gross nor present as a continuous stream. A patchy, irregular and focal distribution of separate tiny air bubbles was present at multiple foci in all three cases, which could be easily differentiated from continuous streaming emphysema found in hangings.

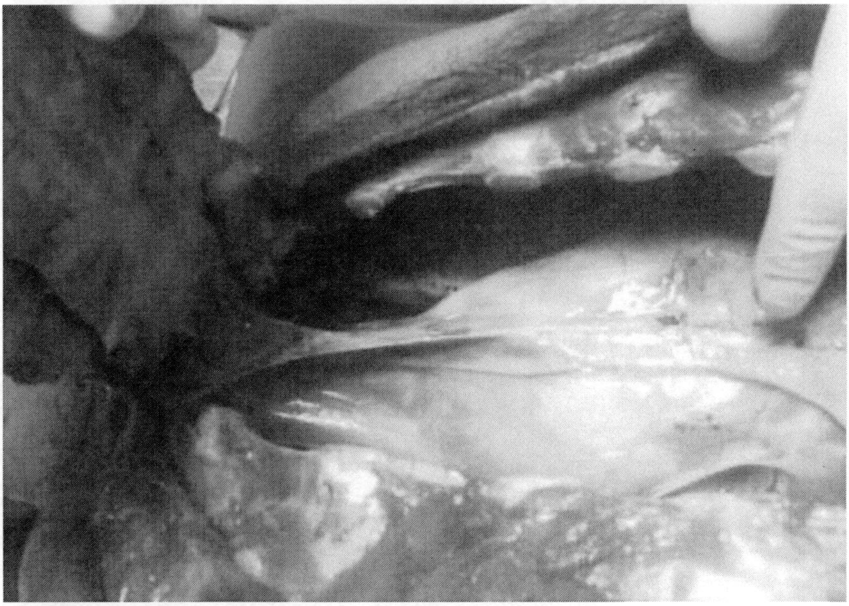

Figure 5: Dissection of the neck and chest.

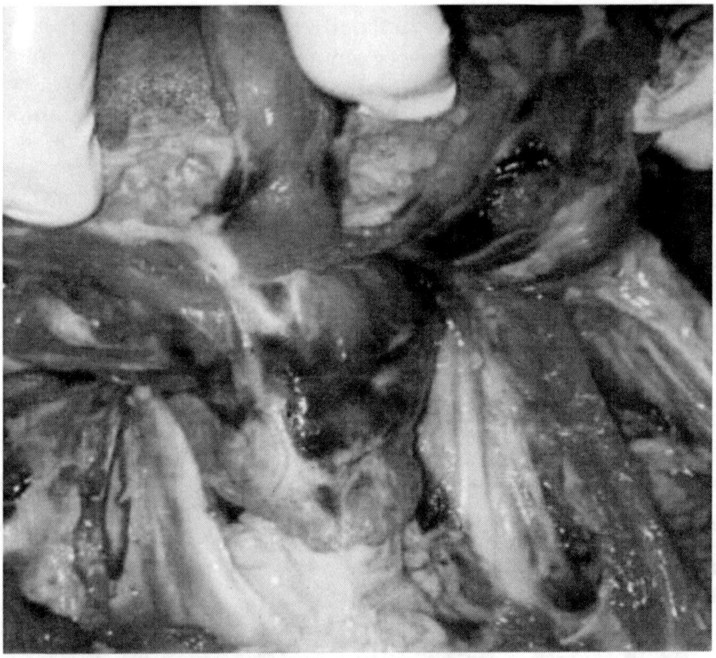

Figure 6: Cervical soft tissue emphysema.

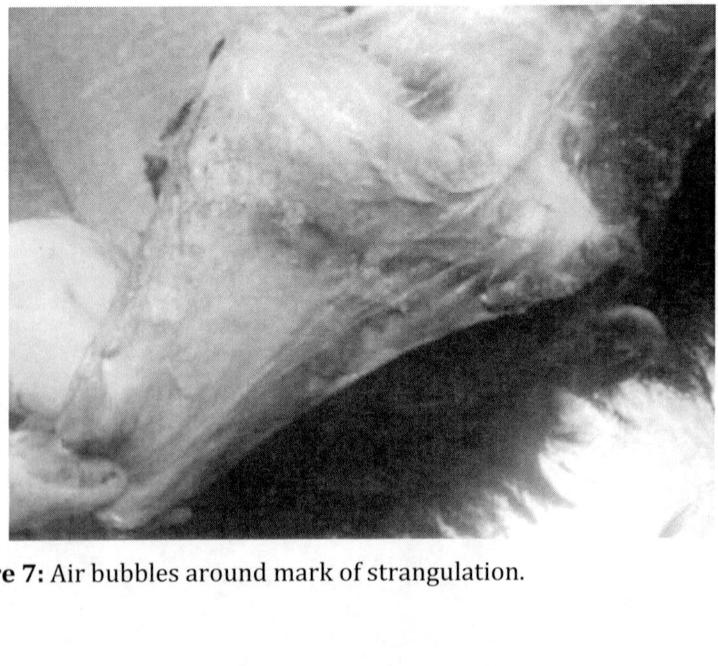

Figure 7: Air bubbles around mark of strangulation.

DISCUSSION

Hussarek and Wolf made the first radiology-based clinical observation of CSTE in cases of attempted strangulation [13]. Canizares et al. reported a case of attempted hanging with subcutaneous emphysema [14]. In the clinical practice pneumomedisatinum and CSTE is commonly observed in cases of attempted strangulations where victims survived, were admitted and underwent radiological examination. The detection of the gas as pneumomedisatinum and/or CSTE has been infrequently demonstrated at autopsy which is evident from the scarce reporting of this finding in deaths due to strangulations [5,6] and in other cases also. In the latter instance, Byard has described a case of death due to chronic obstructive pulmonarydisease with pneumothorax and subcutaneous emphysema in which widespread and massive emphysema was demonstrated at autopsy in the form of air bubbles, all over the upper torso region. The emphysema was dissecting into the dermis and fascia of the muscles such as sternocleidomastoid [15].

A kind of frothy air, soap bubble-like formation in cervical soft tissue can be an artefactual phenomenon resulting from the stretching and exenterating of the neck organs during autopsy. To avoid this and differentiate it from CSTE in cases of hanging, it is important to dissect the neck in layer-wise manner, keeping the traction at minimum while preparing neck muscles and soft tissues [11]. The muscles are cut from their bony attachments and examined on their dorsal surface for any evidence of emphysema that appears as fine air bubbles. The ligature pressure makes a kind of binding for these bubbles so that the emphysema is not appreciable beyond the ligature mark [5].

In the past few years, postmortem imaging has emerged as an effective tool in elucidating the important ante mortem findings prior to the commencement of autopsy. Postmortem multi-slice computed tomography (MSCT) and magnetic resonance imaging (MRI) have been the constant and regular procedures carried out in some developed nations to know the inside story of the corpse prior to dissection [16,17]. The cross sectional radiological imaging methods used in clinical medicine have been implemented in the field of forensic medicine which have revolutionized forensic case work. Although the exact value of postmortem imaging in detecting specific conditions has not been fully established, in specific cases, however, it can be used as a diagnostic tool demonstrating findings that remain undetected during autopsy. Aghayev et al. used postmortem imaging with MSCT and MRI, to demonstrate pneumomediastinum and soft tissue emphysema of the neck in three cases of suicidal hanging. Although the soft tissue emphysema was well appreciated in all three cases on radiological imaging, on autopsy two cases were completely lacking this finding [6]. Similarly, in a case described by Sieswerda- Hoogendoorn et al., pneumomediastinum and soft tissue emphysema were detected with postmortem CT scan in a 3-year-old girl who died due to hanging. However, it was not found subsequently during autopsy [7]. The reasons for these discrepancies might be the variations in dissection techniques and/or inter-individual observation differences. However, these cases demonstrate one of the added values of postmortem imaging, the possibility of demonstrating findings that remain undetected during autopsy.

The underlying mechanism of pneumomediastinum and CSTE in hanging is not completely understood. Two pathophysiological mechanisms have been proposed to explain the presence of gas in the soft tissues. First, a traumatic direct or indirect tear or rupture of lower or upper airways might result from local violence from ligature pressure [6]. The other mechanism involves an increase in intrapulmonary pressure during physical straining causing pulmonary interstitial emphysema [18]. This mechanism was further confirmed by Macklin and Macklin who proposed rupture of terminal alveoli caused by a sudden increase in intra-thoracic pressure due to forceful expiration against a closed glottis during hanging. There is subsequent dissection of air along the visceral spaces around trachea, bronchi, oesophagus and large vessels in the mediastinum and further up in the neck into superficial and/or deep soft tissues. This is the well accepted mechanism at present, also known as Macklin effect [10]. A pressure gradient is formed between the site of alveolar rupture and obstruction zone in the neck, leading to the formation and dissection of air bubbles in the mediastinal and neck soft tissues. Pneumomediastinum and CSTE arise in various conditions because of the Macklin Effect, as a result of an alveolar rupture secondary to high intra-alveolar pressure [8,9]. Both mechanisms are compatible with the finding of air in the visceral space around the trachea, esophagus and big vessels of the neck, the structures serving as guides for air spread. Cases of airway mucosal injuries with complete disruption, including decapitation and transverse laceration of upper airways, have been reported in hanging deaths [19,20]. Gill et al. have reported two cases of crico-tracheal disruption owing to hanging which lead to the development of local subcutaneous emphysema [21].

We did not find mucosal injury to either the upper or the lower airway in any of the cases so the first mechanism is unlikely to be responsible for soft tissue emphysema in this study. The Macklin effect is most likely the responsible mechanism for soft tissue air entry in our cases. These findings are in accordance with previous studies [5,6]. However, recent studies on filmed hangings have suggested that there might not even be fatal airway obstruction in hanging [22]. This largely undermines the effects of complete glottic closure and increased intrathoracic pressure, thereby creating doubts over the true existence of the Macklin effect in hanging. The genuineness of the phenomenon is further questionable based on reports of hanging deaths taking place in individuals with ligature high up in the neck and a permanent tracheostomy opening below [23,24]. However, in the authors' opinion, it is possible that some alternative mechanism of death, in place of respiratory asphyxia, might be operating in the cases lacking this finding. So the pneumomedisatinum and CSTE might not be exclusively found in all cases of hanging. As there has been no mentioning of the knot site and typical-atypical hanging in the videofilmed hanging cases, an exact mechanism of death in those cases cannot be explained, which is frequently dependent on the mode of ligature application and knot site. The involvement of neck vital structures in hanging is also dependent on the amount of weight force acting on the neck which is multifactorial in origin and is different for individual vital neck structures [12]. Furthermore it cannot be emphasized enough that death in hanging is primarily caused by compression of the blood vessels of the neck, thereby causing an insufficient amount of oxygenated blood reaching the brain. Blockage or compression of the air passages is not necessary to cause death in hanging [24]. Findings of cerebral anaemia and hypoxia cannot be seen at autopsy as the death in hanging occurs in too short time interval to make these changes apparent. Although if the victim survives, ischaemic infarcts in the classical watershed zones of the brain may be demonstrated grossly as well as histologically [25]. Therefore, it is not often possible to comment with certainty which mechanism might be predominantly operating in a particular case. Also, unlike postmortem imaging, autopsy does not allow for the perfect detection of the track followed by the gas molecules along chest and neck organs in such cases. Soft and solid neck structures preparation by layers only allows for the visualization of CSTE, but does not explain its origin clearly. Therefore, the patho-physiology of the finding remains unclear most of the time.

Cases of non-traumatic pneumomediastinum and cervical emphysema have been extensively reported in the living individuals including cases of inhalational drug abuse [1], postoperative vomiting [3] and also as a spontaneous phenomenon [26]. A case of spontaneous pneumomediatinum in a trombonist has been described by Ito et al. The authors proposed the mechanism as repetitive overinflation of lungs with alveolar tenderness and high intra-alveolar pressure causing alveolar rupture [27]. Histological demonstration of pulmonary micro-morphology was made by Grellner and Madea in 106 cases of fatal strangulation and hanging. The authors recorded perivascular and intra-alveolar hemorrhages, local dystelectasis and focal emphysema as frequent histological findings in such cases [28].

Numerous vital findings have been described in connection with hanging. Important among these are hemorrhagic zone between ligature marks, periosteal hemorrhages at the clavicular origin of sternocleidomastoid muscle, hemorrhages into the back and auxiliary muscles of respiration, bowel wall hemorrhages, Simon's bleedings and carotid intimal tears (Amussat's sign). These vital reactions are not compulsory in hanging, and they can also be produced postmortem [29]. Along with these, a few controversial findings of doubtful significance have been described in the literature [30,31]. In contrast, pneumomediastinum and cervical emphysema are not the products of circulation, rather they result from attempted breathing against closed glottis due to mechanical obstruction; therefore, they are more reliable.

None of our five cases of deaths due to ligature and manual strangulation revealed any air bubbles in either mediastinum or higher cervical soft tissues. This may be explained by the fact that the mechanism of death predominantly involved in such cases is either jugular venous and/or carotid artery occlusion or sudden vagus induced cardiac arrest. Furthermore the amount of pressure required to compress the trachea and/or to achieve glottis closure is usually not achieved from manual means of strangulation. This is in contrast to hanging where comparably more neck constriction exists making tracheal and glottis closure relatively easier [12]. At first instance, the results and the existing mechanisms make the finding of emphysema quite specific for hanging, especially for cases where the potential confusion exists between hanging and strangulation. But we cannot firmly conclude this finding for cases of strangulation due to the small size of our sample. A larger study involving deaths from homicidal asphyxial means is required for such a firm opinion.

In other cases of mechanical asphyxial deaths, such as smothering, choking, hangings with soft broad ligatures etc., where the obvious injuries to the superficial as well as deep neck structures may be absent, presence of cervical soft tissue emphysema and pneumomediastinum may act as a useful vital parameter in suggesting fatal airway obstruction.

CONCLUSION

The finding of pneumomediastinum and CSTE at autopsy can be important and reliable vital evidence in cases of hanging, especially when the circumstances are non-explanatory. However, it is important to rule out any evidence of incipient putrefaction, cardiopulmonary resuscitation or already existing cardiac or pulmonary pathology before making an opinion of hanging from this finding. Whenever possible, the vitality of the sign should be considered in conjunction with other known vital signs of hanging. Furthermore a layer by layer and bloodless dissection of the neck should be strictly followed in such cases, as artefacts may be introduced at any stage if proper precautions are not taken, which may spoil the vital findings significantly. The underlying mechanism of the soft tissue emphysema is not completely understood, so the interpretation must be made with caution, based on sound medical grounds.

REFERENCES

1. Miller WE, Spiekerman RE, Hepper NG (1972) Pneumomediastinum resulting from performing Valsalva maneuvers during marihuana smoking. Chest 62: 233-234.
2. Maunder RJ, Pierson DJ, Hudson LD (1984) Subcutaneous and mediastinal emphysema. Pathophysiology, diagnosis, and management. Arch Int Med 144: 1447-1453.
3. Bremner WG, Kumar CM (1993) Delayed surgical emphysema, pneumomediastinum and bilateral pneumothoraces after postoperative vomiting. Br J Anaesth 71: 296-297.
4. Akinyemi RO, Ogah OS, Akisanya CO, Timeyin AO, Akande KO, et al. (2007) Pneumomediastinum and Subcutaneous Emphysema Complicating Acute Exacerbation of Bronchial Asthma. Ann Ib Postgrad Med 5: 78-79.
5. Nikolić S,Zivković V,Babić D,Juković F (2012) Cervical soft tissue emphysema in hanging--a prospective autopsy study. J Forensic Sci 57: 132-135.
6. Aghayev E,Yen K,Sonnenschein M,Jackowski C,Thali M,et al. (2005) Pneumomediastinum and soft tissue emphysema of the neck in postmortem CT and MRI; a new vital sign in hanging? Forensic Sci Int153: 181-188.
7. Sieswerda-Hoogendoorn T,Strik AS,Hilgersom NF,Soerdjbalie-Maikoe V,Van Rijn RR (2014) Pneumomediastinum and soft tissue emphysema in pediatric hanging. J Forensic Sci59: 559-563.
8. Wintermark M, Schnyder P (2001) The Macklin effect-a frequent etiology for pneumomediastinum in severe blunt chest trauma. Chest 120: 543-547.
9. Sakai M, Murayama S, Gibo M, Akamine T, Nagata O (2006) Frequent cause of the Macklin effect in spontaneous pneumomodiastinum-demonstration by multidetector-row computed tomography. J Comput Assist Tomogr 30: 92-94.
10. Zylak CM, Standen JR, Barnes GR, Zylak CJ (2000) Pneumomediastinum revisited. Radiographics 20: 1043-1057.
11. Vanezis P (1993) Post mortem techniques in the evaluation of neck injury. J Clin Pathol 46: 500-506.
12. Saukko P, Knight B (2016) Knight's Forensic Pathology. (4th edn), CRC Press, Taylor & Francis Group, Boca Raton, London, New York.
13. Hussarek M, Wolf G (1971) Subcutaneous emphysema of neck and larynx following attempted strangulation. Z Rechtsmed 68: 41-44.
14. Canizares MA, Arnau A, Fortea A, Zarzuela V, Martinez-Vallina P, et al. (2000) Hyoid fracture and traumatic subcutaneous cervical emphysema from an attempted hanging. Apropos a case. Arch Bronconeumol 36: 52-54.
15. Byard RW (2016) Gross and microscopic features of extensive subcutaneous emphysema. Forensic Sci Med Pathol 12: 372-375.
16. Thali MJ,Yen K,Schweitzer W,Vock P,Boesch C,et al. (2003) Virtopsy, a new imaging horizon in forensic pathology: virtual autopsy by postmortem multislice computed tomography (MSCT) and magnetic resonance imaging (MRI)--a feasibility study. J Forensic Sci48: 386-403.
17. Thali MJ,Yen K,Plattner T,Schweitzer W,Vock P,et al. (2002) Charred body: virtual autopsy with multi-slice computed tomography and magnetic resonance imaging. J Forensic Sci 47: 1326-1331.
18. Joannides M, Tsoulos GD (1930) The etiology of interstitial and mediastinal emphysema. Arch Surg 121: 333-339.

19. Tattoli L, Buschmann CT, Tsokos M (2014) Remarkable findings in suicidal hanging. Forensic Sci Med Pathol 10: 639-642.
20. Hejna P, Bohnert M (2012) Decapitation in suicidal hanging–vital reaction patterns. J Forensic Sci 58: S270-S277.
21. Gill SS, Singh CM, Eggleston FC (1977) Cricotracheal disruption owing to strangulation. A report of two cases with successful surgical repair. J Thorac Cardiovasc Surg 73: 948-950.
22. Sauvageau A, Laharpe R, King D, Dowling G, Andrews S, et al. (2011) Agonal sequences in 14 filmed hangings with comments on the role of the type of suspension, ischemic habituation, and ethanol intoxication on the timing of agonal responses. Am J Forensic Med Pathol 32: 104-107.
23. Cooke CT, Cadden GA, Hilton JMN (1989) Unusual hanging deaths. Am J Forensic Med Pathol January 9: 277-282.
24. DiMaio VJ, DiMaio D (2001) Forensic Pathology. Practical aspects of criminal and forensic investigations. (2nd edn), CRC Press, Boca Raton London, NewYork Washington, USA.
25. Oehmichen M, Auer RN, Konig HG (2006) Forensic Neuropathology and Associated Neurology. (1st edn), Springer, Berlin Heidelberg, NewYork, US.
26. Abolnik I, Lossos IS, Breuer R (1991) Spontaneous pneumomediastinum. A report of 25 cases. Chest 100: 93-95.
27. Ito S, Takada Y, Tanaka A, Ozeki N, Yazaki Y (1990) A case of spontaneous pneumomediastinum in a trombonist. Kokyu to junkan. Respiration & circulation 37: 1359-1362.
28. Grellner W, Madea B (1994) Pulmonary micrornorphology in fatal strangulations. Forensic Sci Int 67: 109-125.
29. Puschel K, Turk E, Lach H (2004) Asphyxia-related deaths. Forensic Sci Int 144: 211-214.
30. Zhu BL, Ishida K, Fujita MQ, Maeda H (2000) Immunohistochemical investigation of pulmonary surfactant in fatal mechanical asphyxia. Int J Legal Med 113: 268-271.
31. Fieguth A, Albrecht UV, Bertolini J, Kleemann J (2003) Intracartilaginous haemorrhagic lesions in strangulation? Int J Legal Med 117: 10-13.

CHAPTER 4

Rapid Detection of Fentanyl, Fentanyl Analogues, and Opioids for on-Site or Laboratory Based Drug Seizure Screening using Thermal Desorption DART-MS and Ion Mobility Spectrometry

Akira Namera [1]• Maho Kawamura[2] • Akihiro Nakamoto[2] • Takeshi Saito [3]• Masataka Nagao[1]

[1].Department of Forensic Medicine, Institute of Biomedical and Health SciencesHiroshima UniversityHiroshimaJapan

[2].Forensic Science LaboratoryHiroshima Prefectural Police HeadquartersHiroshimaJapan

ABSTRACT

Fentanyl and fentanyl analogues represent a current and emerging threat in the United States as pure illicit narcotics and in mixtures with heroin. Because of their extreme potency, methods to safely and rapidly detect these compounds are of high interest. This work investigates the use of thermal desorption direct analysis in real time mass spectrometry (TD-DART-MS) and ion mobility spectrometry (IMS) as tools for the rapid and sensitive (nanogram to picograms) detection of fentanyl, 16 fentanyl analogues, and five additional opioids. Competitive ionization studies highlight that detection of these compounds in the presence of heroin is readily achievable, down to 0.1% fentanyl by mass with TD-DART-MS. With IMS, detection of nanogram levels of fentanyl in a binary fentanyl and heroin mixture is possible but can be complicated by decreased resolution in certain commercial instrument models. Modifications to the alarm windows can be used to ensure detection of fentanyl in binary mixtures. Additionally, three complex background matrices (fingerprint residue, dirt, and plasticizers) are shown to have a minimal effect of the detection of these compounds. Wipe sampling of the exterior of bags of questioned powders is shown to be a safe alternative method for field screening and identification, removing the need to handle potentially lethal amounts of material.

1. INTRODUCTION

Opioid abuse is a growing epidemic that creates unique challenges for law enforcement, first responders, and medical personnel. Fentanyl and fentanyl analogues are many times more potent than morphine, making accidental exposure through the seizure of powders and pills life-threatening. The Drug Enforcement Administration (DEA) issued a warning to law enforcement in June of 2016 to exercise extreme caution when handling possible fentanyl-containing materials [1]. The lethal dose of fentanyl can be as low as 2 mg [1], with ingestion, inhalation, and absorption through the skin as possible exposure routes. Carfentanil, commonly used as a large animal tranquilizer, is present in illicit U.S. drug markets, is approximately 100 times more potent than fentanyl, and has been used as a chemical warfare agent in anti-terrorist activities [2]. The Centers for Disease Control (CDC) has issued a health alert on the rise of unintentional overdoses as a result of fentanyl and fentanyl analogues in the U.S. from clandestinely produced and trafficked fentanyl (commonly referred to as non-pharmaceutical fentanyl (NPF)) in the form of counterfeit pills and heroin adulterants [3].

Field testing that requires direct handling of suspected drug powders is discouraged due to safety concerns, as materials must be treated as potential NPF, both in the field and in the lab, until a qualitative analysis can be performed. This extended period of uncertainty about the contents of the suspicious material is undesirable, as it requires extreme vigilance during handling. A rapid, safe screening test for fentanyl and fentanyl analogues that is successful in realistic sample scenarios (i.e. mixtures with heroin) would protect emergency and laboratory personnel and could provide valuable medical treatment information. Current colorimetric techniques require macroscopic amounts of material, making them unsuitable for safe field testing of NPF. Ion mobility spectrometry (IMS) is a field-deployable technology useful for qualitative screening of street drugs and is sensitive to nanogram levels of fentanyl, even in the presence of heroin [4]. The contamination on bag exteriors may be sufficient for detection, providing necessary information regarding the bag's potential contents and what precautions should be taken. What is unknown is the specificity of IMS to detect a wide range of fentanyl analogues, including carfentanil, particularly in the types of mixtures expected in street drugs or when contaminated with dirt and fingerprint residues from the outside of bags.

Another potential field deployable, and laboratory-based, screening technique is thermal desorption direct analysis in real time mass spectrometry (TD-DART-MS), which has been demonstrated for direct analysis of drugs collected onto wipes without the traditional solution-based sample preparation [5]. TD-DART-MS is sensitive to picogram levels of a wide range of illicit drugs [5], and the ability of this technique to detect fentanyl and fentanyl analogues has not been reported to our knowledge. The use of traditional DART-MS has also been extensively reported in literature [6]; [7]; [8]; [9] ; [10]. Additionally, the configuration of the TD-DART-MS system is arranged to minimize potential exposure to the analyst because of aersolization and/or vaporization. TD-DART-MS instruments have potential applications in mobile laboratories, emergency vehicles, and hospitals. The CDC advises two-tiered testing to identify the specific fentanyl compound on which individuals are suspected to have overdosed; the identity of the compound is essential information for understanding and controlling outbreaks [3]. Current guidelines recommend an enzyme-linked immunosorbent assay (ELISA) screen for fentanyl followed by gas chromatography/mass spectrometry (GC/MS) analysis. As NPF concentrations in blood can be quite low, a wipe-based technique such as TD-DART-MS targeting solid trace contamination on the individual or their belongings may be a more effective approach. TD-DART-MS may also be useful in emergency medicine, providing a rapid identification of the specific NPF to make informed choices about treatment. The common antidote for opioid-related overdoses, naloxone, may not be effective in single doses against the more potent NPFs such as carfentanil and other synthetic opioids such as U-47700 [11].

This study provides some of the necessary data needed to support the use of IMS and TD-DART-MS for applications in the law enforcement, emergency, and medical communities. Fentanyl, sixteen different fentanyl analogues, and five opioids (heroin, U-47700, buprenorphine, methadone, and naloxone, which is found in combination with buprenorphine) were evaluated for detection by IMS and TD-DART-MS, to determine spectral characteristics and limits of detection. Selected compounds were evaluated for changes in sensitivity and selectivity of detection when combined with heroin and/or excipients, and against background collected from sampling the outside of bags, fingerprints, and dirt. The potential for false negatives or false positives for NPF using IMS was evaluated.

2. MATERIALS & METHODS

2.1. Materials

A total of 22 compounds were analyzed, as listed in Table 1. These compounds included fentanyl, fentanyl analogues, and five additional commonly abused opioids. All fentanyl analogues were purchased as hydrochloride salts from Cayman Chemical (Ann Arbor, MI, USA)* in powdered form or, when possible, as 1 mg mL^{-1} methanolic solutions. Fentanyl was purchased in powder form as fentanyl citrate (US Pharmacopeia, Rockville, MD, USA) and the remaining opioids were purchased as 1 mg mL^{-1}methanolic solutions from Cerilliant (Round Rock, TX, USA). Powdered samples were dissolved gravimetrically in methanol (Chromasolv® grade, Sigma-Aldrich, St. Louis, MO, USA), and all samples were further diluted in methanol. Solutions were then pipetted or inkjet printed [12] ; [13] (additional parameters in the Supplemental information) onto *meta*-aramid fiber wipes (Nomex®, Smiths Detection, Hempstead, UK) for IMS analysis, or polytetrafluoroethylene (PTFE)-coated fiberglass wipes (DSA Detection, North Andover, MA, USA) for TD-DART-MS analysis. The five excipients examined (acetaminophen, caffeine, mannitol, quinine, and procaine) were purchased in powder form (Sigma-Aldrich) and dissolved in methanol. Materials used as background simulants include an artificial fingerprint material [14] and a dirt simulant (NIST SRM 1944). Polyethylene glycol – 600 (PEG-600) diluted in methanol was used as the mass tuning compound for the mass spectrometer of the TD-DART-MS system.

Table 1. Representative data for all compounds analyzed by IMS and TD-DART-MS. For IMS sensitivities, values with an asterisk (*) are the calculated LOD_{90} value. The average uncertainty in the K_0 value is ±0.0008.

Compound			IMS		TD-DART-MS			
Name	Molecular Weight (Da)	Molecular Formula	K_0-value	Sensitivity (ng)	Base Peak +30 V (*m/z*)	Base Peak +90 V (*m/z*)	LOD_{90} (ng)	90% Upper Confidence Limit (ng)
Fentanyls								
Acetyl Fentanyl	322.204	$C_{21}H_{26}N_2O$	1.0797	5.0	323.212 $[M+H]^+$	105.070 $[C_8H_9]^+$	0.222	0.452
Acryl Fentanyl	334.204	$C_{22}H_{26}N_2O$	1.0569	1.0	335.212 $[M+H]^+$	105.070 $[C_8H_9]^+$	0.145	0.274
Benzyl Fentanyl	322.204	$C_{21}H_{26}N_2O$	1.0817	1.0	323.212 $[M+H]^+$	91.055 $[C_7H_7]^+$	0.0864	0.152
Butyryl Fentanyl	350.236	$C_{23}H_{30}N_2O$	1.0179	5.0	351.244 $[M+H]^+$	188.144 $[C_{13}H_{18}N]^+$	0.140	0.259
Carfentanil	394.226	$C_{24}N_{30}N_2O_3$	0.9712	5.0	395.233 $[M+H]^+$	113.084 $[C_6H_{11}NO]^+$	0.197	0.392
Cyclopentyl Fentanyl	376.251	$C_{25}H_{32}N_2O$	0.9757	1.0	377.259 $[M+H]^+$	188.144 $[C_{13}H_{18}N]^+$	0.165	0.318
Despropionyl Fentanyl	298.184	$C_{19}H_{23}FN_2$	1.1437	1.0	299.192 $[M+H]^+$	105.070 $[C_8H_9]^+$	0.160	0.306
Fentanyl	336.220	$C_{22}H_{28}N_2O$	1.0489	1.401*	337.228 $[M+H]^+$	105.070 $[C_8H_9]^+$	0.142	0.268
ortho-Fluorobutyryl Fentanyl	368.226	$C_{23}H_{29}FN_2O$	1.0112	5.0	369.234 $[M+H]^+$	105.070 $[C_8H_9]^+$	0.296	0.657
para-Fluoroisobutyryl Fentanyl	368.226	$C_{23}H_{29}FN_2O$	1.0020	1.0	369.234 $[M+H]^+$	105.070 $[C_8H_9]^+$	0.351	0.836
Furanyl Fentanyl	374.199	$C_{24}H_{26}N_2O_2$	1.0003	5.0	375.207 $[M+H]^+$	105.070 $[C_8H_9]^+$	0.199	0.397
β-Hydroxythiofentanyl	358.171	$C_{20}H_{26}N_2O_2S$	1.0291	10.0	359.179 $[M+H]^+$	192.085 $[C_{11}H_{14}NS]^+$	0.292	0.647
Isobutyryl Fentanyl	350.236	$C_{23}H_{30}N_2O$	1.0195	5.0	351.244 $[M+H]^+$	188.144 $[C_{13}H_{18}N]^+$	0.183	0.357
4-Methoxy Butyryl Fentanyl	380.246	$C_{24}H_{32}N_2O_2$	0.9595	10.0	381.254 $[M+H]^+$	188.144 $[C_{13}H_{18}N]^+$	0.101	0.180
4-Methoxy Fentanyl	366.231	$C_{23}H_{30}N_2O_2$	0.9861	5.0	367.238 $[M+H]^+$	188.144 $[C_{13}H_{18}N]^+$	0.262	0.563
trans-3-Methyl Fentanyl	350.236	$C_{23}H_{30}N_2O$	1.0217	5.0	351.244 $[M+H]^+$	202.160 $[C_{14}H_{20}N]^+$	0.144	0.273
Valeryl Fentanyl	364.251	$C_{24}H_{32}N_2O$	0.9866	5.0	365.259 $[M+H]^+$	105.070 $[C_8H_9]^+$	0.0807	0.142
Other Opioids								
Buprenorphine	467.304	$C_{29}H_{41}NO_4$	0.9050	3.857*	468.311 $[M+H]^+$	450.301 $[M-OH]^+$	0.473	1.12
Heroin	369.158	$C_{21}H_{23}NO_5$	1.0386	1.080*	370.165 $[M+H]^+$	370.165 $[M+H]^+$	2.55	4.11
Methadone	309.209	$C_{21}H_{27}NO$	1.1004	0.846*	310.217 $[M+H]^+$	265.159 $[M-C_2H_6N]^+$	0.0529	0.0887
Naloxone	327.147	$C_{19}H_{21}NO_4$	1.1425	0.895*	328.155 $[M+H]^+$	310.144 $[M-OH]^+$	0.0930	0.165
U-47700	328.111	$C_{16}H_{22}Cl_2N_2O$	1.0910	0.5	329.119 $[M+H]^+$	172.956 $[C_7H_3Cl_2O]^+$	0.308	0.677

2.2. TD-DART-MS

A thermal desorption direct analysis in real time mass spectrometer (TD-DART-MS) configuration was used in this work and included a thermal desorption unit independent of the DART ionization source. The details of the configuration are discussed elsewhere [5], but briefly, the system consists of a traditional DART source and Vapur® interface (IonSense, Saugus, MA, USA) with a glass T-junction installed in between the source and the interface. A thermal desorption unit (Morpho Detection, Newark, CA, USA) was mounted to the T-junction, perpendicular to the source and interface. Parameters for this portion of the system included 400 °C DART gas stream temperature, 255 °C thermal desorption unit temperature, and Vapur® flow rate of 3.8 L min^{-1} (±0.1 L min^{-1}). After evaporation of the solvent, prepared PTFE wipes were inserted into the thermal desorption unit. The TD-DART system was interfaced to a JEOL JMS-T100LP time-of-flight mass spectrometer (JEOL USA, Peabody, MA, USA). Relevant mass spectrometer parameters included: positive ion detection,+400 V peaks voltage, 100 °C orifice temperature,+5 V orifice 2 and ring lens voltage, 2300 V detector voltage, and a scan range of 60–700 *m/z* at 1 scan s^{-1}. The orifice 1 voltage was switched between +20 V, +30 V, +60 V, and +90 V to evaluate fragmentation spectra. Unless otherwise stated, the data presented was collected using a +30 V orifice 1 voltage. The instrument was tuned using PEG-600 in the range of 70–700 *m/z*.

2.3. IMS

A dual tube IMS (Ionscan 500 DT, Smiths Detection) was used for this work. Prepared meta-aramid wipes containing dried deposits were inserted directly into the thermal desorber unit of the instrument. Tube 1 was configured to detect positive ions (narcotics), and was used exclusively for the data presented herein. The data from tube 2, which was set to detect negative ions (explosives), was not used. The instrument configuration included a desorber temperature of 245 °C, a drift tube temperature of 244 °C, and an inlet temperature of 265 °C. Data was collected for a total of 8 s by sampling full spectra at 30 discrete intervals over that time. A detection algorithm was used to analyze each plasmagram in the group of 30 for peaks within defined detection windows, as the shape and position of peaks can change during the collection time. Most of the compounds evaluated in this study have not been studied previously by IMS, therefore reduced mobility (K_0) values were measured for the most prominent peaks from replicate samples using instrument firmware for peak fitting that references an internal calibrant of nicotinamide ($K_0 = 1.8600$). Nicotinamide also serves as the reactant ion peak (RIP), which is a constant background spectral component. For selected compounds, measured K_0 values were used to create detection windows in the instrument library, using a minimum peak intensity detection threshold of 50 counts to trigger an alarm. In addition to the newly added NPF alarms, a large suite of drug alarms was left active, which included: heroin, methamphetamine, amphetamine, cocaine, 3,4-methylenedioxymethamphetamine (MDMA), hydrocodone, oxycodone, and morphine.

3. RESULTS & DISCUSSION

3.1. Representative response and sensitivity

To minimize the risk of exposure for first responders and forensic scientists, it is crucial that detection of fentanyl and fentanyl analogues be completed with a high degree of specificity and sensitivity, while facilitating sampling from the residue on the outside of a container, not the bulk powder. To evaluate the abilities of the TD-DART-MS and IMS instruments to meet these requirements, representative spectra and sensitivities of the 22 compounds were obtained. Representative spectra using TD-DART-MS were obtained by pipetting 100 ng deposits onto the PTFE wipes and directly inserting them into the thermal desorption unit. All compounds were analyzed at four orifice 1 voltages (+20 V, +30 V, +60 V, and +90 V) to establish fragmentation pathways and be consistent with data compiled in the NIST DART Forensics Library [15]. All compounds readily produced signals differentiable from the background, an example of which is shown in Figure 1 (additional spectra Figs. S1–S17). As has been exhibited with other narcotics [5], responses of the opioids and fentanyl analogues were dominated by protonated molecular ions at low orifice 1 voltages of +20 V and +30 V (Figure 1 and Table 1), with increasing fragmentation observed at the higher (+60 V and +90 V) orifice 1 voltages. Fentanyl and fentanyl analogues followed similar patterns of fragmentation, with the nitrogen-carbon bond being the predominant fragmentation point. At the low fragmentation voltage (+30V) there were three sets of fentanyl analogues that produced identical protonated molecular ion peaks: 1) acetyl and benzyl, 2) o-fluorobutyryl and p-fluoroisobutyryl, and 3) butyryl, isobutyryl, and trans-3-methyl fentanyl. Analysis of these compounds at the higher fragmentation voltages revealed that isomeric compounds (i.e. butyryl and isobutyryl fentanyl) were unable to be differentiated because of identical fragment ions. If isomeric information was required, it may be able to be achieved using an IMS-MS configuration. Non-isomeric compounds, however were differentiated based on their fragmentation spectra (i.e. acetyl fentanyl has a base peak of 105.070 *m/z* at 90 V while benzyl fentanyl had a base peak of 91.055 *m/z*). Fragmentation patterns of other opioids were more unique: the loss of a hydroxide ion was commonly observed for naloxone and buprenorphine; the loss of the nitrogen containing end group was observed with methadone and heroin; and fragmentation across the carbon-nitrogen bond in U-47700 was observed (Table 1, Figure S17).

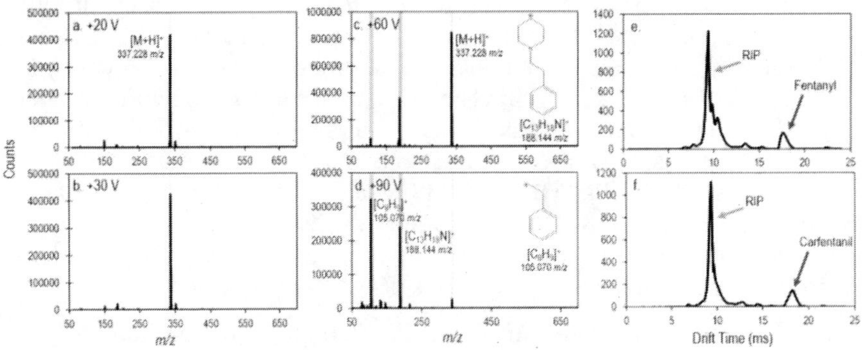

Figure 1. (a–d) Representative TD-DART-MS mass spectra of a 100 ng deposit of fentanyl at the four orifice 1 voltages studied. Representative IMS plasmagrams of 10 ng fentanyl (e) and carfentanil (f) deposits. The reactant ion peak (RIP) is also identified.

In addition to identifying the representative spectra, the instrument sensitivities for detection of these compounds were also established. Limits of detection (LOD_{90}) were determined in accordance with ASTM E2677 (Standard Test Method for Determining Limits of Detection in Explosive Trace Detectors) [16] in which four levels around a presumptive LOD are analyzed (10 replicates/level) in addition to 10 replicates of the blank. From the ASTM Standard a LOD_{90} corresponds to the lowest mass which can be detected off of a wipe with a 90% probability of true detection.

The levels of all compounds except heroin evaluated for TD-DART-MS were (0, 0.5, 1.0, 5.0, and 10.0) ng wipe^{-1}, while heroin analyzed at (0, 1.0, 2.0, 10.0, and 20.0) ng wipe^{-1}. Table 1 shows the calculated LOD_{90} for the compounds examined. For all compounds, with the exception of heroin and buprenorphine, sub ng wipe^{-1} detection limits were obtained, with most fentanyl analogues detectable at or below a few hundred picograms wipe^{-1}.

Determination of representative IMS responses were completed by analyzing 1 ng to 100 ng of the compounds deposited individually on meta-aramid wipes. In most cases, the spectra were simple combinations of the RIP and one distinct peak for the compound, as represented by fentanyl and carfentanil in Figure 1e and f. Shapes and locations of additional fentanyl analogue peaks are shown in Figure S18. The primary peak detected for each compound is likely to be a product ion of M^+ or $[M+H]^+$, and the reduced mobilities are linearly correlated with molecular weight (Figure S18). There are five pairs of fentanyl analogues that cannot be differentiated on this instrument on the basis of K_0 (differences of less than ±0.002). These pairs are: 1) acetyl and benzyl, 2) p-fluoroisobutyryl and furanyl, 3) butyryl and isobutyryl, 4) isobutyryl and trans-3-methyl, and 5) 4-methoxy and valeryl fentanyl. In addition to the fentanyl analogue overlaps, despropionyl fentanyl and naloxone could not be differentiated on the basis of K_0. Since IMS is based on cross-sectional area in addition to molecular mass, the technique was able to differentiate one set of isomeric compounds that the TD-DART-MS could not (o-fluorobutyryl and p-fluoroisobutyryl). Of the 22 compounds listed in Table 1, none have overlaps in K_0 with the illicit drugs already active in the instrument library, and no false positive alarms for these drugs were observed.

LOD_{90} values were determined for a small subset of compounds on IMS (fentanyl, buprenorphine, heroin, methadone, and naloxone), as shown in Table 1, with sensitivities roughly equal to those obtained using TD-DART-MS. For the remaining compounds approximate sensitivities are reported, and were calculated by finding the lowest deposited mass that produced a peak capable of being fit to a Gaussian function, with a height of at least 50 counts, across three replicates. These values are also shown in Table 1. For all compounds analyzed by TD-DART-MS and IMS, sensitivities were at or below 10 ng wipe^{-1}, highlighting the possibility of detection by swiping the outside of a bag, or other surface, instead of needing to sample bulk powder.

3.2. Competitive ionization in binary mixtures

Though it is useful to demonstrate low level detection of these compounds, it is more important to evaluate whether or not these compounds can be detected in the presence of other drugs or excipients. Since fentanyl is commonly encountered in the United States as a minor component in a mixture with heroin [17], it is necessary to establish whether or not the heroin or any excipients will inhibit detection of fentanyl. In order to evaluate whether detection of fentanyl would be possible in more realistic settings, a series of experiments were carried out to establish whether signal suppression would occur in the presence of heroin or common adulterants (mannitol, acetaminophen, quinine, procaine, and caffeine). The excipients investigated were chosen based on data suggesting they are common in heroin [18].

To evaluate the effect of competitive ionization in binary mixtures using TD-DART-MS, increasing amounts of heroin, or an excipient, were added to 1 ng wipe^{-1} deposits of fentanyl. Levels of the "competing compound" (heroin or the excipient) were 0 ng wipe^{-1}(just fentanyl analyzed), 1 ng wipe^{-1}, 10 ng wipe^{-1}, 25 ng wipe^{-1}, 50 ng wipe^{-1}, 100 ng wipe^{-1}, and 1000 ng wipe^{-1} to obtain competing compound: fentanyl ratios of 0:1, 1:1, 10:1, 25:1, 50:1, 100:1, and 1000:1. Once analyzed, the extracted ion chronograms of the fentanyl molecular ion (337.228 m/z) were used to calculate peak areas. The response of fentanyl in the binary mixture relative to pure fentanyl was then calculated by plotting the peak area in the mixture relative to peak area of pure fentanyl (Figure 2).

Figure 2 highlights the effects of three of the compounds – heroin, mannitol, and acetaminophen – on the fentanyl signal. In these plots, the dotted line represents the response of 1 ng wipe^{-1} deposits of fentanyl. Points above the dotted line indicate the competing compound enhanced the response of fentanyl while points that fall below the dotted line indicate compounds that caused suppression of the fentanyl signal. Perhaps the most important of the three plots is the response of fentanyl in the presence of heroin (Figure 2a). No decrease was observed in the peak areas for 1 ng wipe^{-1} fentanyl deposits in the presence of up to 1000 ng wipe^{-1} of heroin (a 0.1% fentanyl mixture by mass). The stability of the fentanyl signal is promising as it shows that fentanyl can be detected in a heroin-fentanyl mixture regardless of the relative percentages of the two compounds. This data also indicates that the proton affinity for fentanyl is likely higher than that of heroin. More interesting than the heroin mixture response is that of

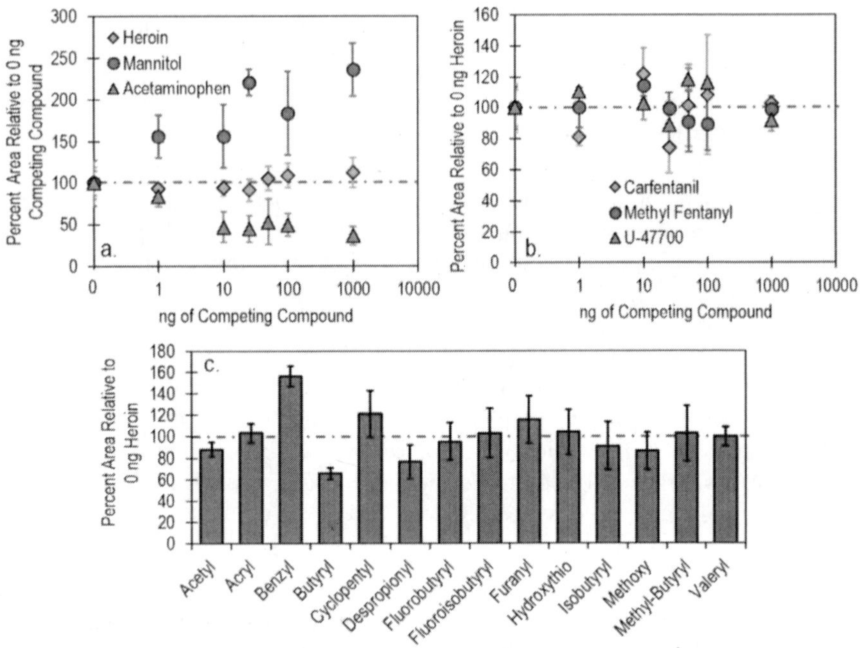

Figure 2. (a) Competitive ionization studies using TD-DART-MS of fentanyl in the presence of increasing amounts of heroin, mannitol, and acetaminophen. (b) Competitive ionization studies of carfentanil, methyl fentanyl, and U-47700 in the presence of increasing amounts of heroin. (c) Competitive ionization study of the remaining fentanyl analogues in the presence of 100 ng heroin. Error bars represent the standard deviation of 5 (a and b) or 3 (c) replicate measurements.

fentanyl in the presence of mannitol (Figure 2a). With the addition of mannitol, a consistent increase in the fentanyl peak area was observed (up to a factor of 2.5 increase at 1000 ng wipe^{-1} of mannitol). This increase may be attributed to either additional proton donation from mannitol, which has a higher propensity to form deprotonated molecular ions than protonated molecular ions [19] or the neutralization of other charge accepting species that could increase available charge for fentanyl. Converse to mannitol, acetaminophen (Figure 2a) caused a reduction in fentanyl response beyond a 1:1 mixture, indicating competitive ionization was occurring. Inhibition of the fentanyl signal did increase with increasing amounts of acetaminophen though the signal was never completely quenched. At a 1000:1 acetaminophen: fentanyl ratio, fentanyl was still readily detectable (36% peak area relative to pure fentanyl) indicating that detection of fentanyl in a mixture containing acetaminophen would likely not pose any issues.

Addition of quinine and procaine exhibited trends similar to heroin (Figure S19) with slight decreases observed at the 1000:1 level. Caffeine exhibited a trend similar to acetaminophen, though a less dramatic reduction in signal was observed (50% peak area relative to pure fentanyl at the 1000:1 level).

A second study was completed to evaluate whether or not the more potent fentanyl analogues and opioids (carfentanil, and methyl fentanyl, and U-47700) would have similar responses to fentanyl in the presence of heroin. Figure 2b depicts the response of these three compounds to increasing amounts of heroin. The response from all compounds remained fairly constant across the range of heroin masses, with 91–102% relative response at the 1000:1 ratio, indicating competitive ionization is not a factor for these compounds regardless of the relative amounts. It is important to note that these experiments were completed using trace levels of material so direct analysis of bulk material may cause different effects. The final study evaluating competitive ionization in binary mixtures was completed by analyzing the remaining fentanyl analogues at a 100:1 ratio of heroin: analogue (100 ng wipe^{-1} heroin and 1 ng wipe^{-1} of the analogue), and comparing the response to the signal of the pure analogue. Figure 2c shows the relative response of all the analogues in the 100:1 ratio, further highlighting that competitive ionization of any of the compounds with heroin would likely not cause a false negative detection. Butyryl fentanyl was the only compound which exhibited an appreciable loss of signal (35% reduction). The similar behavior between fentanyl and the studied analogues suggests that analogues not examined in this work may exhibit the same behavior in the presence of heroin or excipients.

To evaluate competitive ionization effects in the IMS, a parallel study was completed with 5 ng wipe^{-1} used as the constant mass of fentanyl and (0 ng wipe^{-1}, 5 ng wipe^{-1}, 12.5 ng wipe^{-1}, 25 ng wipe^{-1}, 50 ng wipe^{-1}, 250 ng wipe^{-1}, and 500 ng wipe^{-1}) as the masses of competing compounds. This provided ratios of competing compound: fentanyl of 0:1, 1:1, 2.5:1, 5:1, 10:1, 50:1, and 100:1. The response of fentanyl, measured as the maximum peak height during analysis, showed little suppression in the presence of acetaminophen (Figure 3a), while mannitol provided an enhancement in the fentanyl signal, similar to what was observed of caffeine using TD-DART-MS. Caffeine (Figure S19) also induced minimal suppression for all but the highest level (100:1), where an approximately 50% reduction in the fentanyl signal was observed. Quinine enhanced ionization of fentanyl, with an increase in fentanyl signal observed for all levels (Figure S19). Procaine proved to be the most problematic excipient, causing little suppression of the fentanyl signal in a 1:1 mixture but complete suppression at higher ratios. Rapid suppression of the fentanyl peak is accompanied by an increasing procaine peak intensity, indicating that procaine has a higher proton affinity and can consume all available charge. This may mean that additional high proton affinity compounds (such as cocaine) could cause similar reductions in fentanyl signals on the IMS when present in a mixture.

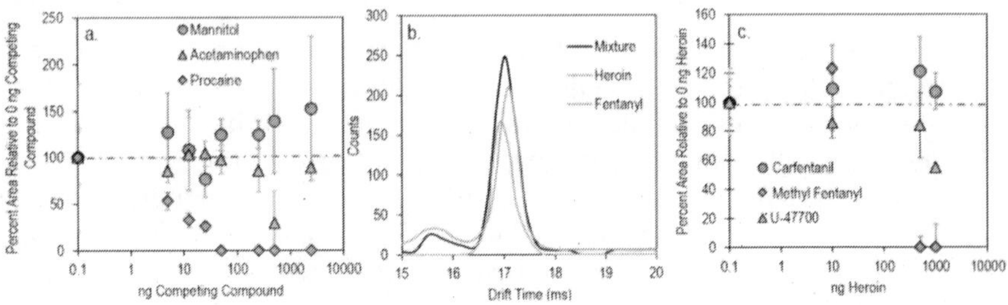

Figure 3. (a) Competitive ionization studies using IMS of fentanyl in the presence of increasing amounts of procaine, mannitol, and acetaminophen. (b) Enlarged overlaid IMS plasmagram windows of the signal for pure heroin, pure fentanyl, and the combination of the two compounds. (c) Competitive ionization studies of select fentanyl analogues in the presence of increasing amounts of heroin. Error bars represent the standard deviation of 3 replicate measurements.

Evaluation of IMS competitive ionization effects in fentanyl – heroin mixtures was complicated by resolution limitations of this instrument. A previous study using an earlier model of instrument (Ionscan 400b, Smiths Detection), reported separation of fentanyl and heroin peaks, with detection of both compounds over a wide range of binary compositions [4]. The instrument used in the current study, although a later model, has a lower resolution, and the peaks for the two compounds are not resolved when analyzing mixtures (Figure 3b). A "combination" peak lies between heroin and fentanyl and has a reduced mobility value of 1.0445 Because alarm windows are typically programed with a window of ±0.003, this combination peak may fall outside of the alarm window of both heroin and fentanyl. This limitation can be addressed by creating a new alarm window for the combination peak. By using three alarm windows (heroin, fentanyl, combination), 100% positive alarms for fentanyl in all heroin: fentanyl combinations (0:1 to 50:1) for 10 ng wipe^{-1} fentanyl were obtained. The alarm for fentanyl was defined as an alarm for either fentanyl and/or the combination peak. Without the combination peak, there were alarms for almost all samples, but typically only for either heroin or fentanyl. Creating additional alarm windows to capture non-resolved and shifted peaks is a workable, but not ideal, solution to the problem. The previous version of the instrument shows that the resolution problems are not a fundamental limitation of IMS, but simply a result of changes in design due to market forces [4]. It is possible to have a system that could resolve fentanyl and heroin combination samples, leading to clear indicators of fentanyl in the presence of heroin.

The same problem with non-resolved peaks occurs for mixtures of heroin with other fentanyl analogues such as methyl fentanyl, o-fluorobutyryl and p-fluoroisobutyryl fentanyl, which are within ±0.037 of the K_0 value of heroin. The combination peaks that form from these 3 analogues with heroin cannot be reliably differentiated from each other, or from the fentanyl:heroin combination peak. However, a combination peak can be a reliable indicator of the presence of one of the 4 fentanyl compounds in a mixture with heroin. Carfentanil, benzyl, furanyl, acetyl, and valeryl fentanyl have K_0 values that are further away from heroin, and can be resolved for the entire range of studied compositions (10 ng $wipe^{-1}$ fentanyl analogue combined with 0 ng $wipe^{-1}$, 10 ng $wipe^{-1}$, 500 ng $wipe^{-1}$, and 1000 ng $wipe^{-1}$ of heroin). For these resolved mixtures, ion suppression could be evaluated by monitoring peak intensity, with carfentanil maintaining its intensity even at the highest additions of heroin (Figure 3c). Detection of U-47700 was also possible across the entire range, and minimal suppression of signal was seen until the highest addition of heroin.

A similar competitive ionization study was completed with buprenorphine and naloxone, the two active ingredients in the drug suboxone®, which are present at a 4:1 ratio. To evaluate the competitive ionization that may occur in this mixture, pure buprenorphine and naloxone, as well as a binary mixture, were analyzed using 10 ng $wipe^{-1}$ and 2.5 ng $wipe^{-1}$ deposits respectively. Integrated peak areas of both compounds, when analyzed by TD-DART-MS, increased in the binary mixture, relative to pure, with a 23% increase in signal observed for buprenorphine and 8.5% increase in signal observed for naloxone. When analyzed by IMS, a slight reduction in naloxone signal was observed (approximately 18%) with a slight increase in buprenorphine (approximately 25%) when compared to the pure compounds. Like the fentanyl-heroin binary mixture, these results are promising as they demonstrate that both compounds would be detected in the drug, at least at these lower mass loadings.

3.3. Detection in background matrices

The final set of experiments aimed to establish whether or not realistic background matrices would hinder the detection of trace levels of these compounds. In scenarios encountered by first responders or forensic scientists, samples would be collected from materials stored under non-ideal conditions, and therefore would likely contain materials such as phthalates and plasticizers from plastic bags, fingerprint residues, and dirt or dust, in addition to heroin or excipients. To evaluate whether background matrices would complicate detection of these compounds, a binary 10:1 heroin: fentanyl mixture in each of three simulated background matrices was analyzed. Background matrices included artificial fingerprint residue, standard reference material (SRM) sediment to simulate dirt, and retrieval of the materials off of a plastic bag.

The first challenge matrix examined was an artificial fingerprint material containing over forty compounds commonly found in fingerprint residue at biologically relevant concentrations [14]. Approximately 3 μg of the fingerprint material, 10 ng of fentanyl, and 100 ng of heroin were deposited directly onto either the PTFE or meta-aramid wipes and directly inserted into the TD-DART-MS or IMS, respectively. The resulting spectra, shown in Figure 4a and b, highlight the detection of both compounds in the presence of a largely varied, and complex, matrix. Analysis by IMS produced an alarm for fentanyl, while one of two replicates also produced an alarm for heroin. Cleardown of the TD-DART-MS system after analysis of this background matrix and all others was on the order of 5–20 s. Analysis of the fingerprint residue did increase the IMS cycle time between samples from 15 s to 120 s. Background spectra for this, and all matrices, can be found in the Supplemental information (Figure S20).

Along with fingerprint residue, it is likely that a surface that is a candidate for bearing trace narcotics will contain dirt or dust. Dirt presents several challenging aspects for trace detection, namely the presence of trace organics, trace inorganics, and particulates which can clog the instrumentation. The effect of dirt on the detection of the binary heroin fentanyl mixture was examined by depositing approximately 50 μL of an aqueous suspension of NIST SRM 1944 (New York/New Jersey Waterway Sediment) onto the wipe along with 100 ng heroin and 10 ng fentanyl followed by direct insertion into the instruments. The aliquot of SRM 1944 was allowed to dry prior to deposition of the heroin and fentanyl. NIST SRM 1944 contains a number of organic and inorganic constituents, as well as particulate with a mean diameter of approximately 150 μm. Figure 4c highlights the detection of both heroin and fentanyl in this matrix using TD-DART-MS, with a strong signal for fentanyl. The relative signal for heroin is reduced in this matrix, compared to the artificial fingerprint, though it is still detectable, with a signal-to-noise ratio of 127. This reduction in signal for heroin was likely caused by enhanced competitive ionization from a number of compounds in the SRM that have high proton affinities. Analysis by IMS (Figure 4d) only produced an alarm for fentanyl, and a reduction in the RIP was also observed.

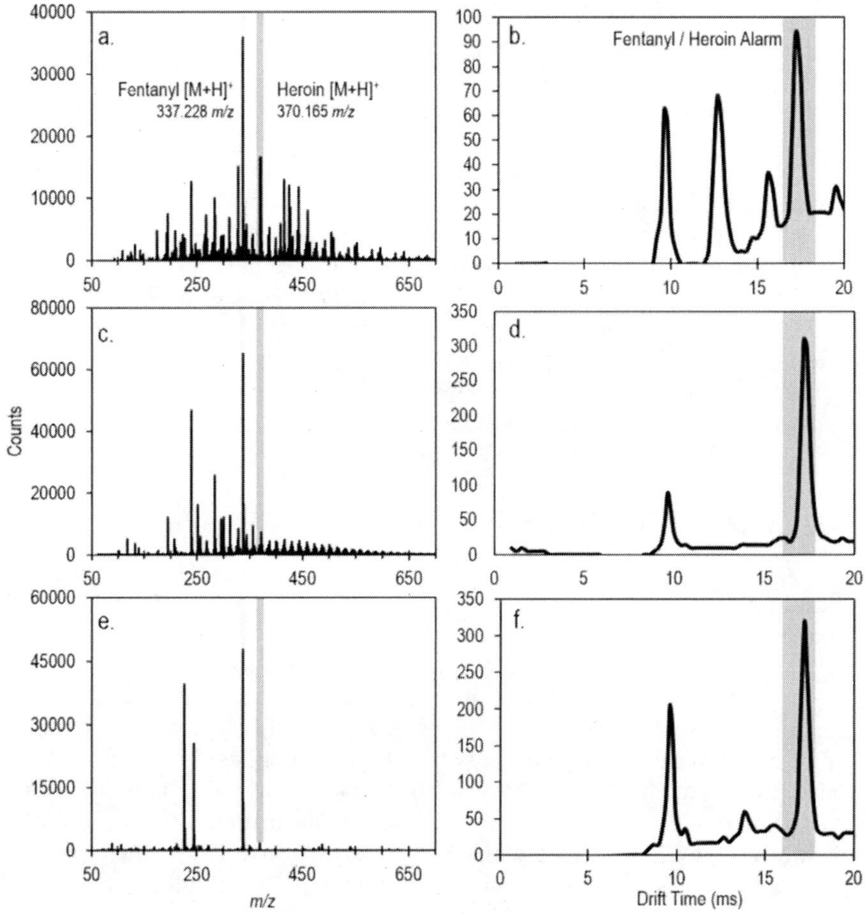

Figure 4. Representative mass spectra (a, c, and e) and averaged IMS spectra (b, d, and f) of fentanyl and heroin in the presence of artificial fingerprint material (a and b), simulant dirt (c and d), and off of a plastic bag (e and f). Spectra are not background subtracted.

The final background matrix intended to investigate real-world wipe collection and the effects of phthalates and plasticizers on detection by collecting the fentanyl and heroin deposit off of a plastic bag using a wipe. To do so, a polyethylene bag, approximately 7.5 cm by 11.5 cm, was handled for several minutes to purposefully contaminate the bag with real fingerprint residue. After handling, for TD-DART-MS, 100 ng fentanyl and 1 µg heroin were deposited. The entirety of the bag was then wiped and inserted directly into the instrument. The higher levels of fentanyl and heroin were chosen because typical collection efficiencies of the PTFE-coated wipes are on the order of single percentages, which allowed for the collection of approximately 5 ng fentanyl and 50 ng heroin [20]. Figure 4e shows the rapid detection of both narcotics by TD-DART-MS after being collected off the plastic bag. Again heroin exhibited a suppression in signal, while the signal for fentanyl was quite strong. For IMS, 50 ng of fentanyl and 500 ng of heroin were used because the meta-aramid wipe has an expected collection efficiency around 20% [20]. Both fentanyl and the combination peak were observed and produced alarms (Figure 4f). Being able to sample rapidly nanogram quantities of material off the outside of a plastic bag, without the need for extraction, is extremely beneficial to first responders and forensic scientists as it provides the opportunity to triage or screen samples without having to open them. This type of approach would minimize risk of exposure to the analyst by reducing the likelihood of coming into contact with powdered material.

4. CONCLUSION

Rapid detection of fentanyl and fentanyl analogues with nanogram to sub-nanogram sensitivity is possible using both TD-DART-MS and IMS. Detection of low levels directly wiped off of a surface can be advantageous to law enforcement, first responders, and forensic scientists as it provides an option for screening an item of evidence without having to come in direct contact with the bulk powder. Both techniques detected fentanyl in the presence of heroin down to 0.1% by weight fentanyl. Lower instrumental resolution in the IMS caused the formation of a heroin and fentanyl combination peak which could be reliably traced to the presence of heroin and fentanyl or a fentanyl analogue. Other adulterants provided varying degrees of signal enhancement or suppression; however, with TD-DART-MS, the fentanyl signal was maintained regardless of excipient and amount present. IMS provided similar results, except with procaine, which consumed all available charge at higher than 1:1 ratios of procaine to fentanyl. Complex matrices, including artificial fingerprint residue, dirt simulant, and a plastic bag, did not prevent detection of fentanyl on either system. The 16 fentanyl analogues behaved similarly to fentanyl, and it is expected comparable results would be obtained in the presence of excipients or background matrices.

While these techniques are not quantitative, they do constitute qualitative tools to help identify and properly mitigate fentanyl and fentanyl analogues in evidence, on persons, and elsewise. Analysis by these techniques is both rapid (less than 10 s), sensitive, and generally discriminative. DART was able to separate all but one pair of isomeric compounds, while IMS had less selectivity in discriminating similar analogues. However, the only identified potential issue for false alarms is the similar K_0 values for naloxone and despropionyl fentanyl. This work highlights the ability of TD-DART-MS and IMS to rapidly detect fentanyl and fentanyl analogues, though prior work already has highlighted their capabilities in the detection of more conventional narcotics. Both tools provide potential avenues for rapid on-site or laboratory-based screenings from mobile units to forensics laboratories to emergency rooms.

DISCLAIMERS

‡Official contribution of the National Institute of Standards and Technology; not subject to copyright in the United States.

*Certain commercial products are identified in order to adequately specify the procedure; this does not imply endorsement or recommendation by National Institute of Standards and Technology, nor does it imply that such products are necessarily the best available for the purpose.

REFERENCES

1. Carfentanil: A Dangerous New Factor in the U.S. Opioid Crisis, 2016
2. J.R. Riches, R.W. Read, R.M. Black, N.J. Cooper, C.M. Timperley, Analysis of clothing and urine from Moscow theatre siege casualties reveals carfentanil and remifentanil use, J. Anal. Toxicol. 36 (2012) 647–656, http://dx.doi.org/ 10.1093/jat/bks078.
3. Influx of Fentanyl-laced Counterfeit Pills and Toxic Fentanyl-related Compounds Further Increases Risk of Fentanyl-related Overdose and Fatalities, Centers for Disease Control and Prevention, 2016. https:// emergency.cdc. gov/han /han00395.asp.
4. J.R. Verkouteren, J.L. Staymates, Reliability of ion mobility spectrometry for qualitative analysis of complex, multicomponent illicit drug samples, Forensic Sci. Int. 206 (2011) 190–196,
5. E. Sisco, T.P. Forbes, M.E. Staymates, G. Gillen, Rapid analysis of trace drugs and metabolites using a thermal desorption DART-MS configuration, Anal. Methods. 8 (2016) 6494–6499,
6. H.J. Kim, E.H. Jee, K.S. Ahn, H.S. Choi, Y.P. Jang, Identification of marker compounds in herbal drugs on TLC with DART-MS, Arch. Pharm. Res. 33 (2010) 1355–1359,
7. A.D. Lesiak, R.A. Musah, R.B. Cody, M.A. Domin, A.J. Dane, J.R.E. Shepard, Direct analysis in real time mass spectrometry (DART-MS) of "bath salt" cathinone drug mixtures, Analyst 138 (2013) 3424,
8. A.D. Lesiak, J.R. Shepard, Recent advances in forensic drug analysis by DARTMS, Bioanalysis 6 (2014) 819–842,
9. A.D. Lesiak, R.A. Musah, M.A. Domin, J.R.E. Shepard, DART-MS as a preliminary screening method for "herbal incense": chemical analysis of synthetic cannabinoids, J. Forensic Sci. 59 (2014) 337–343,

10. J.H. Gross, Direct analysis in real time–a critical review on DART-MS, Anal. Bioanal. Chem. 406 (2014) 63–80,
11. P. Nikolaou, M. Katselou, I. Papoutsis, C. Spiliopoulou, S. Athanaselis, U-47700, an old opioid becomes a recent danger, Forensic Toxicol. 35 (2017) 11–19,
12. R.M. Verkouteren, J.R. Verkouteren, Inkjet metrology: high-accuracy mass measurements of microdroplets produced by a drop-on-demand dispenser, Anal. Chem. 81 (2009) 8577–8584,
13. R.M. Verkouteren, J.R. Verkouteren, Inkjet metrology II: resolved effects of ejection frequency, fluidic pressure, and droplet number on reproducible dropon-demand dispensing, Langmuir 27 (2011) 9644–9653,
14. E. Sisco, J. Staymates, K. Schilling, A chemically relevant artificial fingerprint material for the cross-comparison of mass spectrometry techniques, Can. Soc. Foren. Sci. J. 48 (2015) 200–214,
15. DART Forensics Library, (n.d.). http://chemdata.nist.gov/mass-spc/ms-search/DART_Forensic.html (accessed October 17, 2016).
16. Standard Test Method for Determining Limits of Detection in Explosive Trace Detectors, ASTM International, 2014.
17. Adulterants and diluents in heroin, Amphetamine, and cocaine found on the illicit drug market in Aarhus, Denmark – Research – Aarhus University, Open Foren. Sci. J. 2 (2009) 16–20.
18. C. Cole, L. Jones, J. McVeigh, A. Kicman, Q. Syed, M.A. Bellis, CUT: A Guide to Adulterants, Bulking Agents and other Contaminants Found in Illicit Drugs, (2010). http://www.cph.org.uk/wp-content/uploads/2012/08/cut-a-guide-tothe-adulterants-bulking-agents-and-other-contaminants-found-in-illicitdrugs.pdf
19. E. Sisco, T.P. Forbes, Rapid detection of sugar alcohol precursors and corresponding nitrate ester explosives using direct analysis in real time mass spectrometry, Analyst 140 (2015) 2785–2796,
20. J.R. Verkouteren, J.L. Coleman, R.A. Fletcher, W.J. Smith, G.A. Klouda, G. Gillen, A method to determine collection efficiency of particles by swipe sampling, Meas. Sci. Technol. 19 (2008) 115101,

CHAPTER 5

Simple and Rapid Determination of Myristicin in Human Serum

Andrzej L. Dawidowicz • Michal P. Dybowski

[1].*Faculty of ChemistryMaria Curie Sklodowska UniversityLublinPoland*

ABSTRACT

Myristicin (5-allyl-1-methoxy-2,3-methylenodioxybenzene) is the main component of nutmeg (*Myristica fragrans* Houtt.) essential oil. The increasing use of myristicin as a cheap hallucinogenic intoxicant, frequently causing fatal cases of myristicin poisoning, requires new methods for determination of this compound in blood. This report describes the rapid, simple, and useful procedure for myristicin analysis in human serum, involving myristicin–protein complex degradation before chromatographic analysis. The developed method is characterized by a high recovery (above 99 %), a low detection limit (6.0 ng/g) and good repeatability (average RDS of 2.01 %).

Keywords: Myristicin Human serum SPE Gas chromatography

INTRODUCTION

Myristicin (5-allyl-1-methoxy-2,3-methylenodioxybenzene) is the main component of nutmeg essential oil [1], which exists in the nutmeg nut (*Myristica fragrans* Houtt.) and in mace, the waxy red covering of the nutmeg seed. Readily accessible on local markets, these spices have, at times, been drugs of abuse. On the other hand, nutmeg and mace are used in Asia as traditional medicines treating stomach cramps, diarrhea, and rheumatism. Information currently available on myristicin explains why its effects range from harmful to beneficial. As reported in the literature [2], this alkenylbenzene acts as a serotonin receptor antagonist and as a hallucinogenic compound. Moreover, it induces: neurotoxicity in human neuroblastoma SK-N-SH cells [3]; P450 liver enzyme in several families of rat, including the P450 1A enzyme [4], which is also induced by polyaromatic hydrocarbons, dioxins, and polychlorinated biphenyls; glutathione S-transferase [5] and forms DNA adducts in rodent livers. The ability to form DNA adducts is a common property of most chemical carcinogens and their metabolites [6]. Several tests have shown myristicin to be a weak monamine oxidase (MAO) inhibitor [7]. Acutely toxic doses of myristicin can cause organ damage [8]. Myristicin poisoning can lead to many health problems such as convulsions, delirium, blurred vision, palpitations, nausea, dehydration, general body pain, and others [9]. Those symptoms usually occur 3–6 h after the ingestion of myristicin or foodstuffs containing it, and persist up to 72 h [10]. In recent years, many cases of nutmeg poisoning have been reported, including several fatal myristicin cases [11, 12, 13]. Such poisonings can result not only from the toxic effect of myristicin itself but also from the combined toxic effects of its use with other substances [12]. Because of potential widespread human exposure through foods and beverages and the possibility of adverse effects in diverse populations, myristicin was presented to the Chemical Selection Working Group (CSWG) for review.

Two publications describing myristicin determination in biofluid can be found in the literature [14, 15]. The first concerns the myristicin metabolism in rats, but the results presented in the article were not obtained in a quantitative fashion. As a result, it is difficult to judge the method used for myristicin estimation in the reported study [14]. Moreover, the described method involves a derivatization process, which is frequently inconvenient in GC analysis (the formation of deposits in the liner of GC injector or in chromatographic column). In the second study [15], a method for HPLC determination of myristicin in rat serum is described. The procedure involves deproteinization, separation, and fractionation concentration of myristicin by appropriate switching of columns in the HPLC system. However, it should be stressed that the removal of serum proteins in the HPLC process can lead to the loss of analytes forming analyte–protein complexes. The present report describes the GC procedure for myristicin analysis in human plasma. The procedure involves a protein precipitation process, which generally degrades drug–protein complexes, and solid-phase extraction (SPE) isolation of myristicin from the examined materials. The proposed analytical approach can be considered as a method of choice for the estimation of myristicin in human fluids.

MATERIALS AND METHODS

Chemicals and reagents

Hexane and ethanol, both of analytical grade, were supplied by POCH (Gliwice, Poland). The Sepra C18-E sorbent (50 μm, 65Å) used in the SPE process was purchased from Phenomenex (Torrance, CA, USA). Myristicin, of analytical standard, was supplied by Fluka (Buchs, Switzerland). Blood for the validation procedure and blood samples containing myristicin were obtained from volunteers. Water was purified on a Milli-Q system (Millipore, Bedford, MA, USA).

Sample preparation

Preparation of blood samples

Myristicin-free and myristicin-containing blood samples, obtained from healthy volunteers recruited from our laboratory, were obtained from forearm vein using Sarstedt blood collection systems. Heparin sodium was used as an anticoagulant. Myristicin was administered to the volunteers in the form of an alcoholic drink. Twenty grams of the myristicin solution (5 mg/g) was consumed by each volunteer 1 h before blood sampling.

All volunteers gave consent to participate in the studies and were made aware of the option to withdraw from further participation in the study at any time and for any reason. Each of them signed a consent form for processing of the experimental data obtained in this study.

Standard plasma solutions

To estimate myristicin recovery from plasma in the applied method, eight plasma samples containing 0.01, 0.1, 1, 10, 25, 50, 75, and 100 μg/g, respectively (same concentrations used for equipment calibration—see Chromatographic analysis below), were prepared. Pure plasma samples were spiked with myristicin solutions in ethanol/water mixture (30/70 % v/v). In each case, the volume of the spiking solution was less than 80 µl.

Myristicin isolation by SPE

To eliminate eventual loss of myristicin caused by its binding with plasma proteins, 1.5 ml of ethanol was added to 1 ml of plasma sample in a 4-ml glass vial and vortexed for 5 min. After centrifugation (945 *g* for 15 min), 1 ml of supernatant was mixed with 1.5 ml of water and 2 ml of the obtained solution was transferred into an SPE cartridge containing 0.2 g of SepraC18-E. Before loading, the cartridge was washed with 5 ml of *n*-hexane and then vacuum-dried (ca. 5 min). After removal of the loading solvent from the SPE bed, the remaining components were eluted into a calibrated flask with 2 ml of *n*-hexane and the solution was analyzed by GC. The SPE process was carried out using the SPE vacuum chamber (SPE-12G from J.T.Baker USA) at an eluent velocity of 1 drop per second.

Myristicin isolation by liquid–liquid extraction (LLE)

The mixture of plasma sample (1 ml) and hexane (1 ml) was vigorously shaken for 10 min at 2000 rpm. After phase separation (centrifugation), the hexane layer was subjected to GC analysis.

Chromatographic analysis

The amount of myristicin was estimated using a GC with flame ionization detection (FID) (GC 2010, Shimadzu, Kyoto, Japan). To confirm the peak purity, each sample was also analyzed by GC-mass spectrometry (GC/MS QP2010, Shimadzu, Kyoto, Japan).

For identification of SPE extracts, a ZB5-MS fused-silica capillary column (30 m × 0.25 mm i.d., 0.25 µl film thickness; Phenomenex, USA) was used. Helium (grade 5.0) was used as carrier gas. A 1-µl aliquot of the sample was injected using an AOC-20i type autosampler. The injector's temperature was 310 °C. The following temperature program was applied: 1 min at 50 °C, followed by a linear temperature increase up to 250 °C at 6 °C /min. The mass spectrometer was operated in EI mode at 70 eV; the ion source temperature was 220 °C. The mass spectra were measured in the range 35–360 amu. Qualitative analysis was conducted by comparing the retention indices and MS spectra for the obtained peaks with analogous data from NIST'05 to Adams databases.

Quantification of extracts was performed by injecting 1 μl of the sample using the same autosampler and capillary column. The temperature program during GC-FID separation was the same as for GC-MS.

Myristicin peak identification was carried out by comparing the GC retention index with those from GC-MS and with the retention data for myristicin standard.

The applied GC-FID equipment was calibrated using myristicin standard solutions. The working solutions were obtained by serial dilutions of the stock solution with hexane to obtain the following myristicin concentration: 0.01, 0.1, 1, 10, 25, 50, 75, and 100 μg/g.

RESULTS AND DISCUSSION

Figure 1 presents chromatograms for myristicin analytical standard (Figure 1a), SPE extracts from blank plasma (Figure 1b), and a plasma sample spiked with alcohol solution of myristicin standard (Figure 1c), with the two latter results were obtained according to the presented procedure. Figure 1d shows the chromatogram for the LLE extract from blank plasma, while Figure 1e shows the additional chromatogram for the SPE extract from the plasma of a volunteer who consumed an alcohol solution of myristicin. As appears from the figure, the applied chromatographic conditions allow for sufficient resolution of myristicin from sample matrix peaks, both in the SPE and LLE procedures. However, the use of SPE seems better due to the lack of peaks occurring in the vicinity of myristicin retention in this procedure. In the case of LLE, the extract contains some cholesteric derivatives (Figure 1d), the presence of which requires more frequent cleaning of the GC injector and leads to faster column deterioration. The identity and purity of the myristicin peak were confirmed by the MS spectra and the value of the retention index (1520), which is exactly the same as the literature value [16].

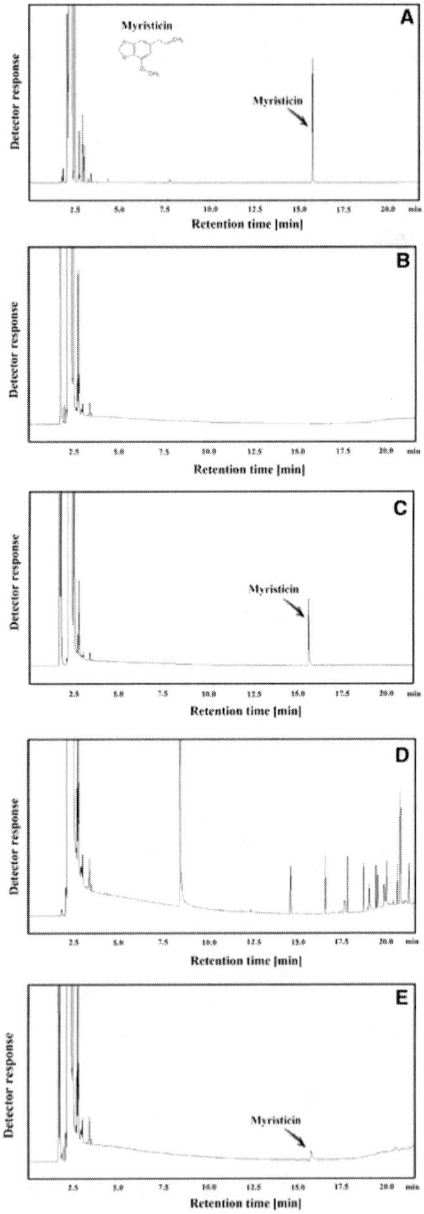

Figure 1. GC chromatograms of: **a** Myristicin analytical standard, **b** SPE extract from blank plasma sample, **c** SPE extract from plasma sample spiked with alcoholic solution of myristicin standard, **d** LLE extract from blank plasma, **e** SPE extract from plasma of a volunteer who consumed alcoholic solution of myristicin

Figure 2 illustrates the calibration plot for myristicin in the concentration range from 10 ng/g to 100.0 µg/g. The resultant calibration curve shows excellent linearity with $R^2 = 0.9998$. The intercept value (48.1) equals 10 % of the detection limit. The obtained value exceeds the FDA requirements of $R^2 \geq 0.999$ from a minimum of five observations [17, 18].

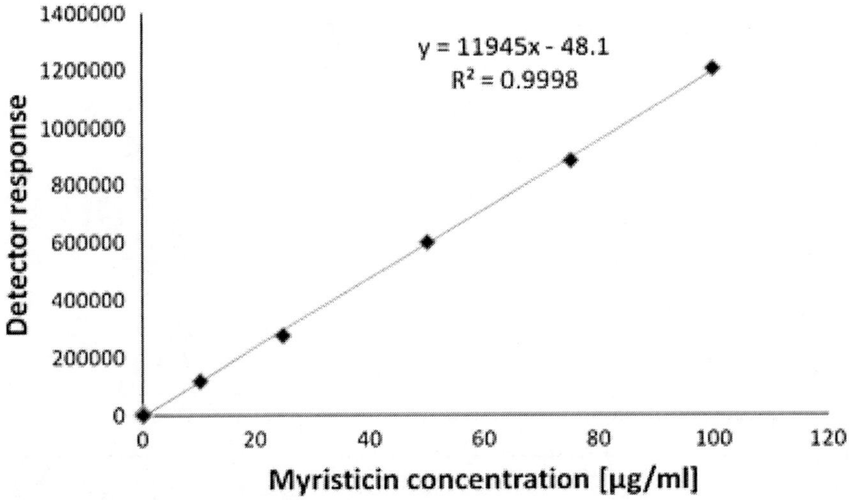

Figure 2. Calibration plot for myristicin: concentration range 0.01–100 µg/g

The precision and accuracy of the SPE step of the applied procedure are shown in Table 1. The presented data were obtained by five successive analyses for each sample spanning the range of the calibration curve. The examined method is characterized by very high SPE recovery (mean recovery 99.08 %) and exhibits excellent precision. The average precision of the SPE step determined as RSD for the samples at low, intermediate, and high myristicin concentrations is 2.01 %.

In order to estimate the limit of detection (LOD) and the limit of quantitation (LOQ), SPE extracts from plasma samples spiked with the myristicin standard were injected into the GC to determine the signal-to-noise (S/N) ratio. The LOD, defined as a S/N ratio of 3 was 6 ng/g, while the LOQ, defined as a S/N ratio of 10, was 20.0 ng/g.

Table 1. Validation data obtained by analysis of spiked samples using the described method (n = 5)

Concentration (µg/g)	Myristicin		
	SD (µg/g)	RSD (%)	Recovery (%)
100	1.684	1.684	99.89
75	1.463	1.951	99.45
50	0.947	1.894	99.84
25	0.461	1.845	99.36
10	0.200	2.001	99.47
1	0.021	2.121	98.47
0.1	0.002	2.241	99.01
0.01	0.0002	2.351	97.15
Average		2.011	99.20

Table 2. Concentration of myristicin in different human serum samples (n = 5)

Test subject	Weight (kg)	Age (years)	Sex	Ingested dose of myristicin alcoholic solution (5 mg/g) (g)	Detected analyte (µg/g)
1	95	24	M	20	17.60
2	82	49	M	20	19.71
3	74	45	F	20	21.70
4	71	49	M	20	22.43
5	55	23	F	20	33.25

M male, *F* female

Table 2 brings together the myristicin concentrations in human plasma samples taken from volunteers who consumed the alcoholic solution of myristicin standard 1 h before sampling. The observed differences in myristicin plasma concentration result from the different metabolisms of the volunteers connected with their age, sex, weight, etc. The estimated myristicin concentrations in human plasma are much higher than the LOQ, although the administered myristicin doses were significantly below the amount believed to cause negative symptoms in humans.

CONCLUSIONS

The analytical procedure of myristicin determination in rat serum described in the literature involves deproteinization, separation, and fractionation concentration by switching of columns in the HPLC system [15]. However, it should be stressed that the removal of serum proteins in the HPLC process can lead to the loss of the analytes forming analyte–protein complexes.

The present report describes the rapid, simple, and useful procedure of myristicin analysis in human serum, involving myristicin–protein complex degradation before chromatographic analysis. The method is characterized by a high recovery (above 99 %) and a low detection limit of 6.0 ng/g, which to date are the most satisfactory values reported in the literature for myristicin determination in plasma [14, 15]. The analysis was achieved by dilution of the plasma alcoholic solution to reach a final alcohol concentration of less than 40 %. The SPE conditions for myristicin isolation from human plasma solutions agree with those elaborated for the isolation of low-molecular oxygen compounds from essential oils [19]. After slight tuning, the described method can be adapted for the determination of myristicin in other biofluids, such as cerebral spinal fluid and urine samples.

OPEN ACCESS

REFERENCES

1. Bakkali F, Averbeck S, Averbeck D, Idaomar M (2008) Biological effects of essential oils—a review. Food Chem Toxicol 46:446–475
2. Barceloux DG (2009) Nutmeg (Myristica fragrans Houtt.). Dis Mon 55:373–379
3. Lee BK, Kim JH, Jung JW, Choi JW, Han ES, Lee SH, Ko KH, Ryu JH (2005) Myristicin-induced neurotoxicity in human neuroblastoma SK-*N*-SH cells. Toxicol Lett 157:49–56
4. Yun CH, Lee HS, Lee HY, Yim SK, Kim KH, Kim E, Yea SS, Guengerich FP (2003) Roles of human liver cytochrome P450 3A4 and 1A2 enzymes in the oxidation of myristicin. Toxicol Lett 137:143–150
5. Zheng GQ, Kenney PM, Lam LKT (1992) Myristicin: a potential cancer chemopreventive agent from parsley leaf oil. J Agric Food Chem 40:107–110
6. Gupta KP, van Golen KL, Putman KL, Randerath K (1993) Formation and persistence of safrole-DNA adducts over a 10,000-fold dose range in mouse liver. Carcinogenesis 14:1517–1521
7. Truitt Jr EB (1979) The pharmacology of myristicin and nutmeg. In: Efron DH, Holmstedt B, Kline NS (eds) Ethnopharmacologic search for psychoactive drugs. Raven press, New york, pp 215–222
8. Bethesda MD (1997) RTECS (registry of toxic effects of chemical substances). Record No.: 13815
9. Demetriades AK, Wallman PD, McGuiness A, Gavalas MC (2005) Low cost, high risk: accidental nutmeg intoxication. Emerg Med J 22:223–225
10. Sangalli BC, Chiang W (2000) Toxicology of nutmeg abuse. Clin Toxicol 38:671–678
11. Forrester MB (2005) Nutmeg intoxication in Texas, 1998–2004. Hum Exp Toxicol 24:563–566
12. Stein U, Greyer H, Hentschel H (2001) Nutmeg (myristicin) poisoning—report on a fatal case and a series of cases recorded by a poison information centre. Forensic Sci Int 118:87–90
13. Weil AT (1996) The use of nutmeg as a psychotropic agent. Bull Narc 4:15–23
14. Lee HS, Jeong TC, Kim JH (1998) In vitro and in vivo metabolism of myristicin in the rat. J Chromatogr B 705:367–372
15. Lee HS, Kim JH, Kim K, Do KS (1998) Determination of myristicin in rat serum samples using microbore high performance liquid chromatography with column-switching. Chromatographia 48:365–368
16. Adams RP (2007) Identification of essential oil components by gas chromatography/mass spectrometry, 4th edn. Allured Publishing Corporation, Carol Stream
17. Center for Drug Evaluation and Research (CDER), US Food and Drug Administration (FDA) (1996) Guide for Industry. Q2B Validation of Analytical Methods Protocol. Metrology. http://www.fda.gov./CDER/guidance/index.htm
18. Center for Drug Evaluation and Research (CDER), US Food and Drug Administration (FDA) (1994) Reviewer Guidance. Validation of Chromatographic Methods. www.fda.gov/CDER
19. Dawidowicz AL, Dybowski MP (2010) SPE isolation of low-molecular oxygen compounds from essential oils. J Sep Sci 33:3213–3220

CHAPTER 6

A Pioneer Case Study on Identification of Infant Rhinoceros Horn

Dinesh Kumar Jha*, Sandeep Kumar Gupta, Nirajan Thapa Kshetry, Raju Panday and Basanta Raj Pokharel

National Forensic Science Laboratory Biology, Khumaltar, Lalitpur, Kathmandu, Nepal

ABSTRACT

Objective: To investigate the authenticity of infant Rhino horn sample from morphometric, microscopic, genetic and phylogenetic studies.

Methods: A suspected rhino horn sample (11 cm × 13 cm) was sent to the laboratory for identification. Relevant morphometric data of sample were taken. Similarly, established methods for microscopic examination of rhino horn were adopted. Using standard universal primer, a part of Cytochrome b (Cyt b) and 12S rRNA was amplified from sample, sequenced and aligned using BLAST of the GenBank (NCBI) and in-house reference database available at Wildlife Institute of India, Dehradun, India for decisive confirmation of species.

Results: Morphometric analysis showed the sample was oval, dome shaped with a bony part which was supported by upper skin. Presence of filamentous units is the microscopic identifying feature of genuine rhino horn. However, microscopic examination showed very unclear filamentous units because of the immature nature of sample. Mitochondrial sequence data revealed the source of suspected sample was of a one-horn Rhinoceros

(Rhinoceros unicornis). Similarly, NJ method for phylogenetic analysis for both Cytb and 12S rRNA identified sequence similarity with Rhinoceros unicornis.

Conclusion: Even though the morphometric and microscopic data do not reveal much about the infant rhino horn, mtDNA sequencing used in this pioneer study can be applied in the wildlife forensic for identification of such parts in future and data used as evidence against the poachers in the court of law.

Keywords: Infantile rhino horn; Forensic science; Filamentous unit; Cyt b; 12S rRNA ; BLAST; Neighbour- joining method

INTRODUCTION

Wildlife and their products represent the third greatest illegal traffic after drugs and arms [1]. Illegal poaching are threats to a wide range of endangered species around the world [2]. Many species are protected both at a national and international level and their conservation requires control of illegal poaching [3]. Concept of "species" is perhaps the most debated subject in evolutionary biology [4]. But accurate identification of the species from confiscated biological material is of paramount importance to the investigation of crime in forensics [5], fixing the accurate wildlife offence [6] and so as to prove the conviction of crime by offenders [7]. The test employed depends on the material seized and the available techniques [3]. Morphology and microscopy are the natural starting points in identification [8]. However, the morphological examination has some limitations especially when material is present as powders, potions, oils [3], examination from old, degraded [9], partial, small parts [7] or processed, tanned, finished wildlife products [10]. Currently, DNA sequencing analysis is the proven and most reliable method used for species identification from confiscated biological samples [7]. A number of case reports have shown the usefulness of DNA analysis for solving cases including endangered species [11-20].

Material suspected of originating from Rhinoceros is frequently seized by forensic organizations especially in parts of Asia where the rhinoceros horns are used traditionally [21]. A suspected rhino horn sample was sent to our laboratory for identification. A variety of methods are available to characterize a genuine rhino horn and its products [22]. However, usual morphological and microscopic techniques were inappropriate because of the nature of suspected sample. In this pioneer article, we report the successful identification of ambiguous infant rhinoceros horn through the combined application of mitochondrial Cytochrome b (Cyt b) and 12S rRNA gene sequence analysis and microscopic examination. The procedure can be used by scientific community and data used as evidence against the poachers to convict the wildlife crime in the court of law.

One-horned rhinoceros (Rhinoceros unicornis) inhabiting primarily in India and Nepal is included in Appendix I of the CITES. This species is also listed in schedule I of the Nepal Government's National Parks and Wildlife Conservation Act 1973 and in Indian Wildlife (Protection) Act 1972. As per provision in the Act, any person found guilty of possessing, selling or buying Rhinoceros horn or other body parts is subject to a fine of NRs. 50,000 to 100,000 (1USD~ 105NRs) or 3-15 years of imprisonment or both. Rhino horn seemed to be world's most valuable substance and thus its illicit trade has been steadily increasing with different purposes including ritual, medicinal and recreational activities [22]. In Nepal, horn bearing nasal part below the three years age of Rhinoceros is locally called Brahma-khaag (Brahma-Hindu creator god, Khaag-Rhino horn) which has notable religious value.

METHODS

Case history

The Department of National Parks and Wildlife Conservation, Babarmahal, Nepal had received a suspicious rhino horn sample. The sample was sent to our laboratory to identify the confiscated samples and if possible species identification also.

Morphometric examination

The overall shape, surface texture and configuration of sample were studied and measurements taken.

Microscopic examination

Thin transverse sections from the apex and margin of the sample skin were randomly taken with the help of a sharp razor blade. The sections were cleansed with xylene, dried, mounted in a clean microscopic slide with DPX and observed under 200X using light microscope (Olympus CX41 RF, Japan). A previously developed methodology was adopted to study the internal microscopic structure of hair [23].

DNA extraction and PCR

Approximately 100 mg skin piece was used for the extraction of DNA using standard protocol [24] in 100 μL extraction volume. The isolated DNA was amplified using the universal PCR primers: mcb398, 5'-TACCATGAGGACAAATATCATTCTG-3' and mcb869, 5'CCTCCTAGTTTGTT AGGGATTGATCG-3' [5] as well as conserved primers: 12S rRNA-F_L1091, GCTTCAAACTGGGATTAGATACCCCAC and 12S rRNA-R_H1478, TGACTGCAGAGGGTGACGGGCGGTGTGT [25] targeting the mitochondrial Cyt b and 12S rRNA gene respectively. The use of universal primers minimizes the effort of researcher, since it can be applied for PCR amplification to the DNA of all species without prior information of the victim species [26].

The PCR reaction mixture was prepared in a dedicated laminar hood. PCR amplification was performed in a Veriti thermal cycler (Applied Biosystems, Singapore) in a final volume of 10 μL containing 1 μL of the extracted DNA, 0.2 μL of forward primer (5 pmol) and 0.2 μL of forward primer (5 pmol), 1 μL of 10X buffer, 0.6 μL of Magnesium chloride (3.0 mM), 1 μL Bovine serum albumin (2 mg/ml), 0.2 μL dNTPs (2.5 mM), 0.2 μL Taq polymerase (5 units/ μL) and 5.6 μL Double distilled water. The PCR conditions were as follows: an initial denaturation at 95°C for 3 min, followed by 35 cycles of denaturation at 95°C for 35 seconds, primer annealing at 53°C for 45 seconds, primer extension at 72°C for 40 seconds. The final extension was at 72°C for 10 min. Negative PCR control was subjected to amplification. The PCR products obtained were electrophoresed on 2% agarose gel, stained with ethidium bromide, and visualized under a UV transilluminator to assess amplification (**Figure 1**).

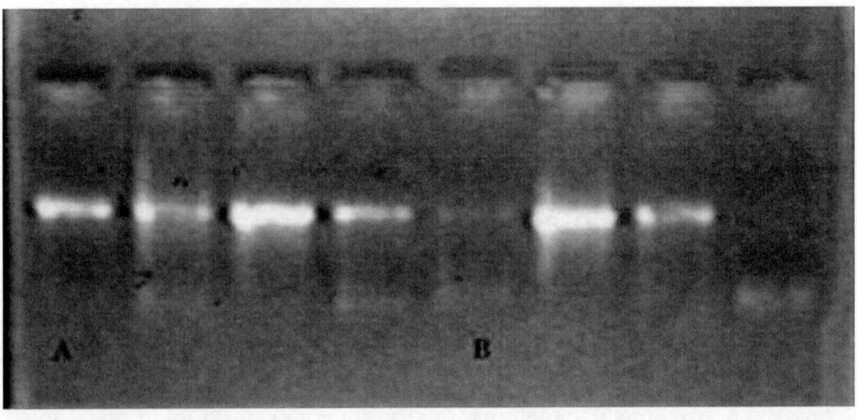

Figure 1: PCR amplification product of *12S rRNA* (Lane A) and *Cyt b* gene (Lane B) electrophored in 2% agarose gel.

Post PCR clean-up and DNA sequencing

The PCR product obtained was treated with Exonuclease I (Exo I) and Shrimp Alkaline Phosphatase (SAP) to clean the unused primers and dNTPs. 1.5 µL of PCR product was directly incubated with 0.5 µL ExoSAP at 37°C for 20 min. followed by inactivation of enzymes at 85°C for 15 min. The PCR product after Exo-I and SAP treatment was cycle sequenced. PCR amplification was performed in a Veriti thermal cycler in a final volume of 10 µL containing 2.0 µL of ExoSAP and PCR products, 1 µL of Big dye 5X sequencing buffer, 0.5 µL of Big dye RR mix v 3.1, 0.2 µL of Primer (Forward only) and 6.3 µL of double distilled water. The PCR conditions were as follows: an initial denaturation at 96°C for 5 sec, followed by 28 cycles of denaturation at 96°C for 10 seconds, primer annealing at 50°C for 5 sec, primer extension at 60°C for 4 min. Sequencing reaction products were purified by Sodium acetate ethanol precipitation method and detected by an ABI 3500xL Genetic Analyzer (Applied Biosystems, USA).

Data analysis

Quality of sequences was determined using Sequencing Analysis software v 6.0 (Applied Biosystem). Further Multiple Sequence Alignments (MSA) was performed by CLUSTAL W algorithm implemented in BIOEDIT version 7.0.9.0 [27]. The obtained consensus sequence was compared with sequences available on public domain at GenBank (http://blast.ncbi.nlm.nih.gov/) using BLAST [28] and in a "in-house database" available in Wildlife Institute of India, Dehradun, India. Phylogenetic trees including all the aligned sequences for the Cyt b and 12S rRNA genes were generated based on Kimura 2 parameter model for nucleotide substitution with 1000 bootstrap values replications using the Neighbor-joining (NJ) method [29] as implemented in MEGA v 5.1 software [30].

RESULTS AND DISCUSSIONS

Morphometry

Suspected sample was oval, dome shaped, bony part supported by upper skin (**Figure 2a** and **2b**). The measured size, height and circumference were 11 × 13, 4.0 and 25 cm respectively. Periphery of the horn was rough while the centre was smooth. Maximum thickness of skin measured was 0.81 cm. Ventral surface had frequent bony elevations (**Figure 1c**). Fewer black, short, coarse hairs were observed at the margins only (**Figure 1b**). Overall, these findings indicated that sample was skull or skull part of a big mammal. Hence, possibility of rhinoceros skull part, especially nasal part of immature ones was not debarred. Rhino horn is anchored to the dermis covering the frontal and nasal bones [31]. Authentic rhinoceros skull was unavailable in the laboratory. Thus, a definite conclusion about the sample was not possible from the morphological data only. Morphological examination has limitations especially when reference sample from known source is unavailable [6]. Identification of the suspected rhinoceros material in the absence of morphological uniqueness is problematic [21]. No any visual evidence of a Rhino horn was observed in the sample. This is probably because a rhinoceros calf is born hornless [32]; pierce the skin at 5 weeks, 39 mm long at 3 months and 100 mm long at 7 months [33].

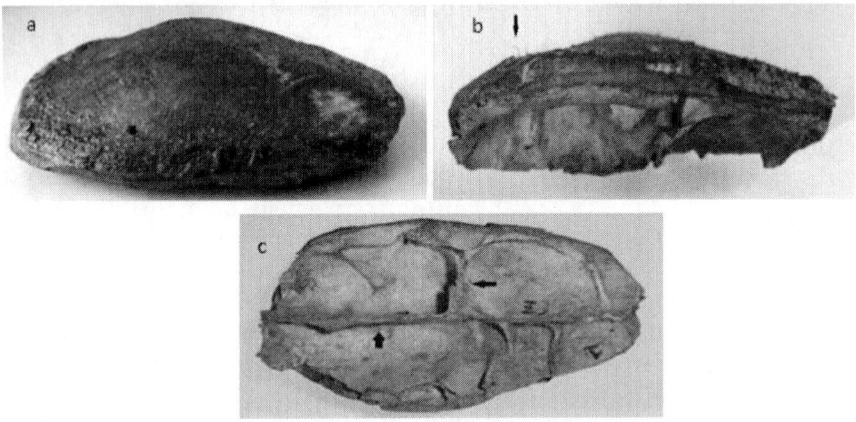

Figure 2: Dorsal (a), Lateral (b) and ventral (c) view of sample. Section site (a), hair (b) and bony elevations (c) indicated by arrow.

Microscopic findings

The occurrence of filamentous units and laminae is the microscopic identifying feature of genuine rhino horn [22]. In microscopic examination, section from margin of sample showed indistinct filamentous units, laminae with larger filamentous spaces (**Figure 3**).

Figure 3: Microscopic structure of section (Filamentous units/ laminae indicated by arrow).

This is probably due to immaturity of horn. Immature horns have greater filamentous units space in comparison to matured rhino horn [22]. Presence of these microscopic identifying features over the skin surface strongly indicated sample as an infant rhinoceros nasal part that bears horn.

Species identification based on hair characteristics has been widely used in dealing the wildlife offence [34]. However, medulla and other specific characteristics were not observed in the hair except profound pigment granules (**Figure 4**). The maximum measured length and diameter of hair was 5 mm and 150 μm respectively.

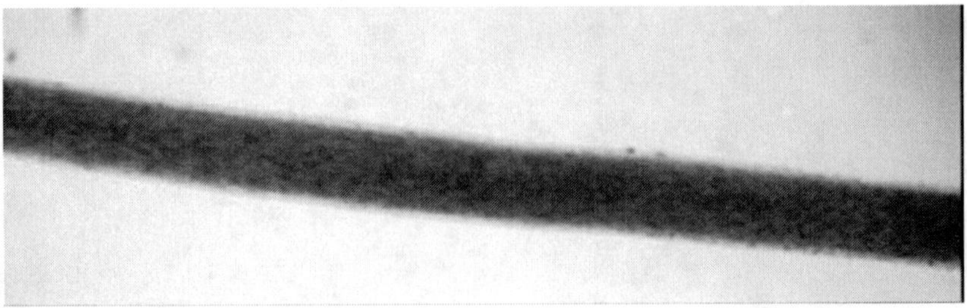

Figure 4: Microscopic configuration of hair (100X magnification).

Genetic analysis

Mitochondrial DNA (mtDNA) is most commonly used in species identification and attracts researchers because of its compact size, high copy number per cell, maternal inheritance, absence of introns and pseudogenes and fast evolutionary rate [35-38]. Cyt b, cytochrome oxidase I (COI), two ribosomal RNA (12S rRNA and 16S rRNA), the control region (D-loop) and subunits of mitochondrial encoded NADH dehydrogenase gene are the widely deployed mitochondrial gene for species identification in wildlife forensic [39].

The central concept in species identification is to match the sequence of the evidence item to a reference sequence, either through DNA sequence similarity searches [40] or by phylogenetic reconstruction [41]. In sequence matching approach, a measure of the similarity between sequences is calculated and the most similar species is attributed to the sample [42,43] A 187 bp consensus sequence of Cyt b was obtained from confiscated sample and a sequence similarity search performed in GenBank provided the highest similarity (100%) with Rhinoceros unicornis (NCBI Accession JN935370) i.e. an Asiatic one-horned Rhinoceros, whereas, it had 87% sequence similarity with Rhinoceros sondaicus (NCBI Accession: AJ245725) followed by other Rhinoceros species/subspecies with a decreasing sequence similarity (**Table 1**). Similarly, 140 bp consensus sequences of 12S rRNA was obtained which also showed highly similarity (100%) to the Rhinoceros unicornis (NCBI Accession: AY739618) with reference sequence (**Table 1**).

Table 1: Details of used species/subspecies and their GenBank Accession number of Cyt b and 12S r RNA gene.

	Cyt b (187 bp)		12S rRNA (140 bp)	
Species/ Subspecies	**Accession Number**	**Similarity (%)**	**Accession Number**	**Similarity (%)**
Rhinoceros unicornis	JN935370.1	100	AY739618.1	100
Rhinoceros sondaicus	AJ245725.1	87	AJ245724.1	98
Rhinoceros sondaicus sondaicus	-	-	AY739620.1	98
Rhinoceros s. annamiticus	-	-	AY739619.1	98
Diceros bicornis minor	-	-	FJ608808.1	98
Coelodonta antiquitatis	GU371439.1	86	-	-
Dicerorhinus sumatrensis	AJ245723.1	85	AJ245722.1	97
Dicerorhinus s. sumatrensis	-	-	AY739616.1	97
Ceratotherium simum simum	FJ619038.1	85	FJ608805.1	96
Ceratotherium simum cottoni	FJ619039.1	84	FJ608806.1	97

Sequence comparisons identified a total of 48 and 8 variable sites for Cyt b and 12S rRNA respectively. This indicates Cyt b sequences of confiscated sample had higher variability with nearby species/ subspecies comparison to 12S rRNA sequence. Appropriate Genetic DNA markers are those which are rather conservative within the species but discriminative between the species [44]. Cyt b shows limited variability within and much greater variation between species [45] thus ideal for species identification [42].

The phylogenetic tree is used to discern the evolutionary relationships between the suspected sample and reference sequences from each possible species [5,46,47]. The position of the suspected sample in the tree allows the closest reference species to be identified as the likely source [7]. There are several different methods for constructing phylogenetic trees (e.g. neighbor-joining, maximum parsimony, maximum likelihood and Bayesian [7,48]. But, there is currently no consensus in wildlife forensic science over which to use [49,50] The neighbor-joining (NJ) phylogenetic tree showed that BLAST analysis and neighbor-joining bootstrap values are similar in Cyt b gene. The NJ trees constructed using the sequences of Cyt b gene showed the similarity of confiscated sample to Rhinoceros unicornis with 100% bootstrap values (**Figure 5a**). The NJ tree made from sequences of 12S rRNA gene also produced strong weight for the genetic similarity of suspected sample with Rhinoceros unicornis (**Figure 5b**).

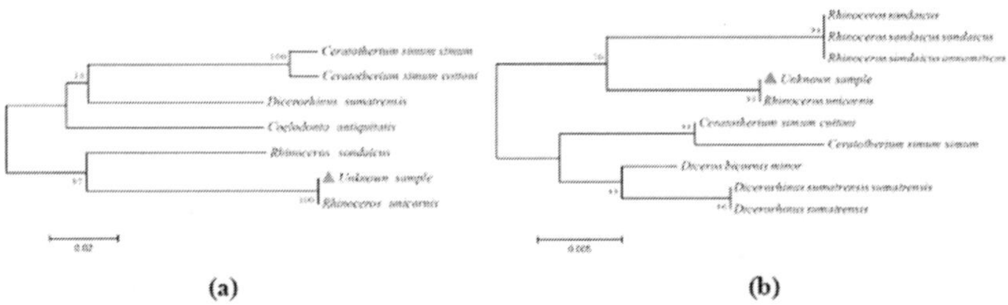

Figure 5: Neighbour-Joining phylogenetic tree based on Cyt b (a) and 12S rRNA (b) showing the relationships of unknown sample with the other related species/subspecies of rhinoceros available in GenBank.

Reference sequence database of different species those are highly in wildlife trades are available at Wildlife Institute of India [51] and such data was used to confirm the BLAST results. Sequences of the 12S rRNA genes was aligned with the reference database of WII and constructed NJ trees showed the similarity of confiscated sample to Rhinoceros unicornis with 100% bootstrap values (**Figure 6**). This is required because The International Society of Forensic Genetics (ISFG) recommends the use of voucher specimen reference DNA sequence, an inhouse and authentic DNA database for species identification for forensic validation [52]. The most commonly used reference databases for comparative species identification searches are the NCBI/EMBL/ DDBJ database collaboration and BOLD [7]. Much of the initial sequence data held on BOLD are unauthenticated, limited regulation and erroneous sequences [53-56]. This can lead to false identifications if the target sample belongs to a previously uncharacterized species [57].

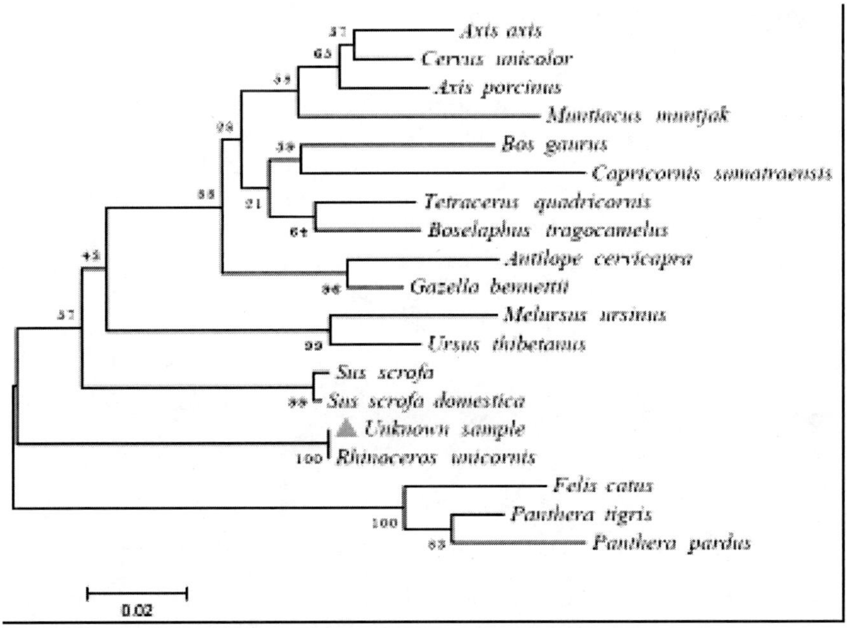

Figure 6: 12S rRNA Neighbour-Joining phylogenetic tree showing the relationships of unknown sample with the other nearby species database available in Wildlife Institute of India.

CONCLUSION

From the sequence alignment and a score table for the case sample based on Cyt b and 12S rRNA, the source of confiscated sample was found to Rhinoceros unicornis. DNA analysis cannot provide information like age [7] or body parts of the organism concerned. A hierarchy of techniques becomes more effective [58]. Thus, on the basis of genetic analysis result integrated with morphological and microscopic findings sample was ultimately confirmed as a nasal part of infantile rhinoceros that bears horn. It is also concluded that infantile rhinoceros horn bears indistinct filamentous unit and laminae with wider filamentous unit space in comparison to mature horn. This is the primary and pioneer work which can be utilized by the scientific community for identification of infantile rhinoceros horn in future.

ACKNOWLEDGEMENTS

The authors are deeply indebted to Dr. Siddhartha B. Bajracharya, National Trust for Nature Conservation, Nepal; Mr. Jiwan Pd. Rijal, Executive Director, National Forensic Science Laboratory, Nepal and Dr. Y.V. Jhala, Wildlife Institute of India, Dehradun, India for their necessary steps to arrange the wildlife DNA forensic training after that genetic examination of sample was made possible. We thank Ajit Kumar, Ved P. Kumar and Madhanraj of WII, India and Nanda Kumar Aryal, National Forensic Science laboratory, Nepal for their leading assistance.

REFERENCES

1. Robert J (2000) Dossier traffic animal: la Mafia a l'assault de la nature. Terre Sauvage 155: 35–50.
2. Karlsson AO, Holmlud G (2007) Identification of mammal species using species-specific DNA pyrosequencing. Forensic Sci Int 173: 16-20.
3. Linacre A, Tobe SS (2011) An overview to the investigative approach to species testing in wildlife forensic science. Investing Genet 2:2.
4. Hey J (2001) The mind of the species problem. Trends Ecol Evol 16: 326-329.
5. Verma SK, Singh L (2003) Novel universal primers establish identity of an enormous number of animal species for forensic application. Molecular Ecology Notes 3: 28-31.
6. Gupta SK (2012) DNA wildlife forensics: Present and future. J Forensic Res 3: 103.
7. Ogden R, Dawnay N, McEwing R (2009) Wildlife DNA forensics-bridging the gap between conservation genetics and law enforcement. Endang Spec Res 9: 179-195.
8. Tobe SS, Linacre AMT (2008) A multiplex assay to identify 18 European mammal species from mixtures using the mitochondrial cytochrome b gene. Electrophoresis 29: 340-347.
9. Gaur A, Singh CS, Sreenivas A, Singh L (2012) DNA-based identification of a snake in a wine bottle using universal primers: A case of mistaken identity. Forensic Sci Int 214: 51-53.
10. Sahajpal V, Goyal SP (2010) Identification of a forensic case using microscopy and forensically informative nucleotide sequencing (FINS): A case study of small Indian civet (Viverricula indica). Sci Justice 50: 94-97.
11. Fang SG, Wan QH (2002) A genetic fingerprinting test for identifying carcasses of protected deer species in China. Biol Conserv 103: 371-373.
12. Verma SK, Prasad K, Nagesh N, Sultana M, Singh L (2003) Was elusive carnivore a panther? DNA typing of faeces reveals the mystery. Forensic Sci Int 137: 16-20
13. Wan QH, Fang SG (2003) Application of species-specific polymerase chain reaction in the forensic identification of tiger species. Forensic Sci Int 131: 75-78.
14. Lorenzini R (2005) DNA forensics and the poaching of wildlife in Italy: a case study. Forensic Sci Int 153: 218-221.
15. Gupta SK, Verma SK, Singh L (2005) Molecular insight into a wildlife crime: the case of a peafowl slaughter. Forensic Sci Int 154: 214-217.
16. An J, Lee MY, Min MS, Lee MH, Lee H (2007) A molecular genetic approach for species identification of mammals and sex determination of birds in a forensic case of poaching from South Korea. Forensic Sci Int 167: 59-61

17. Tsai LC, Huang MT, Hsiao CT, Lin AC, Chen SJ, et al. (2007) Species identification of animal specimens by cytochrome b gene. Forensic Science Journal 6: 63-65.
18. Marin JC, Saucedo CE, Corti P, Gonzalez BA (2009) Application of DNA forensic techniques for identifying poached guanacos (Lama guanicoe) in Chilean Patagonia. J Forensic Sci 54: 1073-1076
19. Caniglia R, Fabbri E, Greco C, Galaverni M, Randi E (2010) Forensic DNA against wildlife poaching: Identification of a serial wolf killing in Italy. Forensic Sci Int: Genetics 4: 334-338.
20. Gupta SK, Kumar A (2014) Molecular identification of man-eating carnivores from scat samples. Conservation Genet Resour 6: 271-274
21. Hsieh HM, Huang LH, Tsai LC, Kuo Y, Meng HH, et al. (2003) Species identification of rhinoceros horns using the cytochrome b gene. Forensic Sci Int 136: 1-11.
22. Jha DK, Kshetry NT, Pokharel BR, Panday R, Aryal NK (2016) A Comparative study of some morphological and microscopic identifying features of genuine Rhino (Rhinoceros unicornis) Horns and Fake Horns. J Forensic Res 6: 315.
23. Rijal JP, Jha DK, Shakya HR (2004) Forensic application of hair comparison, Proceedings of IV National Conference on Science and Technology. pp: 1189-1194.
24. Gupta SK, Kumar A, Hussain SA (2013) Extraction of PCR-amplifiable DNA from a variety of biological samples with uniform success rate. Conservation Genetics Res 5: 215-217.
25. Kocher TD, Thomas WK, Meyer A, Edwards SV, Paabo S, et al. (1989) Dynamics of Mitochondrial DNA Evolution in Animals: Amplification and Sequencing with Conserved Primers. Proceeding of National Academy of Sciences, USA, 86: 6196-6200.
26. Gupta SK, Sharma CP, Singh L (2014) DNA typing established as an unambiguous tool for species identification in a dispute case. Forensic Science Journal 13: 9-14.
27. Hall TA (1999) BioEdit: a user-friendly biological sequence alignment editor and analysis program for windows 95/98/NT. Nucleic Acids Symposium Series 41: 95-98.
28. Altschul SF, Gish W, Miller W, Myers EW, Lipman DJ (1990) Basic local alignment search tool. J Mol Biol 215: 403-410.
29. Saitou N, Nei M (1987) The neighbor-joining method: A new method for reconstructing phylogenetic trees. Mol Biol Evol 4: 406-425.
30. Tamura K, Peterson D, Peterson N, Stecher G, Nei M, et al. (2011) Molecular Evolutionary Genetics Analysis using Maximum Likelihood, Evolutionary Distance, and Maximum Parsimony Methods. Mol Biol Evol 28: 2731-2739.
31. Hieronymus TL, Witmer LM (2004) Rhinoceros horn attachment: anatomy and histology of a dermally influenced bone rugosity. J Morphol 260: 298.
32. Kingdon J, Happold D, Butynski T, Hoffmann M, Happold M, et al. (2013) Carnivores, Pangolins, Equids and Rhinoceroses. In: Mammals of Africa. Bloomsbury Publishing London, UK.
33. Pienaar DJ, Hall-Martin AJ, Hitchnis PM (1991) Horn growth rates of free-ranging white and black rhinoceros. Koedoe 34: 97-105.
34. Bahuguna A, Sahajpal V, Goyal SP, Mukherjee SK, Thakur V (2010) Species identification from guard hair of selected Indian mammals A reference guide. Wildlife Institute of India, India.
35. Brown WM, George Jr. M, Wilson AC (1979) Rapid evolution of animal mitochondrial DNA. Proc Natl Acad Sci USA 76: 1967-1971.
36. Wilson AC, Cann RL, Carr SM, George M, et al. (1985) Mitochondrial DNA and two perspectives on evolutionary genetics. Biol J Linn Soc 26: 375-400.
37. Carracedo A, Bar W, Lincoln P, Mayr W, Morling N, et al. (2000) DNA commission of the international society for forensic genetics: guidelines for mitochondrial DNA typing. Forensic Sci Int 110: 79-85.
38. Pereira SL (2000) Mitochondrial genome organization and vertebrate phylogenetics. Genet Mol Biol 23: 745-752

39. Johnson RN, Wilson-Wilde L, Linacre A (2014) Current and future directions of DNA in wildlife forensic science. Forensic Sci Int Genet 10: 1-11.
40. Altschul S, Madden T, Schaffer A, Zhang J, Zhang Z, et al. (1997) Gapped BLAST and PSI-BLAST: a new generation of protein database search programs. Nucleic Acids Res 25: 3389- 3402.
41. Baker CS, Palumbi SR (1994) Which whales are hunted—a molecular genetic approach to monitoring whaling. Science 265: 1538-1539.
42. Parson W, Pegoraro K, Neiderstatter H, Foger M, Steinlechner M (2000) Species identification by means of cytochrome b gene. Int J Legal Med 114: 23-28.
43. Branicki W, Kupiec T, Pawlowski R (2003) Validation of cytochrome b sequence analysis as a method of species identification. J Forensic Sci 48: 83-87.
44. Paule L, Krajmerová D, Romšáková I, Nedelová L, Slamečka J (2012) Wildlife Forensics genetics-tool for controlling poaching. International symposium on hunting, Modern aspects of sustainable management of game population, Zemun-Belgrade, Serbia pp: 22-24.
45. Caine L, Lima G, Pontes L, Abrantes D, Pereira M, et al. (2006) Species identification by cytochrome b gene: Casework sample. International Congress Series 1288: 145-147.
46. Avise JC (1993) Molecular markers, natural history and evolution. Chapman and Hall, New York, Uk.
47. Roman J, Bowen BW (2003) The mock turtle syndrome: genetic identification of turtle meat purchased in the south-eastern United States of America. Anim Conserv 3: 61-65.
48. Holder M, Lewis PO (2003) Phylogeny estimation: traditional and Bayesian approaches. Nat Rev Genet 4: 275-284.
49. Terol J, Mascarell R, Fernandez-Pedrosa V, Perez-Alonso M (2002) Statistical validation of the identification of tuna species: bootstrap analysis of mitochondrial DNA sequences. J Agric Food Chem 50: 963-969.
50. Wong KL, Wang J, But PPH, Shaw PC (2004) Application of cytochrome b DNA sequences for the authentication of endangered snake species. Forensic Sci Int 139: 49-55.
51. Kumar VP, Kumar D, Goyal SP (2014) Wildlife DNA forensic in curbing illegal wildlife trade: Species identification from seizures. Int J Forensic Sci Pathol 2: 38-42
52. Linacre A, Gusmao L, Hecht W, Hellmann AP, Mayr WR, et al. (2011) ISFG: recommendations regarding the use of non-human (animal) DNA in forensic genetic investigations. Forensic Sci Int Genet 5: 501-505.
53. Dawnay N, Ogden R, McEwing R, Carvalho GR, Thorpe RS (2007) Validation of the barcoding gene COI for use in forensic genetic species identification. Forensic Sci Int 173: 1-6.
54. Groenenberg DSJ, Neubert E, Gittenberger E (2011) Reappraisal of the "Molecular phylogeny of Western Palaearctic Helicidae s.l. (Gastropoda: Stylommato-phora)": When poor science meets GenBank. Mol Phylogenet Evol 61: 914-923.
55. Hassanin A, Bonillo C, Nguyen BX, Cruaud C (2010) Comparisons between mitochondrial genomes of the domestic goat (Capra hircus) reveal the presence of numts and multiple sequencing errors. Mitochondrial DNA 21: 68-76.
56. Bidartondo MI, Bruns TD, Blackwell M, Ryberg M (2008) Preserving accuracy in GenBank. Science 319: 1616.
57. Pereira F, Carneiro J, Amorim A (2008) Identification of species with DNA-based technology: Current progress and challenges. Recent Pat DNA Gene Seq. 2: 187-200.
58. Vipin, Sharma V, Sharma CP, Kumar VP, Goyal SP (2016) Pioneer identification of fake tiger claws using morphometric and DNA-based analysis in wildlife forensics in India. Forensic Sci Int 266: 226-233.

CHAPTER 7

Identification and Characterization of New Designer Drug 4-Fluoro-PV9 and A-PHP in the Seized Materials

Milena Majchrzak[1,3] • Marcin Rojkiewicz[2,3] • Rafał Celin´ski[3] • Piotr Kus´ [2] • Mieczysław Sajewicz[1]

[1].Department of General Chemistry and Chromatography, Institute of ChemistryUniversity of SilesiaKatowicePoland

[2].Department of Organic Synthesis, Institute of ChemistryUniversity of SilesiaKatowicePoland

[3].Toxicology Laboratory ToxLabKatowicePoland

ABSTRACT

In this study, we present identification and physicochemical characterization of new cathinone derivatives, 4-fluoro-PV9 and already known α-PHP in seized materials. Although the disclosure of α-PHP from an illegal product had been reported and characterized to some extent, the data on α-PHP are also presented together with those of 4-fluoro-PV9. The data of characterization for the two compounds were obtained by high-performance liquid chromatography (HPLC)–mass spectrometry and

HPLC–diode array detection, electrospray ionization/ion trap mass spectrometry in MS^2 and MS^3 modes, gas chromatography–mass spectrometry, thermogravimetric analysis, differential scanning calorimetry, Fourier transform infrared spectroscopy, ultraviolet-visible spectroscopy, and nuclear magnetic resonance spectroscopy. To our knowledge, this is the first report for identification and detailed characterization of 4-fluoro-PV9 circulated on the illegal drug market.

Keywords: Designer drugs 4-Fluoro-PV9 α-PHP ESI-MS/MS GC–MS NMR

INTRODUCTION

In the global market of designer drugs, new products continuously appear that contain already known psychoactive compounds and mixtures thereof, but also new compounds, not yet characterized in any way [1, 2, 3]. According to the European Early Warning System, in Europe in 2005 alone, 14 novel psychoactive compounds were identified, and in 2013 the number of such compounds grew to over 80 items [1, 4]. Striking features are a wide spectrum of the marketed compounds and the forms in which they are traded. In Poland, consumables containing psychoactive ingredients are currently marketed as "fireplace kindling", "bidet refreshers", "humidity adsorbents", "driver's charms", or "leech tamers". Among the substances seized on the Polish and European Union drug market, derivatives of piperazine and phenylethylamine are frequently found, along with the synthetic cannabinoids, but the most frequently identified group is that of the cathinone derivatives [1, 2, 3, 5, 6, 7, 8, 9, 10, 11].

Synthetic analogs of cathinone can be divided into four groups; i.e., those containing *N*-alkyl- or alkyl/halogen substituents in any given position of the aromatic ring, those with methylenedioxy-substituted structures also in any given position of the aromatic ring, those with a pyrrolidinyl group at the nitrogen atom, and those with combined substituents of the second and the third groups [12, 13, 14, 15, 16]. Among the most frequently identified cathinones distributed on the Polish market, one can find 3-MMC (2-(methylamino)-1-(3-methylphenyl)-1-propanone), pentedrone (2-(methylamino)-1-phenyl-1-pentanone), 4-MEC (2-(ethylamino)-1-(4-methylphenyl)-1-propanone), MDPV (3,4-methylenodioxypyrovalerone), α-PVP (1-phenyl-2-(1-pyrrolidinyl)-1-pentanone), and MDPBP (3,4-methylenodioxy-α-pyrrolidinobutiophenone) [17, 18, 19, 20, 21].

Due to an increasing number of designer drugs available on the drug market and taking into account the vital possibility of synthesizing and manufacturing structural modifications of already known psychoactive substances, the availability of analytical characteristics of these compounds is inevitable for their efficient and rapid identification [8, 18, 22, 23, 24, 25, 26]. In this study, we present, for the first time to our knowledge, analytical characteristics of 4-fluoro-PV9 (1-(4-fluorophenyl)-2-(pyrrolidin-1-yl)octan-1-one) in seized materials. Although we could detect α-PHP (1-phenyl-2-(1-pyrrolidinyl)-1-hexanone) from another seized material, its disclosure from an illegal product had been reported by Uchiyama et al. [27].

We mainly present and discuss relevant analytical data obtained for 4-fluoro-PV9 and α-PHP by high-performance liquid chromatography–mass spectrometry (HPLC–MS) and high-performance liquid chromatography–diode array detection (HPLC–DAD), ion trap mass spectrometry with electrospray ionization (ESI) in the MS^2 and MS^3 modes, gas chromatography–mass spectrometry (GC–MS), thermogravimetric analysis (TGA), differential scanning calorimetry (DSC), Fourier transform infrared (FTIR) spectroscopy, ultraviolet-visible (UV-VIS) spectroscopy and proton and carbon nuclear magnetic resonance spectroscopy (^1H NMR and ^{13}C NMR).

MATERIALS AND METHODS

Chemicals

In our study, all reagents used were of the HPLC and MS purity grade. Water (Chromasolv), methanol, acetonitrile, 0.1 M formic acid and ammonium formate were purchased from Sigma-Aldrich, Poznań, Poland. For the NMR analysis, deuterated chloroform ($CDCl_3$), also purchased from Sigma-Aldrich, was used.

Sample preparation

The analyzed samples were seized by the police in 2014 in the Silesia region of South Poland and delivered to our laboratory for toxicological examination. Both samples were obtained in powdered form. Sample no. 1 was crumbled pale pink powder, and sample no. 2 was lumped beige powder.

For the purpose of liquid and gas chromatography, 10-mg aliquots of each powder were weighted, dissolved in 1 mL of acetonitrile/methanol (50:50, *v/v*) mixture, and ultrasonicated for 10 min. From each solution, 10 μL were collected and diluted with 990 μL methanol. Finally, the extracts were ready for analysis. For NMR spectroscopic analysis, 10 mg of each sample was dissolved in 0.6 mL $CDCl_3$ and analyzed. For the purpose of FTIR, a 2-mg aliquot of each sample was taken and mixed with KBr, and pellets were formed. The KBr pellets were analyzed. For the purpose of TGA, DSC and UV-VIS spectroscopy, a 5-mg aliquot of each sample was taken for analysis. The TGA and DSC analysis were performed without any further sample treatment, and for the UV-VIS analysis, the samples were dissolved in methanol.

HPLC–MS and HPLC–DAD conditions

In our investigations, the high-performance liquid chromatograph Thermo Surveyor coupled with the mass spectrometer Thermo LCQ DecaXP-Plus with an electrospray ionization source (Thermo Scientific, Warsaw, Poland) was used. The obtained data were processed with use of the Xcalibur and LCQTune programs (Thermo Scientific).

For the HPLC–MS analysis, the Hypersil RP C18 column (150 × 4.6 mm) (Thermo Scientific) was used and analysis was carried out in gradient mode. Mobile phase was composed of solvent A (0.02 M water solution of formic acid/0.05 M water solution of ammonium formate) and solvent B (10 % solvent A/90 % acetonitrile), and the following gradient program was applied: 0–2 min, 95 % (A) + 5 % (B); 2–30 min, 30 % (A) + 70 % (B); 30–32 min, 30 % (A) + 70 % (B); 32–40 min, the solvent ratio was returned to the initial state, i.e., 95 % (A) + 5 % (B). The mobile phase flow rate was 100 μL min^{-1}.

The analytes were electrosprayed in the positive mode (ESI(+)-MS). Fragmentation in the ESI-MS^2 and ESI-MS^3 mode was carried out in the scanning range of m/z 50–500. The source temperature was 250 °C, and the carrier and ionizing gases were nitrogen and helium, respectively.

GC–MS analysis

For GC–MS analysis, the Thermo Trace Ultra chromatograph was used, coupled with the Thermo DSQ mass spectrometer (Thermo Scientific). The analyses were carried out with use of the Rxi®-5Sil MS column (Restek, Bellefonte, PA, USA). The following working parameters were employed: injector temperature, 260 °C; oven temperatures, 100 °C for 2 min, ramp at 20 °C min^{-1} to 260 °C; carrier gas (helium) flow rate, 1.2 mL min^{-1}; MS transfer line temperature, 250 °C; MS source temperature, 250 °C; injection volume, 1 μL, on splitless mode.

The obtained electron ionization (EI) mass spectra were compared to those of an EI-MS library [28] for structural confirmation.

NMR spectroscopy

The NMR spectra were recorded with use of the UltraShield 400 MHz apparatus (Bruker, Bremen, Germany). Deutered chloroform was used as solvent.

Thermogravimetric analysis (TGA) and differential scanning calorimetry (DSC)

TGA was carried out with use of the TGA/DSC1 Mettler–Toledo thermal analyzer (Mettler–Toledo, Greifensee, Switzerland), with a heating rate of 10 °C/min in a stream of nitrogen (60 cm^3 min^{-1}). DSC was performed with a TA-DSC 2010 apparatus (TA Instruments, New Castle, DE, USA) under nitrogen using aluminum sample pans. The DSC experiments were carried out in a nitrogen atmosphere with a temperature range from −70 °C to over the clearing point.

Fourier transform infrared and UV-VIS spectroscopy

The infrared (IR) spectra were acquired with use of the PerkinElmer Spectrum One spectrometer (PerkinElmer, Waltham, MA, USA) using the KBr pellets. The UV-VIS absorption spectra were recorded in the methanol solution using the PerkinElmer Lambda Bio 40 UV-VIS spectrometer.

RESULTS AND DISCUSSION

Identification of the compound contained in sample no. 1

Upon analysis of sample no. 1 by HPLC–MS, the total ion current (TIC) chromatogram (Figure 1a) showed an intense peak at 25.27 min. Otherwise, no marked peaks appeared, showing that sample no. 1 contained almost a single compound with high purity (probably more than 95 %). Figure 1b shows the HPLC–MS spectrum of the intense peak observed in Figure 1a; a base peak appeared at *m/z* 292. Because this peak seemed to be protonated molecular one $[M + H^+]$, the unknown compound may have a molecular weight of 291 Da. In the product ion mass spectra in the MS^2 (tandem) mode using the peak at *m/z* 292 as the precursor ion, a peak at *m/z* 274 appeared (Figure 2a), which indicated the elimination of one water molecule from the protonated molecular ion $[M + H - H_2O]^+$. This transformation is a well-known characteristic of cathinone derivatives. Given that this compound is assumed to be one of the cathinone derivatives, the product ion mass spectrum in the MS^3 mode using the peak at *m/z* 274 as the precursor ion, which neither includes a carbonyl nor hydroxyl group, showed a base peak at *m/z* 203 and a

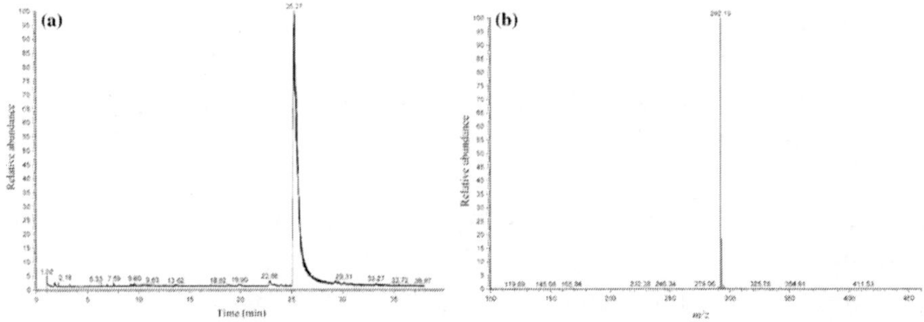

Figure 1. a Total ion current chromatogram (TIC) and **b** the single-stage mass spectrum of the intense peak appearing in the TIC obtained from the compound in sample no. 1, recorded by high-performance liquid chromatography–mass spectrometry (HPLC–MS)

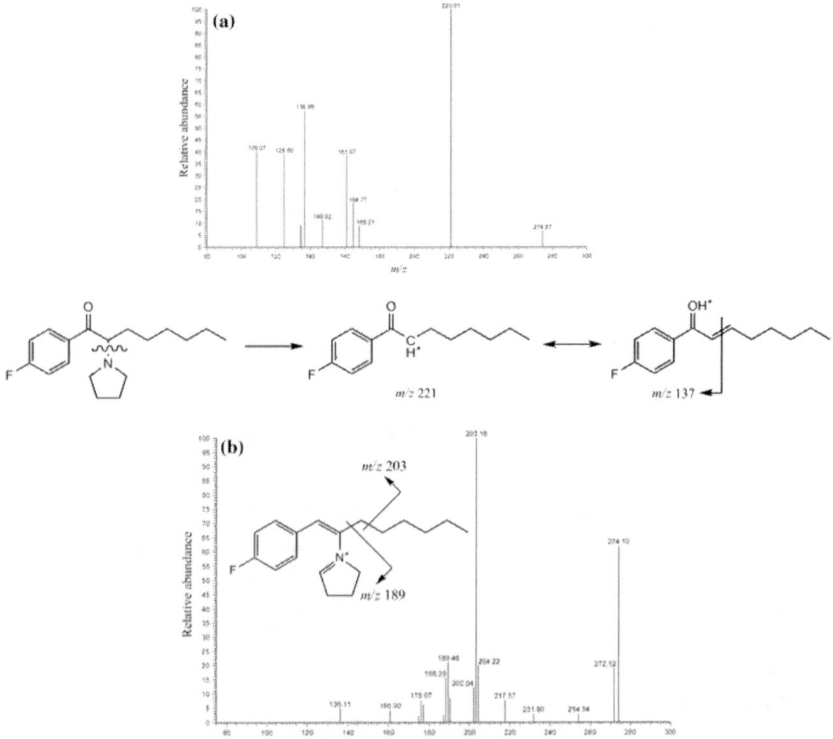

Figure 2. Product ion mass spectra obtained by ion trap mass spectrometry (MS) in the **a** MS^2(tandem) and **b** MS^3 modes for sample no. 1. The precursor ions used for the MS^2 and MS^3 modes were those at *m/z* 292 and 274, respectively

fragment peak at m/z 189 (Figure 2b), suggesting that the cathinone derivative has pyrrolidinyl and fluorophenyl moieties in its structure. The appearance of the base peak at m/z 221 in Figure 2a is easily explicable by removal of the pyrrolidinyl moiety from the molecule.

Unfortunately, in this study, we did not use the high-resolution MS that can give a molecular formula. Instead, we compared the GC–EI-MS spectra to those of an EI-MS spectral library, and fortunately, the EI-MS spectrum obtained from sample no. 1 matched well with that of 4-fluoro-PV9, which had very recently been added to the library. In both spectra, the peak at m/z 168 appeared as the base peak; the fragment peaks at m/z 169, 123, 110, 95, 84 and 55 appeared to be in common (Figure 3).

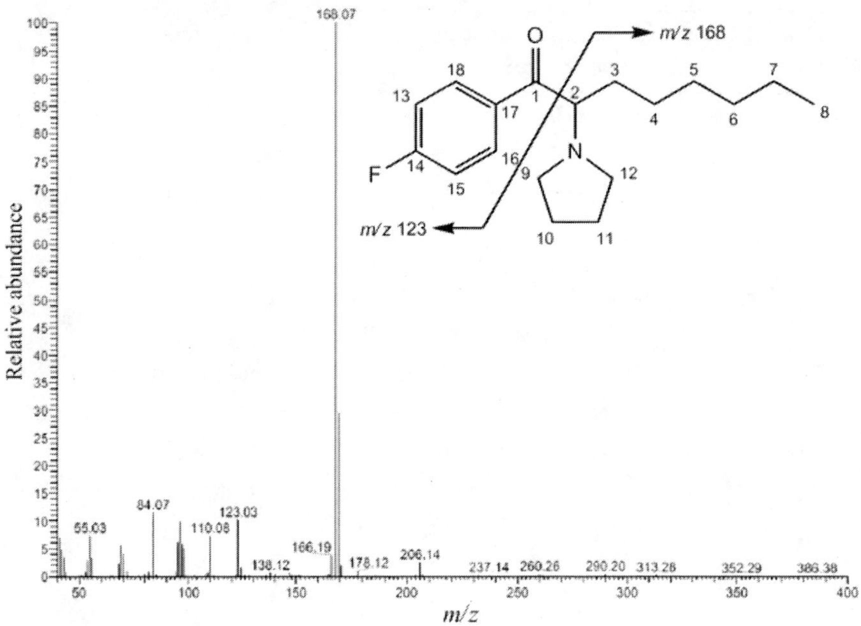

Figure 3. Mass spectrum of the compound contained in sample no. 1, obtained by gas chromatography–mass spectrometry (GC–MS) in the electron ionization (EI) mode

Therefore, it seemed correct that the compound contained in sample no. 1 was fluoro-PV9, but the question of which position of the phenyl ring the fluorine atom was attached remained unclear, because it is well known that such regioisomers give the same or very similar EI-mass spectra.

To finally determine the position of the fluorine atom in the aromatic ring, the NMR spectrum was recorded, as shown in Table 1. The carbon numbering is shown in Figure 4a.

Table 1. ^{1}H and ^{13}C nuclear magnetic resonance (NMR) data for sample no. 1

Carbon atom position	Carbon chemical shifts (ppm)	Proton chemical shifts (ppm)
1	195.20	–
2	64.87	5.50 (dt, 1H)
3	31.13	2.08 (m, 2H)
4, 5, 6, 7	30.82; 29.11; 25.52; 22.36	1.15–1.35 (m, 8H)
8	13.85	0.79 (t, 3H)
9	52.72	3.12 (m, 1H); 3.86 (m, 1H)
10, 11	23.85; 23.66	2.18 (m, 4H)
12	51.23	3.69 (m, 2H)
13, 15	116.51; 116.53	7.22 (m, 2H)
14	165.53; 168.10	–
16, 18	131.88; 131.93; 131.96	8.13 (m, 2H)
17	131.79	–

The carbon atom numbering is shown in Fig. 4a

As a result, confirmation of the anticipated structure was obtained. Namely, in the aromatic range of the ^{1}H NMR spectrum, two signal groups appeared, and witnessed to the presence of the fluorine atom at position *para* of the phenyl ring. Due to the proton interactions with the fluorine atom, two multiplets were observed instead of a "classical" signal arrangement (i.e., the doublets for each group of equivalent protons).

Coupling with the fluorine atom was also visible in the ^{13}C NMR spectrum, where signals originating from carbon atom no. 14, and those from no. 13 and 15 are split to the doublets. Due to the fact that sample no. 1 appeared as a hydrochloride, in the ^{1}H NMR spectrum, a signal is observed at 11.95 ppm, originating from the proton of the quaternary pyrrolidine group. Therefore, it can be concluded that the compound contained in sample no. 1 is 4-fluoro-PV9 (1-(4-fluorophenyl)-2-(pyrrolidin-1-yl)octan-1-one). To our knowledge, this is the first report to identify 4-fluoro-PV9 in an authentic seizure sample.

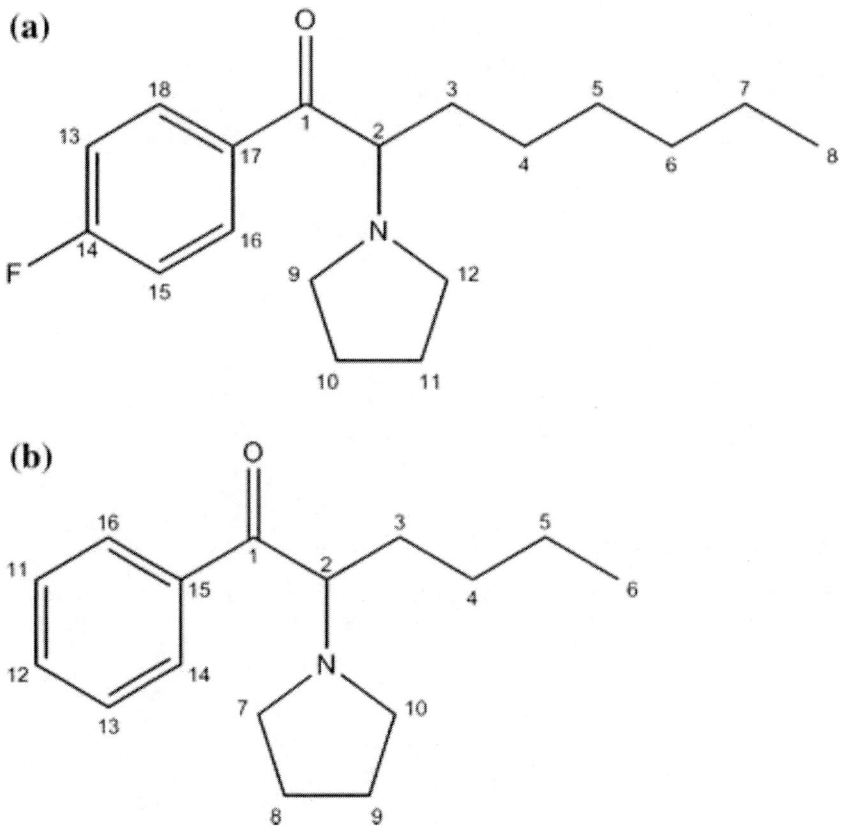

Figure 4. Target compound structures with carbon atom numbering for nuclear magnetic resonance analysis. **a** The compound contained in the sample no. 1; **b** the compound contained in sample no. 2

Identification of the compound contained in sample no. 2

Upon analysis of sample no. 2 by HPLC–MS, the TIC (Figure 5a) showed an intense peak at 18.58 min. Otherwise, no marked peaks appeared, showing that sample no. 2 contained almost a single compound with high purity. Figure 5b shows the HPLC–MS spectrum of the intense peak observed in Figure 5a; a base peak appeared at *m/z* 246. In the product ion mass spectrum in the MS^2 mode (Figure 6a), the peak due to elimination of one water molecule at *m/z* 228 also appeared, suggesting that the compound is also a cathinone derivative, and the peak due to the elimination of the pyrrolidinyl moiety appeared as the base peak at *m/z* 175. The product ion mass spectrum in the MS^3 mode was also recorded using the ion at *m/z* 175 as the precursor ion for further analysis of the probable structure (Figure 6b).

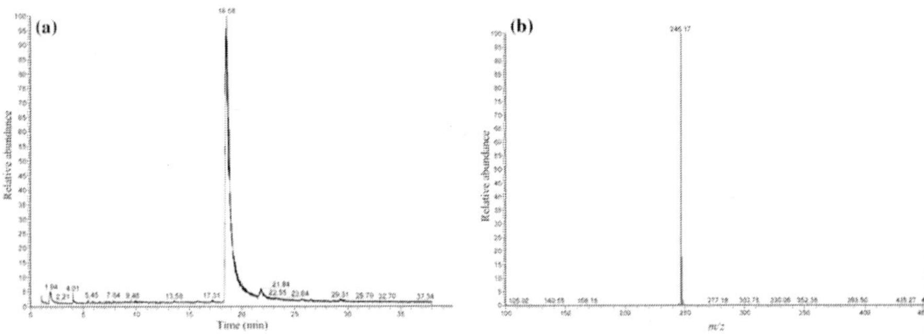

Figure 5. a TIC and **b** the single-stage mass spectrum of the intense peak appearing in the TIC, obtained from the compound in sample no. 2 recorded by HPLC–MS

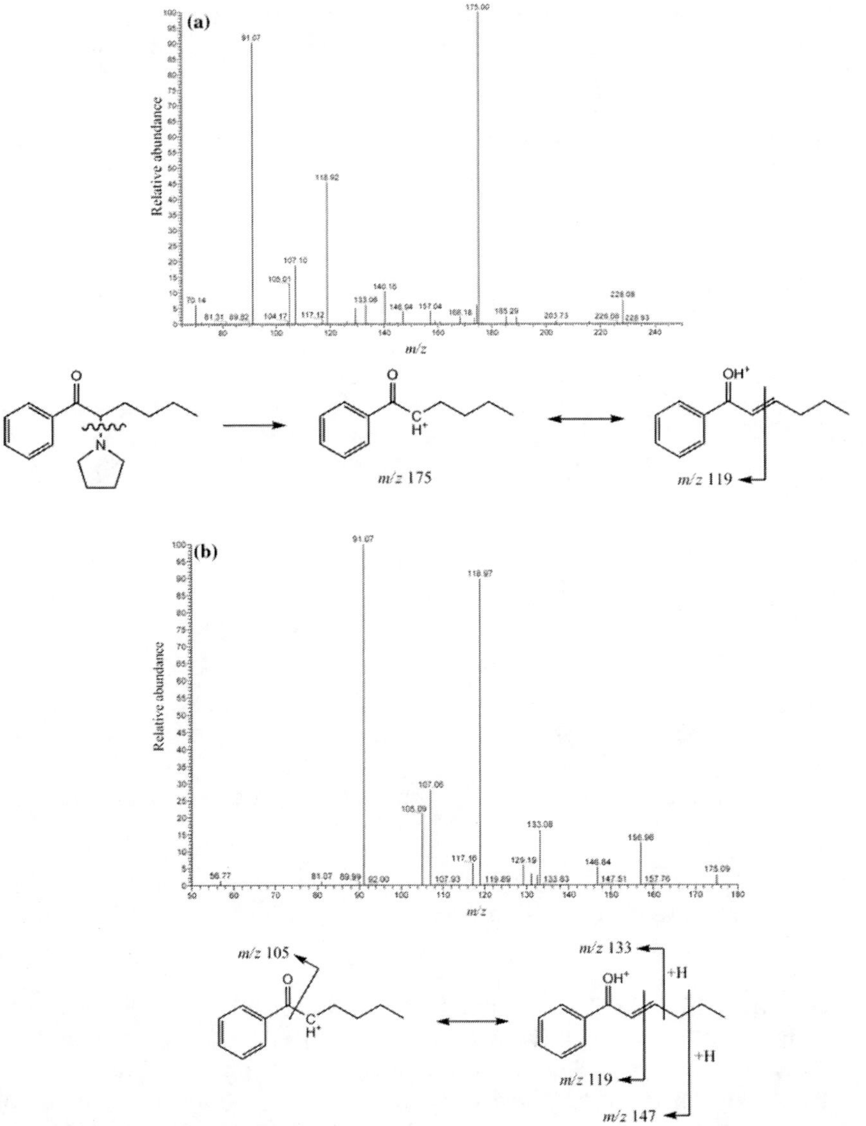

Figure 6. Product ion mass spectra obtained by ion trap MS in **a** MS^2 (tandem) and **b** MS^3 modes for sample no. 2. The precursor ions used for the MS^2 and MS^3 modes were those at *m/z* 246 and 175, respectively

Figure 7 shows the GC–MS spectrum of the compound in sample no. 2, and it was compared to that described in the library [28]. The EI-mass spectrum obtained from sample no. 2 was well matched that of α-PHP (1-phenyl-2-(1-pyrrolidinyl)-1-hexanone). In both spectra, the peak at *m/z* 140 appeared as the base peak; the fragment peaks at *m/z* 141, 105, 96 and 77 appeared in common.

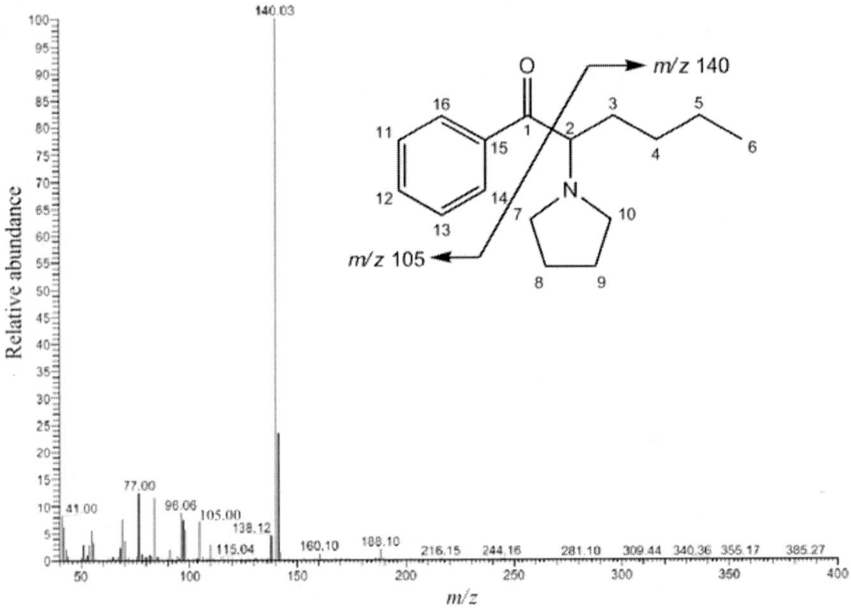

Figure 7. Mass spectrum of the compound contained in sample no. 2, obtained by GC–MS in the EI mode

As shown in Table 2, the data from ^{1}H NMR and ^{13}C NMR analysis of sample no. 2 confirmed the presence of α-PHP (the carbon numbering is shown in Figure 4b). Similar to sample no. 1, the compound in sample no. 2 was also present in the hydrochloride form, which resulted in the appearance of a signal at 12.65 ppm originating from the proton of the quaternary pyrrolidine group.

Table 2. ^1H and ^{13}C NMR data for sample no. 2

Carbon atom position	Carbon chemical shifts (ppm)	Proton chemical shifts (ppm)
1	196.78	–
2	62.87	5.15 (dt, 1H)
3	30.62	2.03 (m, 2H)
4, 5	27.93; 22.50	1.33 (m, 4H)
6	13.48	0.84 (t, 3H)
7	52.90	2.83 (m, 1H); 3.68 (m, 1H)
8, 9	23.93; 23.71	2.21 (m, 4H)
10	49.51	3.84 (m, 2H)
11, 13	129.43	7.58 (t, 2H)
12	135.16	7.73 (t, 1H)
14,16	128.62	7.98 (d, 2H)
15	135.75	–

The carbon atom numbering is shown in Fig. 4b

It should be mentioned that Uchiyama et al. [27] reported the identification of α-PHP in an illegal product in 2014; they reported GC–MS mass spectrum, UV spectrum and ^1H NMR and ^{13}C NMR for α-PHP, which generally agree with the data described in this study.

Other physicochemical analytical results for 4-fluoro-PV9 and α-PHP identified in this study

The UV spectra recorded by HPLC–DAD and UV-VIS spectrometry showed absorption maxima at 254 and 253 nm, respectively, for the compound in sample no. 1 (4-fluoro-PV9); they were 252 and 251 nm, respectively, for that in sample no. 2 (α-PHP) (Figure 8). The symmetrical peaks and the absence of impurity peaks in the spectra show that each compound existed in the materials in high purity.

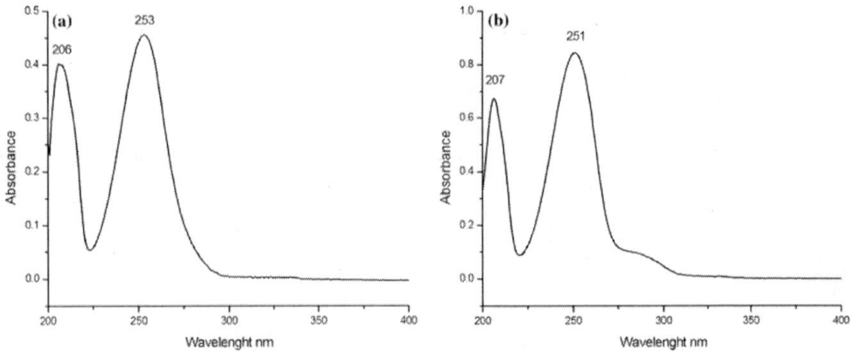

Figure 8. Ultraviolet-visible spectroscopy for the compounds contained in **a** sample no. 1 and **b** sample no. 2

The TGA spectra for both compounds are shown in Figure 9a, b, respectively. Both spectra showed single main transitions corresponding to their decomposition. The decomposition temperatures for those in sample no. 1 and sample no. 2 were 259 and 250 °C, respectively. The only one main transition observed for each compound shows that the purity of each compound was high.

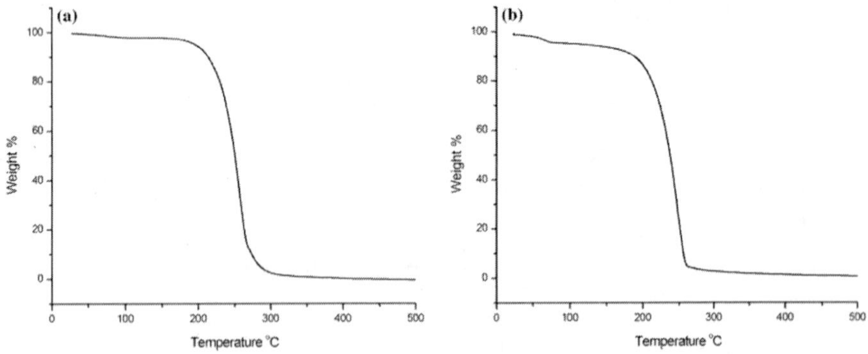

Figure 9. Thermogravimetric spectra for the compounds contained in **a** sample no. 1 and **b** sample no. 2

By DSC analysis, melting points for the compounds in the samples no. 1 and no. 2 were 144 and 65 °C, respectively; those measured in the classic way were 140–142 and 62–64 °C, respectively.

By the FTIR spectroscopy, a strong carbonyl stretch at 1685 cm^{-1}, the aliphatic CH stretching at 2800–3000 cm^{-1} and the amine hydrochloride salt bands at 2476–2752 cm^{-1} were observed for the compound in the sample no. 1; a strong carbonyl stretch at 1678 cm^{-1}, the aliphatic CH stretching at 2800–3000 cm^{-1} and the amine hydrochloride salt bands at 2494–2750 cm^{-1}were observed for the compound in sample no. 2. Due to the structural similarity of both compounds, their IR spectra were very similar, as described above.

CONCLUSIONS

In this study, we experienced the analysis of compounds in two seized materials. We used various analytical tools for their identification and characterization, such as HPLC–MS, HPLC–DAD, ion trap MS in MS^2 and MS^3 modes, GC–MS, TGA, DSC, FTIR spectroscopy, UV-VIS spectroscopy and ^1H and ^{13}C NMR spectroscopy. The compounds contained in sample no. 1 and no. 2 were found to be 4-fluoro-PV9 and α-PHP, respectively. Although α-PHP had been identified and characterized to some extent in 2014, to our knowledge, this is the first report on the identification and characterization of 4-fluoro-PV9. The data presented in this article will be useful for forensic toxicologists who are alert to the emergence of new psychotropic drugs.

COMPLIANCE WITH ETHICAL STANDARDS

Conflict of interest

The authors have no financial or other relations that could lead to a conflict of interest.

Ethical approval

This article does not contain any studies with human participants or animals performed by any of the authors.

REFERENCES

1. Liechti M (2015) Novel psychoactive substances (designer drugs): overview and pharmacology of modulators of monoamine signaling. Swiss Med Wkly 145:w14043

2. Baumann MH, Solis E, Watterson LR, Marusich JA, Fantegrossi WE, Wiley JL (2014) Bath salts, spice, and related designer drugs: the science behind the headlines. J Neurosci 34:15150–15158
3. Weaver MF, Hopper JA, Gunderson EW (2015) Designer drugs 2015: assessment and management. Addict Sci Clin Pract.
4. European Monitoring Centre for Drugs and Drug Addiction (2015) New psychoactive substances in Europe. Publications Office of the European Union, Luxembourg **(An update from the EU Early Warning System)**Google Scholar
5. Arbo MD, Bastos ML, Carmo HF (2012) Piperazine compounds as drugs of abuse. Drug Alcohol Depen 122:174–185
6. Schep LJ, Slaughter RJ, Vale JA, Beasley DMG, Gee P (2011) The clinical toxicology of the designer "party pills" benzylpiperazine and trifluoromethylphenylpiperazine. Clin Toxicol 49:131–141
7. Srisuma S, Bronstein AC, Hoyte CO (2015) NBOMe and 2C substitute phenylethylamine exposures reported to the National Poison Data System. Clin Toxicol 53:624–628
8. Zuba D, Sekuła K, Buczek A (2013) 25C-NBOMe—new potent hallucinogenic substance identified on the drug market. Forensic Sci Int 227:7–14
9. Denooz R, Vanheugen JC, Frederich M, de Tullio P, Charlier C (2013) Identification and structural elucidation of four cannabimimetic compounds (RCS-4, AM-2201, JWH-203 and JWH-210) in seized products. J Anal Toxicol 37:56–63
10. Adamowicz P, Lechowicz W (2015) The influence of synthetic cannabinoid UR-144 on human psychomotor performance—a case report demonstrating road traffic risks. Traffic Inj Prev.
11. Katz DP, Bhattacharya D, Bhattacharya S, Deruiter J, Clark CR, Suppiramaniam V, Dhanasekaran M (2014) Synthetic cathinones: "a khat and mouse game". Toxicol Lett 229:349–356
12. Valente MJ, Guedes de Pinho P, de Lourdes Bastos M, Carvalho F, Carvalho M (2014) Khat and synthetic cathinones: a review. Arch Toxicol 88:15–45
13. Szendrei K (1980) The chemistry of khat. Bull Narc 32:5–35
14. Patel NB (2015) "Natural amphetamine" khat: a cultural tradition or a drug of abuse? Int Rev Neurobiol 120:235–255
15. Brenneisen R, Fisch HU, Koelbing U, Geisshüsler S, Kalix P (1990) Amphetamine-like effects in humans of the khat alkaloid cathinone. Br J Clin Pharmacol 30:825–828
16. Vardakou I, Pistos C, Spiliopoulou C (2011) Drugs for youth via Internet and the example of mephedrone. Toxicol Lett 201:191–195
17. Gibbons S, Zloh M (2010) An analysis of the "legal high" mephedrone. Bioorg Med Chem Lett 20:4135–4139
18. Westphal F, Junge T, Klein B, Fritschi G, Girreser U (2011) Spectroscopic characterization of 3,4-methylenedioxypyrrolidinobutyrophenone: a new designer drug with α-pyrrolidinophenone structure. Forensic Sci Int 209:126–132
19. Kaizaki A, Tanaka S, Numazawa S (2014) New recreational drug 1-phenyl-2-(1-pyrrolidinyl)-1-pentanone (alpha-PVP) activates central nervous system via dopaminergic neuron. J Toxicol Sci 39:1–6
20. Uralets V, Rana S, Morgan S, Ross W (2014) Testing for designer stimulants: metabolic profiles of 16 synthetic cathinones excreted free in human urine. J Anal Toxicol 38:233–241
21. Jankovics P, Váradi A, Tölgyesi L, Lohner S, Németh-Palotás J, Koszegi-Szalai H (2011) Identification and characterization of the new designer drug 4′-methylethcathinone (4-MEC) and elaboration of a novel liquid chromatography–tandem mass spectrometry (LC–MS/MS) screening method for seven different methcathinone analogs. Forensic Sci Int 210:213–220
22. Ibáñez M, Sancho JV, Bijlsma L, Van Nuijs ALN, Covaci A, Hernández F (2014) Comprehensive analytical strategies based on high-resolution time-of-flight mass spectrometry to identify new psychoactive substances. Trac-Trend Anal Chem 57:107–117

23. Uchiyama N, Matsuda S, Kawamura M, Shimokawa Y, Kikura-Hanajiri R, Aritake K, Urade Y, Goda Y (2014) Characterization of four new designer drugs, 5-chloro-NNEI, NNEI indazole analog, α-PHPP and α-POP, with 11 newly distributed designer drugs in illegal products. Forensic Sci Int 243:1–13
24. Fornal E (2013) Identification of substituted cathinones: 3,4-methylenedioxy derivatives by high performance liquid chromatography–quadrupole time of flight mass spectrometry. J Pharm Biomed Anal 81–82:13–19
25. Zuba D (2012) Identification of cathinones and other active components of "legal highs" by mass spectrometric methods. Trac-Trend Anal Chem 32:15–30
26. Westphal F, Junge T, Rösner P, Fritschi G, Klein B, Girreser U (2007) Mass spectral and NMR spectral data of two new designer drugs with an α-aminophenone structure: 4′-methyl-α-pyrrolidinohexanophenone and 4′-methyl-α-pyrrolidinobutyrophenone. Forensic Sci Int 169:32–42
27. Uchiyama N, Shimokawa Y, Kawamura M, Kikura-Hanajiri R, Hakamatsuka T (2014) Chemical analysis of a benzofuran derivative, 2-(2-ethylaminopropyl)benzofuran (2-EAPB), eight synthetic cannabinoids, five cathinone derivatives, and five other designer drugs newly detected in illegal products. Forensic Toxicol 32:266–281
28. Cayman Chemical (2015) Cayman spectral library. https://www.caymanchem.com/app/template/SpectralLibrary.vm. Accessed 6 Mar 2015

CHAPTER 8

Spectroscopic Discrimination of Bone Samples from Various Species

Gregory McLaughlin, Igor K. Lednev*

Department of Chemistry, University at Albany, State University of New York, Albany, USA

ABSTRACT

Determining the species of origin of skeletal remains is critical in a forensic and anthropologic context. However, there are very few methods that use a chemical approach to assist in this determination. In this study, Raman spectroscopy was used to discriminate bone samples originating from four different species (bovine, porcine, turkey and chicken). Spectra were obtained using a near infrared laser at 785-nm. All spectra were combined in a single matrix and processed using partial least squares discriminate analysis (PLS-DA) with leave-one-out cross-validation. Three components were found to adequately describe the system. The first two components which contributed over 85% of spectral data was seen to completely separate the four species of origin in a two dimensional scores plot. A 95% confidence interval was draw around score points of each species class with very slight overlap. The first two components were seen to have large contributions from bioapatite and collagen, the main components of bone. This study serves as a preliminary investigation to evaluate the effectiveness of Raman spectroscopy to discriminate the species of origin of bone tissue.

1. INTRODUCTION

Skeletal remains found at a suspected crime scene are of particular interest to forensic investigators. However, there is little forensic importance if the skeletal remains are in fact non-human. The critical determination of whether skeletal remains are human falls on the forensic anthropologist [1]. This is relatively simple task when analyzing a full skeleton or complete bone, but can be very difficult when only fragmentary remains are obtained.

Visual identification methods are well studied and accepted in the scientific community. So much is known about the skeletal features and morphology of human remains that the determination of human origin of a whole bone is considered relatively simple. However, these methods are complicated when only a fragment of bone is in question. Small bone fragments are common in mass disasters such as earth quakes, terrorist attacks, or military battles.

Bone fragments can be devoid of any species specific morphological characteristics. In these cases, a histological analysis is often employed [2]. This approach requires imbedding and cross sectioning of samples and is therefore a destructive process. This type of analysis is considerably destructive because of the processing involved to obtain a cross section of bone.

Metric analysis of Haversian canals in bones has been used with varying success in the verification of human remains. This type of analysis involves measurements the narrow channel in the center of the osteon substructure and comparing to known values and indexes. A recent study by Cattaneo et al. (2009) [3] has shown significant limits with this type of evaluation. Although they reported a success rate of 70% with adult long bones, this type of analysis was seen to be especially problematic in young human. Such samples had an error rates as high as 93% in the case of newborn long bones and 68% for newborn flat bones.

The field of forensic anthropology is highly specialized, making rapid identification practically impossible. There is an interest for a more rapid, high throughput, reliable and non-destructive technique to identify the species of origin of skeletal remains. A Raman spectroscopic approach offers several advantages. Foremost, this technique is non-destructive and highly flexible for other forensic identification tasks. If a robust, automated data treatment process is developed, an examiner can be minimally trained. This type of analysis could eliminate examiner bias because it is not dependent on subjective observations. Most importantly, using a microscopic Raman system, a spectrum can be obtained from an area in the scale of microns. While current methods are hindered when analyzing fragmentary remains, there would be no such impediment using this approach.

It is our hypothesis that a method for species discrimination based on chemical differences in bone can be established. Using Raman spectroscopy, it is possible that a non-destructive and rapid technique to distinguish human remains can provide a favorable alternative to the current methodology. A technique that avoids physical and chemical alterations is attractive in a forensic setting because evidence is suitable for further testing. The main objective of this study is to develop a method based on Raman spectroscopic data that can be used to discriminate species. The proposed methodology is potentially applicable to the fields of anthropology for cultural studies and forensic science for criminal investigations.

Raman spectroscopy is an extremely flexible analytical technique with a variety of forensic applications [4]. This technique is currently being explored for the identification and classification of drugs [5], gunshot residue [6], fibers [7], paint [8-10] and ink [11] samples. When a laser is applied to a sample, light will be scattered resulting in photons with an initial (elastic scattering) or changed energy. Inelastic scattering occurs when photons loose/gain energy to excite/deactivate molecular vibrations in the sample. A Raman spectrum is obtained by collecting and analyzing these inelastic photons. The intensity of the inelastic photons is plotted against the energy in wavenumbers. Chemical compounds will produce a unique and predictable spectrum that is a result of molecular vibrational modes. Raman spectroscopy is advantageous for the field of forensic science because of the limited sample preparation, rapid results and the in-field analysis possibilities with portable Raman instruments [12-14]. The laser light is typically non-destructive so further testing is not affected in any way.

Bone tissue is mainly composed of inorganic mineral imbedded in an organic protein matrix. The inorganic portion, bioapatite, is similar to the natural geologic mineral hydroxyapatite. Bioapatite is a carbonated form of this mineral. The organic portion consists mainly of type I collagen. By weight, the organic and inorganic portions of bone are roughly 30% and 70% respectively. The composition of bone tissue varies significantly between species. Aerssens et al. reports several measurable differences in cortical bone between human, pig, cow, dog, sheep, chicken, and rat [15]. Key interspecies bone composition differences noted were collagen content, dry ash weight and non-collagen protein content.

Virkler and Lednev have recently introduced a new approach based on the combination of Raman spectroscopy and advanced statistical analysis for identification of body fluid traces [4,16-21] for forensic purposes. The approach relies on the ability of Raman spectroscopy to probe and characterize intrinsically heterogeneous body fluid samples. Further research by Virkler and Lednev has led to the development of methodology to discriminate blood samples of different species [22] with the application of Raman spectroscopy. A recent publication in the Journal of Analytical Chemistry reports on this method of identification of species based on the Raman spectroscopic analysis of blood [22]. We have recently reported on potential application of Raman spectroscopy for determining the burial time based on bone remains [23]. Here we report on a preliminary investigation of species differentiation based on Raman spectroscopy of bones. Our hypothesis is based on the literature which demonstrates that vibrational spectroscopic analysis of bone components and mineralized tissue can discriminate species of origin [24-28]. Species discrimination using Raman spectroscopy of animal tusk has been previously investigated by several researchers. Animal tusk, or dentine, is a mineralized tissue very similar to bone. Dentine tends to be more mineralized and have less collagen than bone tissue. Shimoyama et al. (1997) reports that hard and soft mammoth dentine were discriminated using near infrared (NIR) FT-Raman spectroscopy [24]. Brody et al. (2001) report the discrimination of dentine from six mammalian species, with slight overlap, also using FT-Raman spectroscopy [26] and chemometeric processing. Edwards et al. report using a similar approach to discriminate between Asian and African elephant dentine in a forensic setting [28]. Discrimination in these studies is likely due to organic to inorganic ratio differences between species.

Considering the reported success of the studies of biological material conducted for the purposes of species discrimination, we intend to use similar methods for this study. The main purpose of this research is to prove the principle that bone can be spectroscopically discriminated according to species of origin. This has clear forensic applications, but could also be useful in the study of ancient cultures. For this preliminary investigation, various non-human species were used for practicality; a human bone was not requisite at this stage. If the developed methods prove successful, continued research will focus on the discrimination of human bone specimen from other animal bones.

2. EXPERIMENTAL PROCEDURE

2.1. Samples

Chicken, turkey, cow and pig bone samples were collected from a grocery meat market. Each bone was cleaned of muscle, fat and soft tissue before Raman spectroscopic measurements.

2.2. Raman Microscope

A Renishaw in Via Raman microscope with attached Leica microscope with 20X objective was used. The wavelength of the laser used was 785 nm with approximately 11.5 mW power. Data was collected using Wire 2.0 software and analyzed using Grams v7.01. A small area of cortical bone was cut and placed on the lower plate of a Nanonics AFM MultiView 1000 automatic mapping stage. The automatic mapping stage was used to sample an arbitrary square area of cortical bone measuring approximately 75 μm × 75 μm. Within this square area, 36 points (6 × 6 square distribution) were sampled. Scans were set to 35 seconds with five accumulations and with a full range of 3200 - 100 cm^{-1}.

2.3. Data Treatment

GRAMS v7.01 was used to remove cosmic rays and for baseline subtraction. The data was truncated between the range 3190 - 360 cm^{-1}. The spectral data was then imported as a single matrix into MATLAB version 7.9.0. MATLAB PLS_Toolbox 5.5 by Eigenvector Research Inc. was used for pretreatment and partial least squares discriminant analysis (PLS-DA). The data was normalized about the most intense peak (~960 cm^{-1}) and mean centered. The number of components was estimated using the root mean square error of the leave-one-out crossvalidation (RMSECV). Component analysis was performed using PLS-DA and a model was built based on the component loadings. MATLAB software was also used for spectral averaging.

3. RESULTS AND DISCUSSION

3.1. Main Approach

Previous Raman studies of bone tissue have been limited because of strong fluorescence interference [29]. Golcuk et al. (2006) used photobleaching, a potentially destructtive process, prior to acquiring bone Raman spectra at 532-nm excitation [30]. In this study, to reduce the fluorescence interference we utilized a 785-nm near infrared light for excitation. Two, 60 second accumulations provides high quality Raman spectra with the signal-to-noise ratio of 50:1 or better.

Bone tissue is expected to be heterogeneous, meaning that no individual Raman spectrum measured at a single spot could fully represent a specific tissue. For this reason, we utilized a previously developed a mapping method for acquiring Raman spectra from multiple spots for further statistical analysis. In this study, 36 individual spectra were obtained for each species surveyed using the automatic mapping technique. PLS-DA was used because of the high discriminatory potential of this technique [31,32].

3.2. Spectral Analysis

The spectra were averaged for visual comparison and are shown in **Figure 1**. The averaged spectra between turkey, chicken, and pig bone samples are visually very similar. The peak count and position are identical between these species. There are, however, relative peak intensity variations that could be indicative of the variation in the bone tissue composition for various species. For example, the relative intensities of 1075 cm^{-1} and 1042 cm^{-1} Raman peaks vary (shaded portion of **Figure 1**). The Amide I peak at 1660 cm^{-1}, which is attributed to the contribution from the collagen organic component [33], varies in intensity between samples. The intensity a small peak at 1003 cm^{-1} also changes in intensity between these spectra. This peak is assigned to the phenylalanine amino acid breathing mode.

The averaged spectrum from cow is most distinct from the other three as it appears to be missing several peaks, including three peaks in the 800 - 900 cm^{-1} range, attributed to δ CCH at 855 cm^{-1}(collagen), v C-C 875 cm^{-1} (proline) and v C-C 920 cm^{-1} (collagen) vibrational modes [34]. The peak pattern in the CH_2 region at 2900 cm^{-1} for cow is comparably less intense than other species sampled. This is also seen in the Amide I, Amide III and δ NH bands. Excluding this irregularity, the Raman spectra between species are very similar.

3.3. Determining the Number of Components

Leave-one-out cross-validation was performed and the root mean square error of cross-validation (RMSECV) versus the component plot is shown in **Figure 2**. In this plot, a local minimum or a distinct change in slope will occur at the number of components which describe the system. There are such local minimums around n = 2 - 3. Between the four classes, either two or three components should adequately describe the system. It was seen that the first two components, which contributed over 85% of spectral data, were best at discriminating classes and other components did not vary much between classes. For the purposes of this study, a model built using the first two components is therefore sufficient.

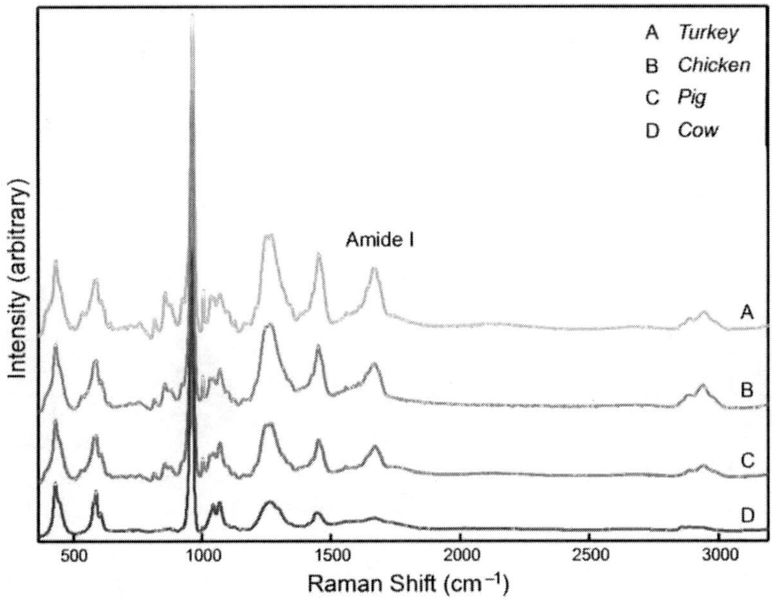

Figure 1. A stack plot showing the combined accumulations of Raman spectra for cow, pig, turkey and chicken. The shaded region is highlights peaks at 1075 and 1042 cm^{-1} which is discussed in Section 3.2. Raman spectra are baseline corrected to aid comparison.

3.4. Component Analysis

PCA is a multivariate data analysis method. This analysis describes a large data set with a small number of extracted components. These components have high variation which serves to highlight patterns in the dataset. This allows relationships in data to be efficiently described and represented by a few components. PLS-DA is an extension of PCA which accounts for class definitions and maximizes the separation of classes. In this case, the classes are the species identity i.e. cow, chicken, pig and turkey.

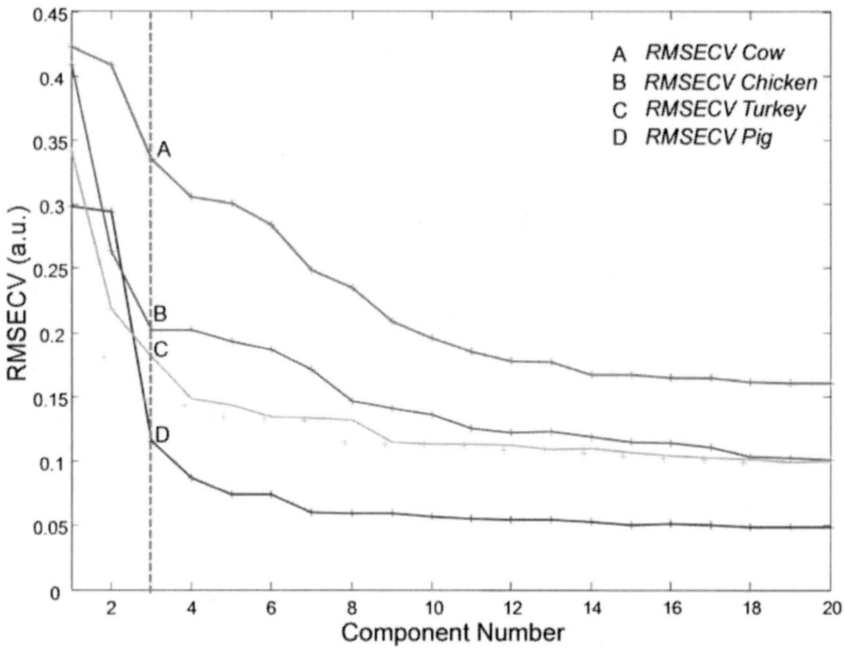

Figure 2. A root mean square error of cross-validation (RMSECV) plot versus the number of components. The break at n = 3 indicates that there are three components.

Component analysis is almost completely automated using MATLAB software and the Eigenvector PLS toolbox. First, the accumulated spectral data is combined into a single matrix. Normalization and mean centering preprocessing is performed on the data matrix to prepare the data for component analysis. When the PLS-DA algorithm is applied to this matrix, a number of components are extracted. The first component accounts for the majority of data variation within the matrix and is usually the most significant for discriminatory purposes. The following components progressively account for less data variance. The shape of a component is called its loading. Next, each individual spectrum is superimposed against the loading of a component to obtain a score. The more a given spectrum has features common to a component, the higher the score. A scores plot is usually displayed in two or three dimensions to graphically display the scores on multiple components. For example, a scores plot in two dimensions usually displays the scores on component one on the x-axis and the scores on component two on the yaxis. The actual numerical score of a given point is arbitrary because the zero, zero (0,0) point in a two dimensional scores plot is positioned at the total average score.

PLS-DA is very useful for discriminating samples that have similar chemical constituents. In the case of bone tissue, chemical constituents are essentially identical between species. All bone tissue is expected to be comprised of mostly bioapatite and type I collagen. However, the percent composition is expected to vary between species. Since the main components of bone, bioapatite and collagen, are detectable with Raman spectroscopy, PLS-DA analysis is appropriate for discriminatory purposes.

The components derived from this type of data analysis can often be chemically informative [35]. It is seen in this study that the components loadings reflect the spectral signatures of collagen and bioapatite. Raman signals from these materials have been previously separated and identified in a modern mouse bone. Freeman et al. (2002) reports the Raman spectrum from both organic and inorganic constituents [29]. A general rule is that the mineral phase of bone has a majority of Raman bands below 1000 cm^{-1} with a distinct sharp peak centered at 960 cm^{-1}. Conversely, the organic phase contributes a majority of peaks towards higher wavenumbers than 1000 cm^{-1} and are generally broad. Component assignments were made by comparing loading patterns with previously published Raman spectrum of isolated bone constituents.

The loadings for components one and two are shown in **Figure 3**. For component one, there are major positive features in the 3000 - 2800 cm^{-1} range as well as a triplet which has peaks centered at around 1260 cm^{-1}, 1450 cm^{-1} and 1670 cm^{-1}. This positive triplet corresponds to the Amide III, δ NH and Amide I Raman bands respectively. Hence, this component is dominated by organic contributions.

Conversely, component two is dominated by inorganic features. The dominant peak centered at 960 cm^{-1} is the v_1 PO_4^{3-} vibrational mode. The two peaks at lower wavenumbers from this, centered around 430 cm^{-1} and 590 cm^{-1} correspond to v_2 PO_4^{3-} and v_4 PO_4^{3-} vibrational modes. The peak at 1070 cm^{-1} corresponds to the carbonate ion (v_3 CO_3^{2-}) vibrational mode. The broad hump centered around 1260 cm^{-1} corresponds to the Amide III vibrational mode.

These component loading features are in agreement with isolated spectra of collagen and bioapatite published by Freeman et al. (2002) Although there is slight mutual contribution, in general component one represents a collagen contribution, while component two is representative of bioapatite.

The first two components were used to create a scores plot to determine if spectra will group with respect to species of origin. A two dimensional component scores plot is shown in **Figure 4**. A 95% confidence interval

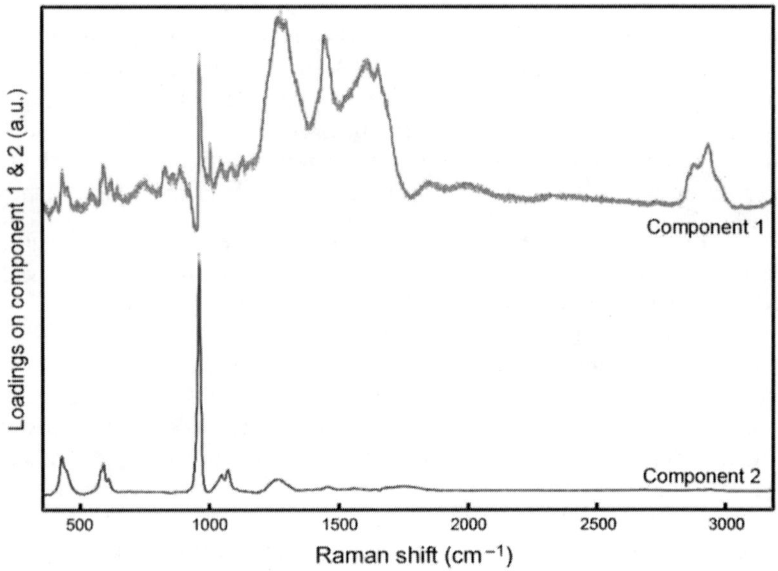

Figure 3. The component loading for spectral component one (top/red) and two (bottom/blue).

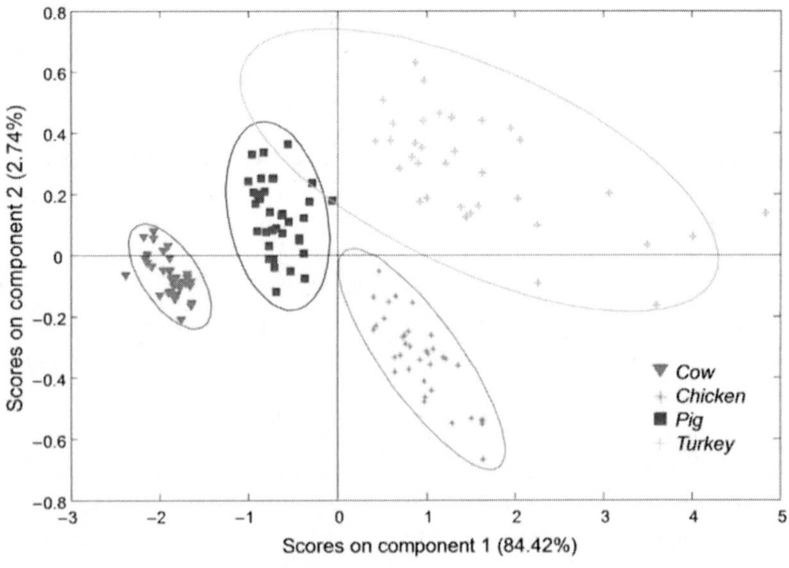

Figure 4. A two dimensional scores plot for cow, chicken, turkey and pig based on the first two spectral components. A 95% confidence ellipsoid encompasses each species class.

ellipsoid is drawn around each class. It is apparent that the classes are completely separated with respect to species of origin and there is only slight group overlap of the ellipsoid representing a 95% confidence interval.

4. CONCLUSION

The goal of this research project was to investigate the potential of Raman spectroscopy to discriminate between bone samples based on species of origin. Previous research demonstrates that the chemical composition of bone is measurably different in various mammalian species [15]. This approach is appropriate since components of bone are easily detected using Raman spectroscopy. PLS-DA was chosen for this purpose because of its powerful discrimination ability. Using this method, spectra from four species were successfully discriminated with no overlap. The model built for discrimination reflects the chemical composition of bone material. The separation of groups is likely due to the differences in bone composition that have been previously reported in different species. These positive results demonstrate that the species of origin determination of bone in a practical setting may be possible. This would be a substantial contribution to forensic anthropology, a field that is currently lacking in spectroscopic analysis.

Gregory McLaughlin, Igor K. Lednev Further research is necessary for a more complete representation of each species. Variables such as specimen age, bone type, and gender are known to affect bone composition and need to be controlled. In addition, degenerative factors that change the composition of bone after death need to be analyzed for this approach to have practical applications. To prove forensic applicability, a human bone also needs to be sampled.

REFERENCES

1. M. Houck and J. Siegel, "Fundamentals of Forensic Science," Academic Press, San Diego, 2006.
2. A. Chamberlain, "Human Remains," University of California Press, Los Angeles, 1994.
3. C. Cattaneo, D. Porta, D. Gibelli and C. Gamba, "Histological Determination of the Human Origin of Bone Fragments," Journal of Forensic Sciences, Vol. 54, No. 3, 2009, pp. 531-533.
4. K. Virkler and I. K. Lednev, "Analysis of Body Fluids for Forensic Purposes: From Laboratory Testing to Non-destructive Rapid Confirmatory Identification at a Crime Scene," Forensic Science International, Vol. 188, No. 1-3, 2009, pp. 1-17.

5. C. M. Hodges and J. Akhavan, "The Use of FourierTransform Raman-Spectroscopy in the Forensic Identification of Illicit Drugs and Explosives," Spectrochimica Acta Part A—Molecular and Biomolecular Spectroscopy, Vol. 46, 1990, pp. 303-307.
6. E. M. A. Ali, H. G. M. Edwards and I. J. Scowen, "In-situ Detection of Single Particles of Explosive on Clothing with Confocal Raman Microscopy," Talanta, Vol. 78, 2009, pp. 1201-1203.
7. G. Jochem and R. J. Lehnert, "On the Potential of Raman Microscopy for the Forensic Analysis of Coloured Textile Fibres," Science & Justice, Vol. 42, No. 2, 2002, pp. 215-221.
8. J. Zieba-Palus and R. Borusiewicz, "Examination of Multilayer Paint Coats by the use of Infrared, Raman and XRF Spectroscopy for Forensic Purposes," Journal of Molecular Structure, Vol. 792-793, 2006, pp. 286-292.
9. S. E. J. Bell, L. A. Fido, S. J. Speers and J. Armstrong, "Raman Spectroscopy for Rapid, One-step Forensic Analysis of Paint," Forensic Science International, Vol. 136, 2003, pp. 354-355.
10. S. E. J. Bell, L. A. Fido, S. J. Speers, W. J. Armstrong and S. Spratt, "Forensic Analysis of Architectural Finishes using Fourier Transform Infrared and Raman Spectroscopy, Part II: White Paint," Applied Spectroscopy, Vol. 59, No. 11, 2005, pp. 1340-1346.
11. J. Mania, B. Trzcinska, M. Kunicki and P. Kooecielniak, "Comparison of the CE method with FTIR and Raman Spectrometry in the Field of Forensic Ink Analysis," Forensic Science International, Vol. 136, 2003, pp. 75-75.
12. B. Eckenrode and E. Bartick, "Portable Raman Spectroscopy Systems for Field Analysis," Forensic Science Communications, Vol. 3, 2001.
13. M. D. Hargreaves, K. Page, T. Munshi, R. Tomsett, G. Lynch and H. G. M. Edwards, "Analysis of Seized Drugs using Portable Raman Spectroscopy in an Airport Environment—A Proof of Principle Study," Journal of Raman Spectroscopy, Vol. 39, No. 7, 2008, pp. 873-880.
14. M. Perez-Alonso, K. Castro, I. Martinez-Arkarazo, M. Angulo, M. A. Olazabal and J. M. Madariaga, "Analysis of Bulk and Inorganic Degradation Products of Stones, Mortars and Wall Paintings by Portable Raman Microprobe Spectroscopy," Analytical and Bioanalytical Chemistry, Vol. 379, No. 1, 2004, pp. 42-50.
15. J. Aerssens, S. Boonen, G. Lowet and J. Dequeker, "Interspecies Differences in Bone Composition, Density, and Quality: Potential Implications for In Vivo Bone Research," Endocrinology, Vol. 139, No. 2, 1998, pp. 663- 670.
16. K. Virkler and I. K. Lednev, "Raman Spectroscopy Offers Great Potential for the Nondestructive Confirmatory Identification of Body Fluids," Forensic Science International, Vol. 181, No. 1-3, 2008, pp. e1-e5.
17. K. Virkler and I. K. Lednev, "Raman Spectroscopic Signature of Semen and its Potential Application to Forensic Body Fluid Identification," Forensic Science International, Vol. 193, No. 1-3, 2009, pp. 56-62.
18. K. Virkler and I. K. Lednev, "Raman Spectroscopic Signature of Blood and its Potential Application to Forensic Body Fluid Identification," Analytical and Bioanalytical Chemistry, Vol. 396, No. 1, 2010, pp. 525-534.
19. K. Virkler and I. K. Lednev, "Forensic Body Fluid Identification: The Raman Spectroscopic Signature of Saliva," Analyst, Vol. 135, 2010, pp. 512-517.
20. V. Sikirzhytski, K. Virkler and I. K. Lednev, "Discriminant Analysis of Raman Spectra for Body Fluid Identification for Forensic Purposes," Sensors, Vol. 10, No. 4, 2010, pp. 2869-2884.
21. A. Sikirzhytskaya, V. Sikirzhytski and I. K. Lednev, "Raman Spectroscopic Signature of Vaginal Fluid and its Potential Application in Forensic Body Fluid Identification," Forensic Science International, 2011, Article in Press.
22. K. Virkler and I. Lednev, "Blood Species Identification for Forensic Purposes Using Raman Spectroscopy Combined with Advanced Statistical Analysis," Analytical Chemistry, Vol. 81, No. 18, 2009, pp. 7773-7777.

23. G. McLaughlin and I. K. Lednev, "Potential Application of Raman Spectroscopy for Determining Burial Duration of Skeletal Remains" Analytical and Bioanalytical Chemistry, Vol. 401, No. 8, 2011, pp. 2511-2518.
24. M. Shimoyama, H. Maeda, H. Sato, T. Ninomiya and Y. Ozaki, "Nondestructive Discrimination of Biological Materials by Near-infrared Fourier Transform Raman Spectroscopy and Chemometrics: Discrimination Among Hard and Soft Ivories of African Elephants and Mammoth Tusks and Prediction of Specific Gravity of the Ivories," Applied Spectroscopy, Vol. 51, No. 8, 1997, pp. 1154- 1158.
25. M. Shimoyama, S. Morimoto, and Y. Ozaki, "Non-destructive Analysis of the Two Subspecies of African Elephants, Mammoth, Hippopotamus, and Sperm Whale Ivories by Visible and Short-wave Near Infrared Spectroscopy and Chemometrics," Analyst, Vol. 129, 2004, pp. 559-563.
26. R. H. Brody, H. G. M. Edwards, and A. M. Pollard, "Chemometric Methods Applied to the Differentiation of Fourier-Transform Raman Spectra of Ivories," Analytica Chimica Acta, Vol. 427, 2001, pp. 223-232.
27. D. M. Hashim, Y. B. C. Man, R. Norakasha, M. Shuhaimi, Y. Salmah, and Z. A. Syahariza, "Potential Use of Fourier Transform Infrared Spectroscopy for Differentiation of Bovine and Porcine Gelatins," Food Chemistry, Vol. 118, No. 3, 2009, pp. 856-860.
28. H. G. M. Edwards, N. F. N. Hassan and N. Arya, "Evaluation of Raman Spectroscopy and Application of Chemometric Methods for the Differentiation of Contemporary Ivory Specimens I: Elephant and Mammalian Species," Journal of Raman Spectroscopy, Vol. 37, No. 1-3, 2006, pp. 353-360.
29. J. J. Freeman and M. J. Silva, "Separation of the Raman Spectral Signatures of Bioapatite and Collagen in Compact Mouse Bone Bleached with Hydrogen Peroxide," Applied Spectroscopy, Vol. 56, No. 6, 2002, pp. 770-775.
30. K. Golcuk, G. S. Mandair, A. F. Callender, N. Sahar, D. H. Kohn and M. D. Morris, "Is Photobleaching Necessary for Raman Imaging of Bone Tissue using a Green Laser?" Biochimica Et Biophysica Acta-Biomembranes, Vol. 1758, No. 7, 2006, pp. 868-873.
31. S. Lars and W. Svante, "Partial Least Squares Analysis with Cross-Validation for the Two-Class Problem: A Monte Carlo Study," Journal of Chemometrics, Vol. 1, 1987, pp. 185-196.
32. M. Barker and W. Rayens, "Partial Least Squares for Discrimination," Journal of Chemometrics, Vol. 17, No. 3, 2003, pp. 166-173.
33. R. Smith and I. Rehman, "Fourier-Transform RamanSpectroscopic Studies of Human Bone," Journal of Materials Science-Materials in Medicine, Vol. 5, 9-10, No. 1994, pp. 775-778.
34. A. Awonusi, M. D. Morris and M. M. Tecklenburg, "Carbonate Assignment and Calibration in the Raman Spectrum of Apatite," Calcified Tissue International, Vol. 81, No. 1, 2007, pp. 46-52.
35. H. Nocairi, E. M. Qannari, E. Vigneau and D. Bertrand, "Discrimination on Latent Components with Respect to Patterns. Application to Multicollinear Data," Computational Statistics & Data Analysis, Vol. 48, No. 1, 2005, pp. 139-147.

CHAPTER 9

A Preliminary Study on Earprint Identification among Young Adults in Malaysia

Lai PS[1]*, Osman K[1] and Ismail NA[2]

[1]*Forensic Scientific Officer, National Institute of Forensic Medicine, Hospital Kuala Lumpur, Malaysia*

[2] *Lecturer of Forensic Science Program, Faculty of Allied Health Sciences, National University of Malaysia, Kuala Lumpur, Malaysia*

ABSTRACT

(74 males and 74 females) ranging in age from 18 to 25 years were taken from Campus Kuala Lumpur of National University of Malaysia (UKMKKL). A total of twelve biometrics based on Iannarelli's method as well as ear length and ear width were used as measurements. Statistical analysis indicated that left and right ears had no significant difference ($p > 0.05$). Gender difference was found to be highly significant for nine out of fourteen ear biometrics ($p < 0.05$). However, there were only five out of fourteen ear biometrics had significant difference between Chinese and Malays ($p < 0.05$). Statistical analysis had also indicated that differences among ages for each ear biometrics were insignificant and no clear correlation between ear biometrics and ages was obtained ($p > 0.05$). Formulae for gender and race estimation were obtained by using Classification and Regression (C & R)

Tree. Formulae of ear length had the highest accuracy (78.7%) for gender estimation.

However, the accuracy was achieved up to 92.9% when the combination of ear biometrics was used. In race estimation, only one ear biometric was able to form a formula with a single deviation point. In fact, most of the ear biometrics did not provide any useful formulae in race estimation if compared to gender estimation. In conclusion, this study demonstrated that ear biometrics played an important role in gender estimation so as to enable ear print to be accepted as strong evidence in the court of law in Malaysia.

INTRODUCTION

Ear print identification is a matching process between the ear prints from crime scene and the ear prints from database or suspects (Champod 2009) [1]. Individualization of the ear prints is established by finding the significant similarities of the individual characteristics to preclude the probability of coincidence matching. Ear biometric becomes identification method based on biological and physiological characteristics with high stability, universality, uniqueness, non-invasiveness and extractability (Singh & Purkait 2009) [2]. The ear length of a normal adult is ranging from 55.0 mm to 65.0 mm while the ear width is ranging from 50% to 60% of its length (Jung & Jung 2003; Kalcioglu., *et al.* 2003) [3,4]. However, the increase in ear dimension at different progressing of age periods is basically due to growth, gravitational effects and diminutive tensile strength of connective tissue (Liu 2008) [5]. Although this might affect the chance of finding a print in a database after a long period of time, it is unlikely to make the ears less unique. In addition, the natural changes in the ears are small enough and suitable to be used in forensic practice and legal prosecution (Meijerman., *et al.* 2006a) [6]. The unique ear print has been used in forensic medicine and criminal investigations as an identification proof, for example, in airplane crashes or crime scenes. It has been used as tamper-proof complementary evidence to the fingerprint and DNA evidence because the courts of law required two or more types of evidence to place a suspect at a scene of crime (Nanni & Lumini 2009) [7]. The excellent evidential value of ear prints is identified with the higher level of inter-person variation and lower level of intra-person variation (Meijerman., *et al.* 2006a) [6].

In 1989, the first manual method was used in Iannarelli's first study in which ten thousands of human ears were randomly selected in California.

Uniqueness of ear characteristics was proved based on measuring the twelve distances between the specific points of the ears (Choras 2008) [8]. Iannarelli's second study was regarding the differences of physiological ear features between identical and non-identical twins. The identical twins were proved to have similar but not identical ears physiological features especially in the concha and lobule areas (Rahman., *et al.* 2007) [9]. This uniqueness held true even in cases of triplets twins and quadruplets twins (Kieckhoefer., *et al.* 2006; Hurley., *et al.* 2007) [10,11].

Extraction methods of ear biometrics can be divided into photography methods and lifting methods. Photography methods used include camera-based photography, lateral cephalometric radiography, magnetic resonance imaging or computed tomography (Lugt 2000; Wang., *et al.* 2010) [12,13]. While lifting methods used include inkless impression kit, aluminium powder used with black gelatine lifter or lumibox mechanism (Kieckhoefer., *et al.* 2006; Meijerman., *et al.* 2006c; Alberink & Ruifrok 2007) [10,14,15]. On the other hands, analysis methods of ear biometrics can also divided into quantitative measurements and qualitative ear structures. Quantitative measurements used include twelve Iannarelli manual measurements, caliper measurements on length and width of ear, lobule, concha and tragus. There were some semi-automation and full automation methods mainly used for matching purpose, not for gender and race estimation, such as computed tomography, Neural network, force field transformation, geometric, iterative closest point etc. (Barut & Aktunc 2006; Hurley., *et al.* 2007; Purkait & Singh 2007; Sforza., *et al.* 2009; Wang., *et al.* 2010) [11,13,16-18]. While qualitative ear structures include the ear shape, lobule shape, presence of auricular tubercle and minutiaes (Barut & Aktunc 2006; Meijerman 2006; Chattopadhyay & Bhatia 2009; Singh & Purkait 2009) [2,16,19,20].

The purpose of this study was firstly to determine the level of significant difference and correlation between right and left ear biometrics as well as between age and ear biometrics. Secondly, this study was also to develop formulae to estimate the gender and race of an unknown individual by using ear biometrics since there was no similar study available for Malaysia population. The formulae could only be used in the same population in which the formulae developed from, which in this case a Malaysian population.

MATERIALS AND METHODS

A total of 148 young adults, 74 subjects for each gender, from Campus Kuala Lumpur of National University of Malaysia (UKMKKL) with age ranging from 18 to 25 years old were selected. Two major ethnic chosen were

Malays and Chinese following the ratio of target population with 100 (67.6%) and 48 (32.4%) subjects respectively. The subjects must be Malaysian young adults of pure Malays and Chinese who owns Malaysian identification card called "Mykad". Subjects were confirmed healthy without any type of anomaly or congenital syndrome as well as never undergone any surgical procedure that could affect ear measurements (Singh & Purkait 2009) [2]. Any attachments to the ear such as earrings were needed to be taken off before photos were captured because it could generally extend the length of ear lobe and further influence the ear length (Sharma., *et al.* 2007) [21].

This study was approved by UKM Research Ethics Committee. Intelectual Property Rights (IPRs) Agreement and Consent Form were signed by subjects prior to the study. Subjects were then required to sit down with the head positioned in Frankfurt horizontal plane, namely a horizontal line connecting both inferior orbitale and tragion. Subjects were subsequently required to paste their ear slowly and lightly on the glass with fixed scale of 150 mm ± 1 mm (Lynn Peavey Company, US) as portrayed in Figure 1. Two photos of right ears and two photos of left ears were taken with Canon 350D Digital Single-Lens Reflex camera (Canon Inc., Japan) perpendicular to the subjects in enough lighting condition from 8:00 pm to 10:00 pm to avoid any diurnal changes. All photos of ears were printed out through Canon MP146 printer (Canon Inc., Vietnam) after normalizing accordingly with Iannarelli method.

The measurements involved were the twelve ear biometrics based on Iannarelli method, ear length and ear width. Measurements were taken using stainless steel ruler of 150 mm ± 0.5 mm (Penguin, China) in millimetre (mm) with one decimal place. One observer was only involved for the measurements so as to prevent any inter-observer error. They were repeatedly measured for three times in which their average values were used. Similar instruments and protocols were applied throughout the measurements in order to act as quality assurance in this study.

Biometric A (Iannarelli measurements) were obtained by putting Iannarelli Inscribed transparent card with eight lines that inclined in similar angle of 45o on top of the ear photos as portrayed in Figure 2 (Hurley *et al.* 2007) [11]. All measurements were named as Biometric A1, Biometric A2 and further until Biometric A12 such as indicated in Table 1. Ear length (Biometric EL) was the maximum distance of vertical line between top end of superior helix rim and bottom end of inferior lobule of ear. While ear width (Biometric EW) was the maximum distance between the most anterior end of helix rim and most posterior end of helix rim where a horizontal line, that cross perpendicularly with the line of ear length, was formed.

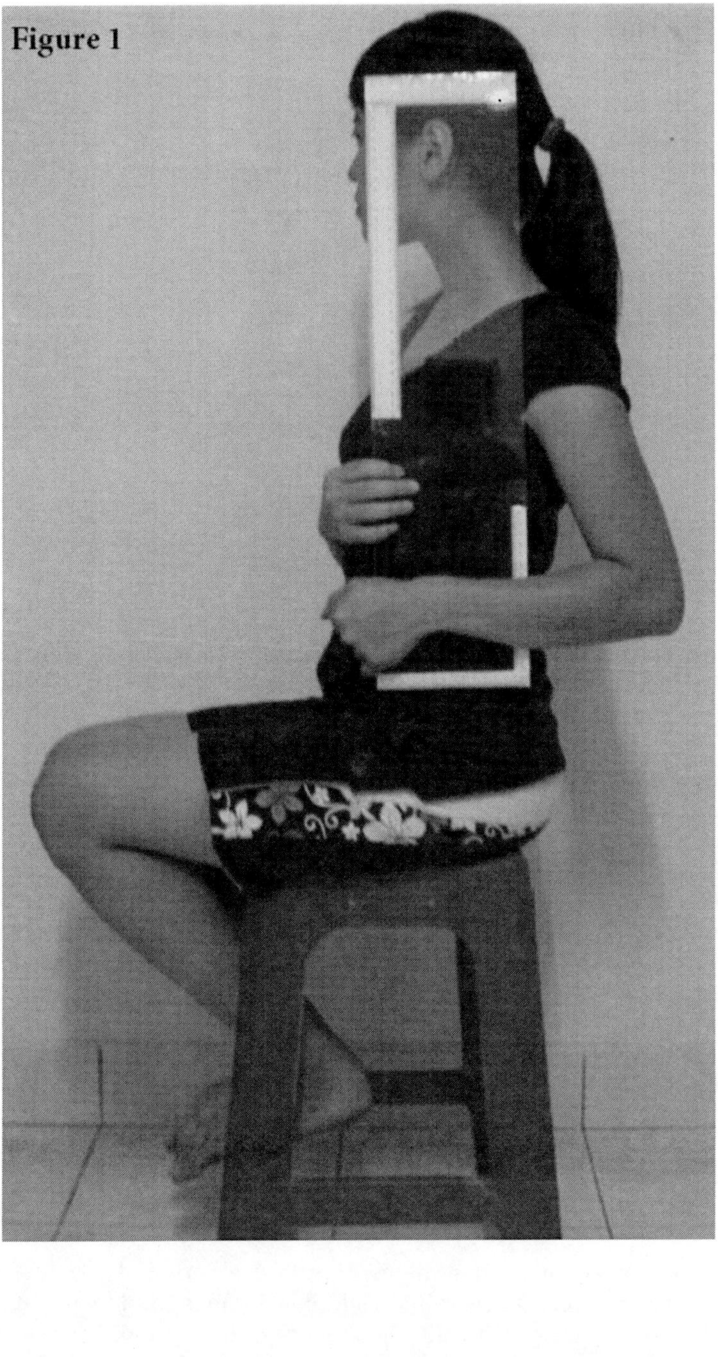

Figure 1

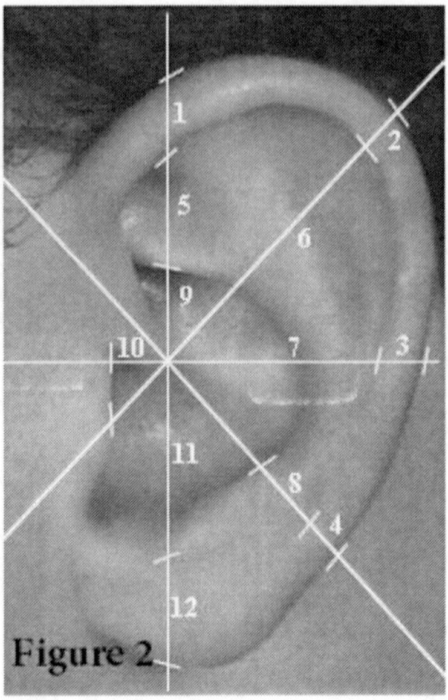

Figure 2

All biometrics were recorded as numerical data in tabular form and then analysed using Predictive Analytics Software (PASW) 18.0. Kolmogorov-Smirnov and Shapir-Wilks tests were carried out to determine the distribution normality. Independent groups t-test and one-way between-groups ANOVA test were used to determine whether there were any differences in biometrics. The p-value is the level of marginal significance within a statistical hypothesis test representing the probability of the occurrence of a given event. The significance level in all tests was $p < 0.05$. The numerical data from PASW 18.0 was then transferred into Statistical Package for Social Sciences (SPSS) Clementine 11.1. Classification and Correlation (C & R) Tree Modeling was used to estimate gender and race by forming formulae. The overall samples were randomly divided into 40% of training group, 40% of test group and 20% of validation group. The R-value denoted the correlation coefficient or power of the relationship between variables.

RESULTS

Table 1 showed the mean values, standard errors and p-values for each biometrics between left ear and right ear. By using Independent groups *t*-test, there was no statistically significance of bilateral differences for all of the ear biometrics where their significant values were $p > 0.05$. Among all the biometrics, Biometric A9 was closely to have significant differences between left and right ears with $p > 0.05$ in which 7.6 ± 0.1 mm and 7.9 ± 0.1 mm respectively. In addition, correlation analyses for the bilateral symmetry between left and right ears were all significant with $p < 0.01$ and r value ranging from 0.503 to 0.920. The highest correlation of bilateral symmetry involved the ear length with $r = 0.920$.

Comparative statistics with the mean values, standard errors, standard deviations and p- values for each biometrics between genders were tabulated in Table 2. By using independent groups t-test, there were statistically significance of gender differences for nine of the biometrics including biometrics A3, A4, A5, A6, A7, A11, EL and EW with $p < 0.01$ while biometrics A9 with $p < 0.05$. However, the other biometrics has no significant difference between genders. Classification formulae for gender were tabulated in Table 3. The accuracies of gender estimation following descending order were 78.7% for EL, 77.4% for biometric A6, 72.6% for biometric A5, 71.0% for biometric A11, 64.9% for biometric A4, 65.5% for EW, 63.2% for biometric A3 and 58.8% for biometrics A7. The best combination formulae were discovered for gender estimation with very high accuracy up to 92.9% as listed in Table 4. The best combination was the combination of five biometrics including A4, A6, A11, EL and EW.

Table 1: Bilateral difference in ear biometrics (mm)

Ear Biometrics	Left ear (n = 148)		T-test			
	Mean	Standard Error	Mean	Standard Error	p	R *p < 0.05
A1	5.0	0.1	5.1	0.1	0.541	0.683*
A2	4.5	0.1	4.4	0.1	0.336	0.631*
A3	4.2	0.1	4.1	0.1	0.341	0.503*
A4	4.9	0.1	4.9	0.1	0.941	0.691*
A5	14.6	0.2	14.6	0.2	0.847	0.742*
A6	29.2	0.3	29.9	0.3	0.068	0.805*
A7	15.9	0.2	16.2	0.2	0.168	0.817*
A8	6.3	0.1	6.5	0.1	0.143	0.551*
A9	7.6	0.1	7.9	0.1	0.058	0.552*
A10	7.0	0.1	7.0	0.1	0.754	0.763*
A11	12.2	0.2	12.6	0.2	0.151	0.791*
A12	17.5	0.2	17.6	0.2	0.717	0.856*
EL	58.9	0.4	59.4	0.4	0.385	0.920*
EW	31.6	0.2	31.8	0.2	0.482	0.751*

Table 2: Descriptive statistics for all biometrics (mm) in gender

Ear Biometrics	Female (n = 148)				Male (n=148)				T-test
	Minimum	Maximum	Mean	Standard Error	Minimum	Maximum	Mean	Standard Error	P
A1	3.5	7.0	5.0	0.1	3.3	6.5	5.1	0.1	0.738
A2	3.0	6.3	4.4	0.1	3.0	6.0	4.5	0.1	0.259
A3	2.5	6.0	4.0	0.0	3.0	6.0	4.4	0.1	< 0.01
A4	2.5	7.0	4.7	0.1	3.3	6.8	5.1	0.1	< 0.01
A5	7.5	18.5	13.5	0.2	10.0	21.0	15.7	0.2	< 0.01
A6	20.5	35.5	27.6	0.2	25.8	38.0	31.5	0.2	< 0.01
A7	10.0	20.0	15.6	0.2	11.5	21.3	16.5	0.2	< 0.01
A8	3.3	9.5	6.3	0.1	3.8	10.0	6.5	0.1	0.288
A9	4.0	11.3	7.6	0.1	5.3	11.3	7.9	0.1	< 0.05
A10	4.8	11.8	7.1	0.1	4.0	9.3	6.9	0.1	0.078
A11	7.0	16.3	11.5	0.2	7.5	18.5	13.3	0.2	< 0.01
A12	13.3	23.0	17.3	0.2	12.3	25.3	17.9	0.2	0.053
EL	46.0	65.5	56.4	0.3	53.0	72.5	61.9	0.3	< 0.01
EW	24.5	35.5	30.7	0.2	28.0	38.0	32.7	0.2	< 0.01

Table 3: Classification formulae based on gender

Ear Biometrics	Deviation Point	Category	Accuracy (%)	Overall of Accuracy (%)
A3	≤ 4.40 mm	Female	68.8 – 88.6	63.2
	> 4.40 mm	Male	36.8 – 53.4	
A4	≤ 4.65 mm	Female	52.0 – 63.6	64.9
	> 4.65 mm	Male	71.9 – 82.1	
A5	≤ 15.15 mm	Female	76.2 – 79.2	72.6
	> 15.15 mm	Male	64.2 – 75.0	
A6	≤ 29.15 mm	Female	66.1 – 73.3	77.4
	> 29.15 mm	Male	78.1 – 86.5	
A7	≤ 14.65 mm	Female	21.8 – 48.4	58.8
	> 14.65 mm	Male	78.3 – 93.8	
A11	≤ 13.40 mm	Female	80.7 – 91.7	71.0
	> 13.40 mm	Male	53.0 – 65.0	
EL	≤ 58.15 mm	Female	69.1 – 73.5	78.7
	> 58.15 mm	Male	80.0 – 89.2	
EW	≤ 30.40 mm	Female	38.3 – 41.0	65.5
	> 30.40 mm	Male	88.0 – 92.9	

Table 4: Best classification formulae based on gender

Combination	Ear Biometrics	Accuracy (%)		Overall of Accuracy (%)	Performance Evaluation
		Female	Male		
Two	A6 dan A4	75.0 – 88.1	87.9 – 92.5	87.8	0.565
Three	A6, A11 dan A4	93.2 – 98.2	82.8 – 94.3	91.9	0.610
Four	A6, A11, A4 dan A7	90.6 – 96.5	83.8 – 96.2	91.2	0.610
Four	A6, EW, A4 dan A7	90.6 – 98.3	83.8 – 87.9	91.2	0.610
Five	EL, A6, A11, EW dan A4	87.5 – 98.3	91.4 – 94.6	92.9	0.620
Six	EL, A6, A5, A11, A3 dan A7	86.0 – 93.8	91.4 – 100.0	92.6	0.617
Six	EL, A6, A11, EW, A4 dan A7	84.4 – 96.6	89.7 – 97.3	92.6	0.617
Seven	EL, A6, A5, A11, A4, A3 dan A7	86.0 – 93.8	91.4 – 100.0	92.6	0.617
Eight	EL, A6, A5, A11, EW, A4, A3 dan A7	84.7 – 93.8	91.4 – 97.3	90.5	0.595

Comparative statistics with the mean values, standard errors, standard deviations and p- values for each biometrics between races were shown in Table 5. By using independent groups *t*-test, there were statistically significance of gender differences for only six of the biometrics including biometrics A2, A4, A8, A10, A11 and A12 with $p < 0.01$ while biometric A2 with $p < 0.05$. However, most of the biometrics has no significant difference between races. Classification formulae for races were tabulated in Table 6. The race estimation using biometrics A4, A8, A10, A11 and A12 were all with moderately high accuracy of 67.9%, 62.5%, 68.2%, 69.3% and 69.3%. However, there were extremely large differences of accuracy between Chinese and Malays for all those biometrics mentioned. Biometric A12 was the only one that successfully providing the formula with single deviation point for race estimation. In fact, most of the biometrics did not provide any useful formulae compared to gender estimation.

Table 5: Descriptive statistics for all biometrics (mm) in races

Ear Biometrics	Chinese (n = 48)				Malays (n = 100)				T-test
	Minimum	Maximum	Mean	Standard Error	Minimum	Maximum	Mean	Standard Error	p
A1	3.8	6.3	5.0	0.1	3.3	7.0	5.1	0.1	0.738
A2	3.0	5.8	4.3	0.1	3.0	6.3	4.5	0.1	0.259
A3	3.0	6.0	4.2	0.1	2.5	6.0	4.1	0.0	< 0.01
A4	3.5	6.5	4.7	0.1	2.5	7.0	5.0	0.1	< 0.01
A5	11.0	19.3	14.7	0.2	7.5	21.0	14.6	0.2	< 0.01
A6	22.0	38.0	30.0	0.3	20.5	37.8	29.4	0.2	< 0.01
A7	11.5	19.8	15.9	0.2	10.0	21.3	16.1	0.1	< 0.01
A8	4.0	10.0	6.0	0.1	3.3	9.5	6.5	0.1	0.288
A9	5.0	11.3	7.6	0.1	4.0	11.3	7.8	0.1	< 0.05
A10	4.0	9.0	6.6	0.1	5.0	11.8	7.2	0.1	0.078
A11	8.3	17.0	13.0	0.2	7.0	18.5	12.1	0.2	< 0.01
A12	12.3	23.5	17.1	0.2	13.5	25.3	17.8	0.2	0.053
EL	53.0	66.0	59.5	0.3	46.0	72.5	59.0	0.4	< 0.01
EW	27.3	35.5	31.9	0.2	24.5	38.0	31.6	0.2	< 0.01

Table 6: Classification formulae based on races

Ear Biometrics	Deviation Point (mm)	Category	Accuracy (%)	Overall of Accuracy (%)
A4	Multiple points	Chinese	4.9 – 21.7	67.9
	Multiple points	Malays	89.7 – 95.2	
A8	Multiple points	Chinese	36.8 – 47.4	62.5
	Multiple points	Malays	69.1 – 77.0	
A10	Multiple points	Chinese	0.0 – 4.7	68.2
	Multiple points	Malays	100.0	
A11	Multiple points	Chinese	0.0 – 15.4	69.3
	Multiple points	Malays	96.3 – 100.0	
A12	≤ 13.40 mm	Chinese	2.9 – 11.5	69.3
	> 13.40 mm	Malays	100.0	

Table 7: Descriptive statistics for all biometrics (mm) in ages

	19 (N = 58)	20 (N = 60)	21 (N = 56)	22 (N = 66)	23 (N = 56)	ANOVA p	Correlation r *p < 0.05
A1	5.0 ± 0.1	5.2 ± 0.1	5.2 ± 0.1	4.9 ± 0.1	4.9 ± 0.0	0.056	-0.082
A2	4.4 ± 0.1	4.7 ± 0.1	4.4 ± 0.1	4.4 ± 0.1	4.4 ± 0.1	0.075	-0.097
A3	4.3 ± 0.1	4.3 ± 0.1	4.1 ± 0.1	4.1 ± 0.1	4.0 ± 0.1	0.093	-0.145*
A4	5.0 ± 0.1	5.1 ± 0.1	5.0 ± 0.1	4.7 ± 0.1	4.8 ± 0.1	0.057	-0.157*
A5	14.6 ± 0.3	14.7 ± 0.3	14.6 ± 0.3	14.5 ± 0.3	14.8 ± 0.3	0.965	0.015
A6	30.1 ± 0.5	29.6 ± 0.4	29.4 ± 0.5	29.1 ± 0.4	29.8 ± 0.4	0.456	-0.048
A7	16.4 ± 0.2	15.9 ± 0.3	16.1 ± 0.3	15.7 ± 0.2	16.3 ± 0.3	0.307	-0.021
A8	6.2 ± 0.1	6.4 ± 0.2	6.4 ± 0.2	6.2 ± 0.2	6.7 ± 0.2	0.260	0.083
A9	7.5 ± 0.1	7.7 ± 0.2	7.6 ± 0.2	7.8 ± 0.2	8.0 ± 0.2	0.349	0.108
A10	6.7 ± 0.1	7.1 ± 0.2	7.3 ± 0.2	7.0 ± 0.1	6.9 ± 0.1	0.109	0.018
A11	12.8 ± 0.3	12.4 ± 0.3	12.2 ± 0.3	12.2 ± 0.2	12.4 ± 0.3	0.578	-0.051
A12	17.1 ± 0.3	18.1 ± 0.3	17.3 ± 0.3	17.4 ± 0.3	18.0 ± 0.3	0.095	0.050
EL	58.8 ± 0.6	60.2 ± 0.6	58.3 ± 0.6	58.6 ± 0.7	59.9 ± 0.7	0.123	0.014
EW	31.8 ± 0.3	31.7 ± 0.3	31.5 ± 0.4	31.5 ± 0.3	32.0 ± 0.3	0.691	0.024

Table 7 showed the comparative statistics with the mean values, standard errors, standard deviations and p-values for each biometrics among ages. Sample size was composed of 58 ears with age of 19, 60 ears with age of 20,

56 ears with age of 21, 66 ears with age of 22 and 56 ears with age of 23. By using one-way between-groups ANOVA, there was no statistically significance of age differences for all the ear biometrics with $p > 0.05$. In addition, correlation analyses showed no relationship between biometrics and age. There was an exception for biometrics A3 and A4 with $p < 0.01$, yet, the relationships were reverse and too low with $r = -0.145$ and $r = -0.157$ respectively. The only closely significant positive relationship could be seen on biometric A9 where $r = 0.108$, $p = 0.062$.

DISCUSSION

There was no study regarding ear biometrics being done based on Malaysian population throughout the decades. In this study, the results found that ear biometric was statistically insignificant for bilateral differences. This showed that left ear and right ear of young adult in Malaysia were generally similar in term of measurements, shapes and anatomical structures. However, it could be differentiated between left and right ears with other descriptive method such as anatomical positioning of the ear.

Results of this study was supported by most of the previous studies in which bilateral differences were not statistically significant for ear length and ear width where the differences were all less than 1 mm although significant levels might be different (Barut & Aktunc 2006; Bozkır., *et al.* 2006; Purkait & Singh 2007; Sforza., *et al.* 2009; Yamada 2009; Wang., *et al.* 2010) [13,16-18,22,23]. This bilateral insignificance showed that intra-person variation was lower than inter-person variation. There might be some exceptions that caused by different population, gender, sample size and range of ages used in different studies. In fact, pediatric more regularly having asymmetry in ear biometrics if compared to adulthood while male more often to show asymmetry in ear biometrics if compared to female (Alexander., *et al.* 2010; Wang., *et al.* 2010) [24,13].

Ear symmetry and ear dimension were affected by human karyotype especially on chromosomes 5q and 22q11 in which controlling the development of first and second branchial arches responsible for ear growth (Chattopadhyay & Bhatia 2009; Sforza., *et al.* 2009) [18,20]. In addition, environmental stress from para-genetic factors such as cigarette smoking and drinking alcohol during pregnancy also contributed to the fluctuating asymmetry of ear biometrics. Ear length was proven to have lowest level of fluctuating asymmetry among other parts of body (Zadzinska 2007) [25]. The correlation coefficient for bilateral symmetry for ear length in present study was as high as previous studies such as Alexander., *et al.* (2010) and Liu (2008) [5,24]. Nevertheless, bilateral comparison could still be done with qualitative methods based on ear structures without considering symmetry level of ear biometrics (Singh & Purkait 2009) [ ].

On the other hand, gender difference was statistically significant, $p < 0.05$ for nine of the ear biometrics. The significant difference between gender was supported by the study under Karl Marx-Leipzig University (Lugt 2009) [26]. These biometric were all higher in males than females and being proven by previous studies (Barut & Aktunc 2006; Liu 2008; Sforza., *et al.* 2009; Yamada 2009; Wang., *et al.* 2010) [5,13,16,18,23]. One of the main factors was due to the difference of puberty period between genders. The age of puberty in male was 1 year later than female in which allowing them to have extra time for ear growth (Rutty., *et al.* 2005) [27]. Another factor was regarding the sex dimorphism process that affected by genetic materials possessed in different gender. This sex dimorphism process has been started since postnatal growth and causing different ear growth between genders (Sforza., *et al.* 2009) [18].

In addition, full ear length and ear width were achieved during different age between genders. For example, males have at least extra one-year or two-year time for vertical ear growth while at most four-year time for horizontal ear growth if compared to females. Other than genetic factor, males and females react in differently towards the environmental disturbances. For example, females with two chromosomes X react in slower manner and lower intensity towards environmental stress in growth (Zadzinska 2007) [25]. Furthermore, the difference in ear biometric between genders was also caused by ear expansion that appeared to be off to a slower start in females as compared with males. For example, ear expansion from age range of 12 – 15 years old to 19 – 30 years old caused an increase of ear length and ear width in respectively about 4.1 mm and 1.8 mm for males and in about 0.9 mm and 0.8 mm. This showed that increase in ear length was much higher than increase in ear width and this difference was in different rate between males and females (Meijerman., *et al.* 2007) [28].

When previous studies were taken into account, C & R Tree was used for the first time in gender estimation. It could be the easiest and useful method as it possessed only an average of estimation risk in which less than 0.300 for gender estimation. The result showed that ear length gave better prediction of an unknown gender with 78.7% of overall accuracy if compared to other biometrics. Instead, combination of ear biometrics were encouraged in gender estimation for forensic investigation as it contribute much higher percentage of accuracy up to 92.9%. A validity test using ten selected subjects was done and the data was tabulated in Table 8. The validation percentage was very accurate with 100% correct for gender estimation. In the future, study of ear biometrics should be focused in formation of these gender estimation formulae so as to enhance its application in forensic.

Table 8: Validity test for gender estimation using ten randomly selected subject

No	Actual Gender	Ear Biometrics (mm) *Valeu of deviation points*								Estimated Gender	Minimum Formula Accuracy
		A3 *4.4*	A4 *4.7*	A5 *15.2*	A6 *29.2*	A7 *14.7*	A11 *13.4*	EL *58.2*	EW *30.4*		
1	F	4.3	4.8	12.3	28.0	16.5	13.3	57.3	29.5	F	84.7%
2	F	4.5	5.0	10.0	23.0	15.0	11.3	49.0	29.5	F	89.8%
3	F	4.0	4.0	12.0	27.3	16.0	14.3	50.5	29.8	F	87.7%
4	M	3.3	3.8	18.0	31.8	18.3	16.0	67.5	33.3	M	84.5%
5	M	3.5	4.3	16.5	33.8	19.5	15.5	60.5	34.0	M	84.5%
6	M	4.3	5.5	15.0	37.3	21.0	15.3	72.5	37.8	M	93.1%
7	M	4.5	5.5	19.0	41.0	20.5	11.8	71.0	36.8	M	87.9%
8	M	3.5	4.0	19.8	33.3	21.5	12.8	62.0	36.5	M	82.8%
9	M	5.3	5.5	14.8	34.3	18.0	12.8	70.5	33.5	M	87.9%
10	M	4.3	5.3	16.3	37.3	24.8	17.3	71.3	39.0	M	89.7%

F = Female and M = Male
A = Iannarelli measurements; EL = Ear lenth and EW = Ear width
≤ deviation points estimated as female
> deviation points estimated as male
Red-coloured values represented majority matching through gender estimation accordingly to deviation

In term of ethnicity, Iannarelli ascertained that growth differences of the ears existed between Whites, Blacks and Asian (Lugt 2009) [26]. There was significant difference ($p < 0.05$) between Malaysian Chinese and Malays based on only five out of fourteen ear biometrics. On the other hand, there was no significant difference of ear length and ear width between Malaysian Malays and Chinese in this study. This was concurring with study of Adamson., *et al.* in year 1965 in which White and Negro populations were not showing any significant difference in ear length and ear width [29]. This could be due to genetic convergence among different races while living in similar environment. This genetic convergence could also occur when there was intermarriage between different races. Although there was no intermarriage within three nearest descendents among the subjects chosen in this study, it cannot be ignored that there was possible occurrence of intermarriage outside from the three known descendants. In addition, Malaysian Malays and Chinese were grouped as Asian Mongoloid in which the genetic background was not in big different because both races were in closely related gene pool.

In fact, the genetic convergence was not complete for race differential, but these tracks were at least visually established (Carter 2010) [30]. Study of

Alexander., *et al.* (2010) in which significant difference between races was proven among males [30,24]. This contradicted with our study and might be due to differences in measurement technique and might represent the true differences between the diverse subjects. Comparison among races could be done by referring to various previous studies such as Alexander *et al.* (2010), Liu (2008), Wang *et al.* (2010), Yamada (2009), Sforza *et al.* (2009) as well as Jung and Jung (2003) [5,13,18,24,23]. The differences between population in these previous studies migth be caused by several factors including races, genetical variation, individual constitution, ages and measurement methods as stated by Bozkir., *et al.* (2006). According to the results of Classification Tree, only biometric A12 was able to form a formula with single deviation point in race estimation. Nevertheless, it gave high risk race estimation of more than 0.300. As a result, these formulae were not good enough in race estimation using the available ear biometrics although complete ear print recovered from crime scene.

This study only involved young adults within the range of 18 and 25 years old in which there was one-year difference for each age. Statistically, there was no significant difference between ages and all the ear biometrics. Hence, it proven that young adults experienced the stabile ear growth as there was no significant increase in ear size. The outcome of the one-way between-groups ANOVA was supported by study of Liu (2008) in which there was no significant difference across age groups from 20 – 39 years olds until 50 – 59 year olds [5]. This fact concurred with the findings of Meijerman., *et al.* (2007) in which the natural changes on ear was too small and not significant within 20 – 50 years old after full ear size was achieved at puberty age [28].

According to Adamson *et al.* in year 1965, 85% from ear size of an adult was achieved on the age of three years old and then full ear size started to be completed during 12 – 14 years old until 98 – 100% of adult's ear size was achieved on the age of 18 – 20 years old depending on gender (Sforza., *et al.* 2009; Wang., *et al.* 2010) [13,18]. Based on the study of Schalbe., *et al.* in year 1891, post maturity growth caused the ear size expansion to occur within the age of 50 years old until 70 - 80 years old (Meijerman., *et al.* 2007; Sforza., *et al.* 2009) [28,18,31]. Ear expansion was also contributed by loss of elastin and resilience, diminutive tensile strength of connective tissues as well as alteration in intracellular tissues. Gravitational forces and reduction in number of cartilage cells per unit area have contributed to the lengthening of lobule and hence can result in increasing ear length of 11.0 mm in males and 13.0 mm in females by the age of 80 years (Rutty., *et al.* 2005) [27]. While increasing ear width was caused by expansion of extracellular matrix of cartilage that in fact pushing helical rim posteriorly instead of conchal fossa (Meijerman *et al.* 2007; Sforza *et al.* 2009) [28,18].

Nevertheless, there were several limitations that should not be neglected in this study. Firstly, the formulae formed for gender estimation were only applicable for Malaysian young adults in two main races except other minorities. According to interview with former Malaysia Crime Scene Investigation Chief Officer named Sir Amidon Anan (2011), the limitations in this study also included flexibility of ear shape, variation of pressure applied on glass surface, subjectivity of practitioner as well as immature discipline [29]. In addition, the factors affecting ear biometrics included quality of surface containing ear prints, lifting methods used, hair occlusion as well as the absence of formal standard protocol of ear print recovering especially in Malaysia that hence causing the limited use of ear identification in forensic practical. It was important to enhance study on ear biometrics and update the database of ear prints from time to time for comparison in forensic practical especially for court cases.

CONCLUSION

In conclusion, there was no significant bilateral difference in ear biometrics but with good correlation of left and right ears. In addition, only six out of fourteen ear biometrics showed significant but not uniform difference between Malays and Chinese in Malaysia population. Generally, there was also absence of significant difference and correlation between ear biometrics across ages. On the other hand, ear biometrics has provided a good reliability in estimation of gender in forensic investigation. Ear length showed a better estimation of gender using formula formed by C & R Tree. Combination of ear biometrics was required to increase the accuracy in estimation of gender. Hence, it could be very useful and accepted as strong evidence in court of law in Malaysia.

ACKNOWLEDGEMENT

Gratefully thanks to supervisors and friends for guidance and assistance throughout the study.

REFERENCES

1. Champod C, In Li SZ, Jain AK (2009) Encyclopedia of Biometrics. 248-53.
2. Singh P, Purkait R (2009) https://www.ncbi.nlm.nih.gov/pubmed/19748090.
3. Jung HS, Jung HS (2003) Surveying the dimensions and characteristics of Korean ears for the ergonomic design of ear-related products. Int J Ind Ergon 31: 361-73.

4. Kalcioglu MT, Miman MC, Toplu Y, Yakinci C, Ozturan O (2003) Anthropometric growth study of normal human auricle. Int J Pediatr Otorhinolaryngol 67: 1169-77.
5. Liu BS (2008) Incorporating anthropometry into design of ear-related products. Appl Ergon 39: 115-21.
6. Meijerman L, Sholl S, Conti FD, Giacon M, Lugt CVD, et al. (2004) Exploratory study on classification and individualisation of earprints. Forensic Sci Int 140: 91-9.
7. Nanni L, Lumini A (2009) Fusion of color spaces for ear authentication. Pattern Recognit 42: 1906-13.
8. Choras M (2008) Perspective methods of human identification: ear biometrics. Opto-Electronics Review 16: 85-96.
9. Rahman M, Islam R, Bhuiyan NI, Ahmed B, Islam A (2007) Person identification using ear biometrics. Int J Comput Internet Manag 1582: 1-8.
10. Kieckhoefer H, Ingleby M, Lucas G (2006) Monitoring the physical formation of earprints: optical and pressure mapping evidence. Measurement 39: 918-35.
11. Hurley DJ, Arbab-Zavar B, Nixon MS, (2007) The ear as a biometric In Flynn P Jain AK & Ross AA (editor) Handbook of Biometrics. 131-50.
12. Lugt CVD (2000) Ear prints In Siegel JA, Saukko PJ, Kunupfer GC (editor) Encyclopedia of Forensic Sciences US: Academic Press.
13. Wang B, Dong Y, Zhao Y, Bai S, Wu G (2010) Computed tomography measurement of the auricle in Han population of north. China J Plast Reconstr Aesthetic Surg 64: 34-40.
14. Meijerman L, Thean A, Maat GJR (2006) Earprints in forensic investigations. Forensic Sci Med Pathol 1: 247-56.
15. Alberink I, Ruifrok A, (2007) Performance of the FearID earprint identification system. Forensic Sci Int 166: 145-54.
16. Barut C, Aktunc E (2006) Anthropometric measurements of the external ear in a group of Turkish primary school students. Aesthetic Plast Surg 30: 255-9.
17. Purkait R, Singh P (2007) Anthropometry of the normal human auricle: a study of adult Indian men. Aesthetic Plast Surg 31: 372-9.
18. Sforza C, Grandi G, Binelli M, Tommasi DG, Rosati R, et al. (2009) Age-and sex-related changes in the normal human ear. Forensic Sci Int 187: 110.
19. Meijerman L (2006) Inter- and Intra Individual Variation in Earprints Leiden: Barge's Anthropologica.
20. Chattopadhyay PK, Bhatia S (2009) Morphological examination of ear: A study of an Indian population. Leg Med 11: S190-3.
21. Sharma A, Sidhu NK, Sharma MK, Kapoor K, Singh B (2007) Morphometric study of ear lobule in northwest Indian male subjects. Anat Sci Int 82: 98-104.
22. Bozkır MG, Karakaş P, Yavuz M, Dere F (2006) Morphometry of the external ear in our adult population. Aesthetic Plast Surg 30: 81-5.
23. Divya Kalra1 (2015) Anthrometric measurements of external ear: An in vivo study. Int J EnhancRes Med Den Care 2. 10-6.
24. Alexander KS, Stott DJ, Sivakumar B, Kang N, et al. (2010) A morphometric study of the human ear. J Plast Reconstr Aesthet Surg 64: 41-7.
25. Marriott I, Huet-Hudson YM (2006) Sexual dimorphism in innate immune responses to infectious organisms. Immunol Res 34: 177-92.
26. Lugt CVD (2009) Ear identification. Crime & Clues 5: 21-8.
27. Rutty GN, Abbas A, Crossling D (2005) Could earprint identification be computerised? An illustrated proof of concept paper. Int J Leg Med 119: 335-43.
28. Meijerman L, Lugt CVD, Maat GJR (2007) Cross-sectional anthropometric study of the external ear. J Forensic Sci 52: 287-93.
29. Amidon A (2011) Limitations of ear print identification Interview 3 Mac 2011.
30. Carter LR (2010) Long-run relationships in differential US mortality forecasts by race and sex: Tests for co-integration. Ageing in Adv Industrial States 8: 47-75.
31. Meijerman L, Thean A, Lugt CVD, Munster RVM, Antwerpen GV, et al. (2006) Individualization of earprints. Forensic Sci Med Pathol 2: 39-49.

CHAPTER 10

Collagen III (Reticulin) – A Location Marker of Pre-Sinusoid Arteriole Sphincters in the Villous System of Human Placenta

Gheorghe S Drăgoi[1,2*], Elena Pătraşcu[2], Ileana Marinescu[2], Elena Dincă[2] And Petru Răzvan Melinte[2]

[1]Romanian Academy of Medical Sciences, Bucharest, Romania

[2]University of Medicine and Pharmacy of Craiova, Romania

ABSTRACT

Blood micro-circulation within the placental villous system still raises a large number of problems of functional anatomy concerning the angio-structure of terminal villi, the communication between arterioles and venules of the umbilical vascular system, the spatial extension of the "extra-vascular contractile system" and, last but not least, the existence of the pre-sinusoid arteriole sphincters. The authors make a micro-anatomic analysis of the spatial distribution of the collagen fibres type III (reticulin) as markers of the pre-sinusoid arteriole sphincters. The study was performed on 20 human placentae of gestation age between 25 and 35 weeks. After paraffin embedding, 10 serial sections were obtained from each paraffin block. They were stained in Hematoxilin Eosin, or silver-impregnated according to Gömöri method. We had two objectives: to evaluate the spatial distribution of post arteriole sinusoid blood vessels within the terminal villi in the placenta, on the one hand, and the visualization of the location of a presinusoid arteriole sphincter system, on the other. The authors believe that umbilical arterioles and venules are connected "serially" at the level of terminal villi through arteriole-venule sinusoid blood vessels. The location of the trajectory of collagen type III (reticulin) bundles enveloping pre-sinusoid arterioles are markers of the existence of a pre-sinusoid arteriole sphincter at this level.

Keywords: Collagen type III (reticulin); Pre-sinusoid arteriole sphincter; Terminal villi; Mature intermediate villi; Sinusoid blood vessels

INTRODUCTION

The angioarchitecture of placental villi is a fascinating field of the functional anatomy of the foetus – placenta system. Although the foetus placenta blood circulation is anatomically autonomous from that of the mother, they are functionally interdependent. The hemodynamics in the placenta villous system still raises a large number of functional anatomy problems, i.e.

- What does the vascular structure of the human placenta terminal villi look like? Is there a capillary micro-circulation through a sinusoid, looping blood vessel that directly joins the arteriole and the venule of the umbilical vascular system?
- How is spatial communication between arterioles and venules at the level of the terminal villi realized: "serially" or in "parallel"? What are their functional influences on the blood flow?
- Can we view the blood circulation in the terminal villi as an "extraordinary arteriole – venule system"?
- Is there a spatial extension of the structures of the "extravascular contractile system" in the mature intermediate and/or the terminal villi?
- What are the relationships between the collagen III (reticulin) fibres and the contractile elements within the biodynamic structures of placenta villi?

In the course of the history of functional anatomy of the human placenta villouse system, the concept of "extravascular contractile structure" has evolved. Initially the cells similar to those in the smooth muscle were defined [1-4] and later these contractile elements were integrated in the "extra-vascular contractile system" at the level of stem villi, associated with the extra-cellular matrix of villose stroma [5,6]. The correlations between smooth muscle cells and the elastic fibres with the collagen fibres in a myofibroelastic anatomical-functional unit are well known [6].

The presence of pre-capillary sphincters was proved in many studies of micro-anatomy and physiology. Chamber and Zweifach [7], working on rabbit mesentery, identified the functional involvement of meta-arterioles in the hemodynamics in the capillary network. In his study on micro-circulation at the level of cat mesentery, Zweifach [8,9] has confirmed, anatomically and functionally, the presence of precapillary sphincters.

The next question was whether pre-capillary sphincters and metaarterioles are universal components of tissular microcirculations [10,11]. A large number of physiology, pharmacology and microanatomy research have proved the existence of these structures in: striated somatic muscle [12-15]; integumentum commune [16,17] ; subcutaneous fat tissue [18]; gastric mucosa [19]; liver [20]; heart [21] and in the small intestine submucosa [22].

We started from the assumption that, at the level of terminal villi there is a communication system between the arterioles and umbilical venules through sinusoid blood vessels. To control the blood flow within the terminal villi the presence of a pre-sinusoid arteriole sphincter system regulating the flow is required.

In the present paper, the authors aim at visualizing the pre-sinusoid arteriole sphincter location by impregnating with silver the collagen type III fibre (reticulin) bundles. The object of the paper concerns the spatial distribution of the sinusoid blood vessels within the terminal villosities on the one hand, and the distribution of the collagen fibres type III as markers of the location of the pre-sinusoid arteriole sphincters, on the other.

MATERIALS AND METHOD

The study was conducted on human biologic material, respecting the norms of the Code of Medical Ethics and the Code for Scientific Research Ethics. The micro-anatomic study of the placental structures was performed on 20 human placentae (gestation age 25 to 35 weeks) processed shortly after deliverance inside Hospital Filantropia of Craiova, Departments Obstetric-Gynecologie and Pathology. They were washed in physiological saline solution 9% and immersed in buffered formaldehyde 5%, pH 7.4. Three days after embedding tissular fragments of 2/2 cm are taken from the para-central and marginal zones of the placenta discus. After re-embedding in histological paraffin, 10 serial sections (for 5 μm) were cut from each paraffin block. The sections were either colored in Hematoxilin- Eosin for general histo-topographic orientation in order to evaluate the angioarchitecture of placenta villosities, or silver-impregnated according to Gömöri method in order to visualize the collagen fibres type III (reticulin) bundles. Micro-photography was performed on microscope Nikon 80 i by means of Sight DS-Fi1 Hight Definition Color Camera Head and processed in Software NIS – Elements Advanced Research

RESULTS

Based on the microanatomic analysis of the serial sections of placenta, the authors had two objectives in view:

- Evaluation of the spatial distribution of post-arteriole sinusoid blood vessels in the terminal villi;
- Visualization of the location through an argirophilic collagen marker, at the level of the arterioles existing in mature intermediate villi, of a pre-sinusoid arteriole sphincter system.

Micro-anatomic analysis of the spatial distribution of arteriole sinusoids in terminal villosities

The examination in X10 objective and X40 objective of the placenta terminal villi connected to the mature intermediate villi enabled the authors to evaluate their form and angiostructure. The form is determined by the intravillouse, vascular and stromal anatomic structures on the one hand, and their biunivocal interconnections. The location of intravillouse blood vessels is variable: marginal in cortex villi, central in the centrum villi (**Figures 1C,1D, 1F** and **1G**).

The external surface of the terminal villi, lined with syncytiotrophoblasts is bosselated by prominances reflecting the presence of marginal sinusoid blood vessels. The form of terminal villi is similar to that of a bunch of grapes (**Figures 1A-1D,1F** and **1G**). Marginal blood vessels compress the syncytiotrophoblast and participate in the formation of the syncytial – vascular membrane (**Figure 1H**). It is difficult to establish - in micro-anatomic sections - the appurtnance of intravillouse sinusoid blood vessels to the advehent/in-coming arteriole revehent/out-going venous systems.

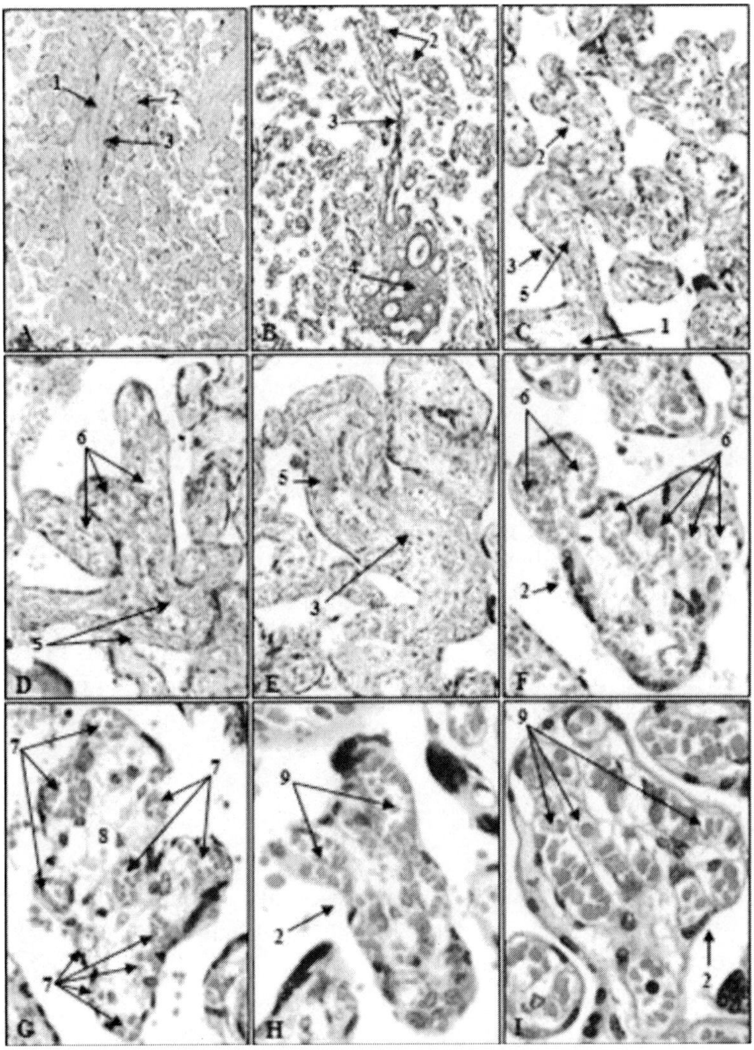

Figure 1: Spatial distribution of pre-sinusoid arterioles into mature intermediate villi and of sinusoid blood vessels into terminal villi of human placenta. 1: Stem villi; 2: Terminal villi; 3: Mature intermediate villi; 4: Villous tree; 5: Pre-sinusoid arteriole within the mature intermediate villi, longitudinal section; 6: Sinusoid blood vessel, within the terminal villi, oblique section; 7: Sinusoid blood vessel within terminal villi, transverse section; 8: Villous stroma; 9: Loop – trajectory blood vessel within the terminal villi, longitudinal section; 10: Syncytial vascular membrane. Paraffin sections, Hematoxylin Eosin stain (A and C – I); Reduced Silver Nitrat Gömöri Method (B). Magnifier: Large image X10 (A,B), X140 (CE); X420 (F–I).

Micro - anatomic analysis of the location and trajectory of collagen fibres type III (reticulin) bundles at the level of presinusoid arterioles in intermediary mature villosities

The examination in X10 objective and X40 objective made us easily identify pre-sinusoid arterioles from their origins in the trunkular arteries of stem villi to the terminal villi. Crossing the mature intermediate villi, pre-sinusoid penetrate the terminal villi where they are continuous with the arteriole sinusoids, and through the latter with the in-coming sinusoids of the post-sinusoid venules, without possible micro-anatomic personalization (**Figures 2A-2D**). The sinusoid trajectory of pre-sinusoid arterioles and of arteriole and venule sinusoid vessels is clearer on the serial sections in varied planes of the terminal villi and/or mature intermediate villi: frontal, saggital, paramedial and/or horizontal (**Figures 2A,2B,2E,2G** and **2H**).

The examination in X20, X40, X60 obiectives of silver-impregnated serial sections - Gömöri method enabled the visualization of the location and trajectory of the bundles of collagen fibres type III (reticulin) at the level of pre-sinusoid arterioles and of the villous stroma (**Figures 2B,2D-2F** and **2H**). At the level of pre-sinusoid arterioles we saw dark-colored, argerophilic strips/bands alternating with clear, nonsilver-impregnated ones. This image suggests that at this level there is a pre-sinusoid arteriole sphincter (**Figures 2B,2D,2F** and **2H**).

DISCUSSION

Based on the previous information, we could attempt a few answers to some of the issues that blood biodynamics in the placenta villous system is confronted with.

Stereo distribution of sinusoid blood vessels in villous microcirculation

Blood micro-circulation in the placental villous system takes place in the absence of a true capillary network between arterioles and venules in the umbilical vessels. It is achieved through sinusoid blood vessels in the terminal villi. However, Bernischke si Kaufmannn [23] use two terms for the intravillous blood vessels: capillaries and sinusoids. They define

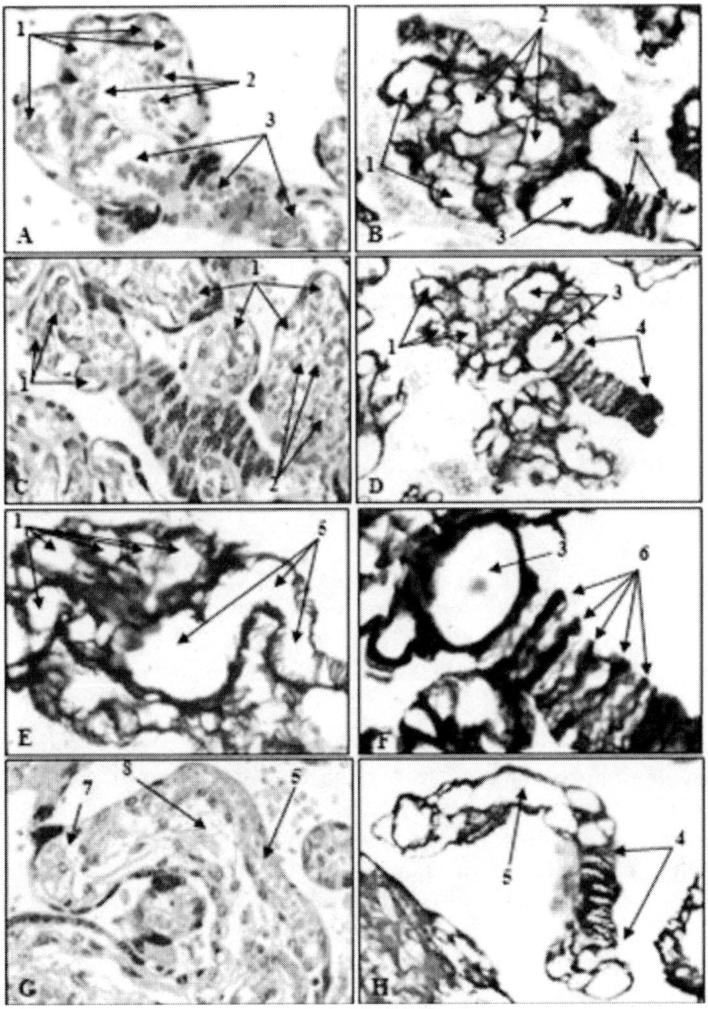

Figure 2: Location and trajectory of bundles of collagen fibres type III (reticulin), at the level of pre-sinusoid arterioles, in mature intermediate villi. 1: Sinusoid blood vessels in the terminal villi cortex; 2: Sinusoid blood vessels in the center of terminal villi; 3: Pre-sinusoid arteriole, in transversal section; 4: Pre-sinusoid arteriolar sphincter; 5: Pre-sinusoid arteriole, in longitudinal section; 6: Bundles of collagen fibres type III (reticulin) of looping trajectory, at the level of pre-sinusoid arteriole; 7: Terminal villi; 8: Stroma in the central part of mature intermediate villi. Paraffin section. Hematoxilin Eosin stain (A,C,G). Reduced Silver Nitrat Gömöri Method (B,D,E,F,H). Magnifier: X140 (A,D), X280 (B,C,H), X420 (E-G).

sinusoids as dilated capillaries of approximately 50 μm. We suggest that umbilical arterioles and venules are "serially" connected through an anastomosis arteriole-venule by means of arteriole-venule sinusoids. The absence of an interposed capillary network between arterioles and venules is replaced by post-arteriole sinusoid blood vessels distributed fanlike. Thus, we can talk about an "extraordinary arteriole-venule sinusoid system". This anatomic characteristic allows the growth of the transversal section of the intravillous vascular bed and, implicitly, a decrease of resistance to the blood flow, according to Laplace law: the flow speed is reverse to the transversal section in the blood vessel. This is the mechanism through which the conditions for the mother-Foetusnutrients and gas exchanges are made possible at the level of the syncytial-vascular membrane.

Pre-sinusoid sphincters in the mature intermediate villi system

Pre-capillary sphincters described by Rhodin (1967) [24] in varied tissues as complete perivascular sleeves have not been identified in human placenta [24]. It is thought that arterioles are continuous with 1-2 capillaries by the gradual decrese of arteriole diameter which are free from smooth muscle cells [25].

The authors have visualized bundles of collagen fibres type III (reticulin) having circumferential and spiral trajectory at the level of the pre-sinusoid arteriole in the mature intermediate villi, this image suggesting the existence of a pre-sinusoid arteriole sphincter at this level. Graf et al. (1994) [5] have described "perivascular contractile sheaths" in the stem villi, also called "extra-vascular contractile system". Ultrastructural studies have suggested the association at this level of smooth muscle cells and subpopulations of microfibroblasts differentited from the perivascular mesenchimal stroma [26,27].

The presence and trajectory of collagen fibres type III (reticulin) we have identified at the level of mature intermediate villi, around the presinusoid arteriole, are markers of the existence of a pre-sinusoid sphincter at this level. It can be integrated in the "extravascular contractile system" as an extension of the latter from the stem villi into the mature intermediate villi.

Allostasic regulation of placental blood flow

"The remarkable sinusoid arteriole" venular system located in villi terminalis ensures the direct transition of the blood from umbilical arteries to the similar vene. Between these huge vascular streams there are sinusoidal capillary loops that should not be assimilated to a true capillary network.

The presence of pre-sinusoid arterial sphincters is a lead to a allostasic, holistic concept of placental blood flow stability by selfregulation [28,29].

We believe that the allostatic regulation of placental blood flow is achieved by changes, multifactorially determined, of the villose functional structures: variations of pre-sinusoid arteriole diameter, and, implicitely, of the sinusoids within the villi terminalis; phenotypical transformations of the villose stroma; villose intra and extra fibrin deposits; the existence of angio and vasculo-genetic processes, and, last but not least, of pathological states diabetes, hypoxia, pre-eclampsia, maternal malnutrition that involve the modulation of vascular resistance and entail restrictions in the foetal intrauterine growth.

CONCLUSION

Pre-sinusoidal arteriolar sphincters are fibro-muscular structures involved in the allostatic self-regulation of placental blood flow stability;

The fascicles of collagen fibres type III (reticulin) visualized at the level of pre-sinusoidal arterioles, by location and their spiral trajectory can be viewed as micro-anatomic markers of the existence of presinusoidal sphincters;

Allostatic regulation of villose blood flow achieved through phenotypical transformations of villose structures as well as through the presence of pre-sinusoidal arteriolar sphincters involves modulation of vascular resistance with consequences upon the intrauterine growth of the foetus.

REFERENCES

1. Iisuda S (1916) Concentration of muscle fibres in the human placenta. Beitr Geburtsh Gynaek 19: 101.
2. Naujoks H (1922) Do anatomical changes of the childish have an ejaculation influence on the time of the bubble jump? Z Geburtsh Gynaek 84: 304.
3. Dubreuil G, Rivière M (1932) Fibromuscular formations of the chorion and villus of the human placenta. CR Soc Biol 111: 170-172.
4. Spanner (1935) Maternal and cognitive circulation of the human placenta and its strombahnen. Z Anat Development 105: 163-242.
5. Graf R, Langer JU, Schöenfelder G, Oeney T, Hartel Schenk S, et al. (1994) The extravascular contractile system in the human placenta. Morphological and immunohistochemical investigations. Anat Embryol 190: 541-548.

6. Graf R, Metejevic D, Schuppan D, Nuedeck H, Shakibaei M, et al. (1997) Molecular anatomy of the perivascular sheath in human placental stem villi : contractil apparatus and its association to the extracellular matrix. Cell Tissue Res 290: 601-607.
7. Chambers R, Zweifach BW (1994) Topography and function of the mesenteric capillary circulation. Am J Anat 75: 173-205.
8. Zweifach BW (1949) Basic mechanisn in peripheral homeostasis. In : Zweifach BW (eds) Factors regulating blood pressure. Transactions of the third conference, New York, Josiac Macy, Jr Fundation pp: 13-52.
9. Zweifach BW, Shorr E, Black MM (1953) The influence of the adrenal cortex on behavior of terminal vascular bed. Ann NY Acad Sci 56: 626-633.
10. Johnson LR (2003) Essential medical physiology. Elsevier Academic Press, MA, USA.
11. Boron WF, Boulpaep EL (2008) Medical Physiology. (2nd edn), Philadedlphia, Elsevier. Sauders, Philadelphia, USA.
12. Mellander S (1966) Comparative effects of acetylcholine, butyl-nor-dynephrine (Vascular), noradrenaline and ethyl-adrainol (Effonti) on resistance, capacitance and precapillary sphincter vessels and capillary filtatration in cat skeletal muscle. Angiologica 3: 77-99.
13. Owen DA, Stuerner E (1971) Effect of dihydroergotamine (DHG) on the capacitance, resistance and precapillary sphincter vessels of denervated cat skeletal muscle. Br J Pharmacol. 42: 655-656.
14. Hillman J, Lundvall J (1981) Classification of beta-adrenoceptors in the microcirculation of sckeletal muscle. Acta Physiol Scand 113: 67-71.
15. Bentzer P. Kongstad L, Grande PO (2001) Capillary filtration coefficient in independent of number of perfused capillaries in cat skeletal muscle. Am J Physiol 280: 2697-2706.
16. Tooke JE (1980) A capillary pressure disturbance in young diabetics. Diabetes 29: 815-819.
17. Widmer RJ, Laurinec JE, Joung MF, Mohiddin MW, Laine GA, et al. (2008) The origin of the biphasic flow response to local heat in skin. Microcirculation 15: 349- 357.
18. Burcher E, Olgart L, Gazelius B (1977) Comparative effects of adrenaline and felypressin (octapressin) on consecutive sections of the vascular bed in canine adipose. Acta Physiol Scand 100: 215-220.
19. Perry MA, Granger DN (1985) Regulation of capillary exchange capacity in the dog stomach. Am J Physiol 248: 437-442.
20. Oda M, Han JY, Yokomori H (2000) Local regulators of hepatic sinusoid microcirculation : recent advances. Clin Hemorrheol Microcirc 23: 85-94.
21. Schneeweiss A, Sherf L, Lehrer E, Lieberman Y, Neufeld HN (1982) Segmental study of the terminal coronary vessels in coarctation of the aorta : a natural model of the effect of coronary hypertension on human coronary circulation. Am J Cardiol 49: 1996-2002.
22. Granger DN, Kvietys PR, Perry MA, Barrowman JA (1987) The microcirculation and intestinal transport. In: Johnson LR (eds) Physiology of the Gastrointestinal Tract. (2nd edn), Raven Press, New York, USA pp: 1671-1697.
23. Rhodin JAG (1967) The ultrastructure of mammalian arterioles and precapillary sphinters. Ultrastruct 18 : 181-223.
24. Benirschke K, Kaufmann P (2000) Pathology of the human placenta, (4th Edn) Springer-Verlag New York, USA pp: 116-148.
25. Kaufmann P, Bruns U, Leiser R Luckhandt M, Winterhager E (1985) The Foetal vascularisation of term placental villi. II. Intermediate and terminal villi. Anat Embryol 173: 203-214.
26. Kohen G, Kertschanska S, Demir R, Kaufmann P (1996) Placental villous stroma as a model system for myofibroblast differentiation. Histochem Cell Biol 105: 415-429.
27. Demir R, Kosanke G, Kertschanska S, Kaufmann P (1997) Classification of human placental stem villi: review of structural and functional aspects. Microsc Res Tech 38: 29-41.

28. McEwen BS, Wingfield JC (2003) The concept of allostasis in biology and biomedicine. Horm Behav 43: 2-15.
29. Sterling P, Allostasis EJ (1988) A new paradigm to explain arousal pathology. In: Fisher S, Reason J (eds) Handbook of life stress, Cognition and health. John Wiley&Sons, New York, USA pp: 629-649.

CHAPTER 11

DNA Fingerprints: Advances in their Forensic Analysis Using Nanotechnology

Mahima Kaushik[1,2*], Swati Mahendru[2], Swati Chaudhary[2] and Shrikant Kukreti[2]

[1]*Cluster Innovation Centre, University of Delhi, Delhi*

[2]*Department of Chemistry, University of Delhi, Delhi*

ABSTRACT

For deciphering the secrets of forensic science, nanotechnology has quite extensively been utilized. Generally, for identifying the fingerprints, a lot of combination of different materials and film assemblies have already been utilized. Since the mode of interaction between the nanoparticles and fingerprint marks, is still not clearly understood, fabricating the nanoparticle assemblies for their identification is quite challenging. Complete identification of fingerprint marks which are generally because of a combination of some kind of proteins and fatty acids, is still a difficult task and is only partially being done with the help of different techniques. Nanotechnology has already shown immense potential in many fields like medicine, molecular biology, genetics, material science etc. and it has also marked promising potential even in forensic analysis studies. This review aims to discuss the details of the process of fingerprint formation, their role in forensic analysis along with the latest advances in the field of nanotechnology for their identification. This information may enhance our understanding about the progress in the forensic analysis of fingerprints, which may further be utilized in solving the puzzle of various criminal cases.

Keywords: DNA Fingerprinting; Forensic science; Applications of nanotechnology in forensic science; Techniques of DNA fingerprinting

INTRODUCTION

Distinctive lines and swirls at the end of our fingers give a unique pattern of ridges and valleys to each individual, known as fingerprints or Frictional Ridge Skin (FRS). Owing to the fact that this pattern is unique to each individual and doesn't change throughout a person's life, it has been used as a biometric identifier by the criminologists and law enforcement officers in forensic investigations since late 19th century [1]. Apart from the scars which originated from any accidents, fingerprints remain same throughout a person's life. Even identical twins with identical DNA have different fingerprints [2]. Fingerprint formation begins, when the baby is still in the womb at about 10th week. Growth in the size of hands, feet etc. after the embryonic volar pads stop growing trigger the cell differentiation on the fingertip skin. Embryonic volar pads are the raised pads on the fingers, palms and feet due to swelling of mesenchymal tissue, a precursor of blood vessels and connective tissues. Flow of amniotic fluid around the foetus as well as position of foetus during the differentiation process creates a microenvironment responsible for the finer details of the fingerprints. Fingerprints are fully formed at about seven months of foetus development [3,4].

The fingerprints can be collected from practically any solid surface such as porous or non-porous surface, adhesives, metals, fabric, human skin and many others. Depending on the residue composition, they can be classified as patent (visible) or latent (invisible) fingerprints. Patent fingerprints are formed, when they are based on blood, ink, paints, etc. Latent fingerprints are impressions of the fingers ridge pattern formed on the surface, when a finger picks eccrine sweat, together with oily substances such as sebum. Eccrine glands are major sweat glands of human body and eccrine fraction of fingerprints mostly consist with inorganic compounds, amino acids, proteins, lipids and miscellaneous compounds such as lactate, urea, creatine, creatinine, glucose and many others. Sebaceous glands are responsible for waterproofing and lubricating the skin, as they secrete oily or waxy matter. So the sebaceous part of fingerprints comprise of fatty acids, phospholipids, wax esters, sterols, squalene and other organic compounds.

This review would briefly update on the fingerprint formation, various techniques utilized for their identification in forensic analysis and the recent most advances in the nanotechnology field.

Various techniques utilized for forensic analysis of fingerprints

The composition of human sweat and sebum is unique for each individual and depends on a number of factors such as individual's age, health condition, diet etc. The composition of fingerprints already deposited on surface can change with time due to a variety of factors, such as evaporation of volatile components of fingerprints, diffusion of fingerprints into porous surface, exposure to light, temperature and humidity, and many more [1,5]. Even though there are a number of optical, chemical and physical techniques available for fingerprint analysis a considerable portion of the latent fingerprints still escape detection, resulting in inaccurate analysis. If appropriately gathered and accurately analysed, fingerprints are impeccable and can serve crime fighters with invaluable knowledge to identify the criminals [6]. Interestingly, fingerprints not only reveal your identity, it can also hint about a person's profession as well as substances touched and secretions from a person's body. Fingertips of a guitar player are rough with calluses due to plucking and brushing the strings, whereas that of a person who deals with a lot of paper work are often smooth [7]. Various physical traits of an individual can be determined by analysing the components of sebum and sweat present in the fingermarks, as the gland secretions are controlled by the body's hormone system Muramoto et al. [8] demonstrated a novel test sample for the spatially resolved quantification of illicit drugs on the surface of a fingerprint using Time-of-Flight Secondary Ion Mass Spectrometry (ToF-SIMS) and Desorption Electrospray Ionization Mass Spectrometry (DESIMS). Using an ink-jet-printed array of known quantities of drugs, researchers calibrated their mass spectrometry techniques to measure specks of the chemicals. Then, using 3-D printed plastic finger and artificial sebum, drug-tainted fingerprints pressed onto paper or silicon were created, to investigate the effect of matrix and substrate effects on analyte signal response [9].

Traditionally, fingerprint analysis was mostly performed by powdering method, for which different types of commercially available powders including magnetic powder, aluminium flake powder, luminescent powder and iron flake powder are used [1]. Techniques such as mass spectroscopy in addition to chromatography are used to identify the molecular ion of explosive residues or prohibited drugs important for forensic detection [10]. Other methods used for fingerprint analysis are Surface Assisted Laser Desorption/ Ionization (SALDI) coupled with mass spectroscopy [11],

Desorption Electrospray Ionization (DESI) mass spectroscopy [12], Matrix Assisted Laser Desorption/ Ionization mass spectroscopy (MALDI) [13], Raman spectroscopy, Infrared spectroscopy [1]. However, with the advent of Nano technological advances in the field of forensics, cutting-edge techniques for fingerprint analysis are being emerging.

Role of nanotechnology in forensic analysis

Nanotechnology has vast application in the field of forensic science. It helps in diagnosing and examining samples at the nano-scale level that was earlier facing difficulty due to the detection limits of machines. Samples including explosives, heavy metals, gunshot residues, DNA fingerprints can be detected using nanotechnology [14]. Nanotechnology is the everyday growing field of scientific research and technology. It can be coined as the technique that works at nanometer scale including choice of materials, analysis, synthesis, measurement through devices, etc. At nanometer scale, materials can be handled at the level of different atoms or molecules. Nanotechnology can also be used to develop productive machines by employing atoms as basic building blocks [15]. At this level, physical as well as chemical properties of a material can be controlled. Many scientific areas such as electronic engineering, biomedical sciences, material sciences and physical sciences have vast application of nanotechnology nowadays. Nanostructures are characterized by greater stability, larger surface area and are considered stronger materials [16]. Different Nanoparticles like gold nanoparticles, Zinc sulfide/ cadmium selenide and various organic-inorganic silica based particles etc. are used to enhance the lucidity of the fingerprints.

Recently, the efficiency of forensic science has increased in various fields. Analysis of DNA by illustration of micro-device or chip-based technologies in forensic science is done by DNA research and development program, whereas chemical and biological defence program involves development of inexpensive or wearable device, which can alert against unexpected biological and chemical threat in a very short time, for the wearer to take powerful constructive measures [17]. Techniques for instance, Scanning Electron Microscopy (SEM), Transmission Electron Microscopy (TEM), Raman Microspectroscopy (Micro-Raman), and Scanning Probe Microscope (SPM) are used for nanomaterial analysis and they have been effectively modified and applied to the forensic research area. Semiconductor nanomaterials (quantum dots), carbon nanotubes, nanoparticles, and peptide nanotubes are few of the examples of nanomaterials, which are used nowadays in the field of environment engineering, medicine, quantum computer, catalysis, consumer products and communication [18].

Identification and evaluation of toxic substances from different crucial samples, which are considered important for forensic analysis such as hair, blood, saliva, palm-print or fingerprint samples, and skeletal remains, can be done using nanotechnology. Recently, illegal drug such as cocaine found in the fingerprint samples are quantified by exploring the spectroscopic properties of nanoparticles. Detection limit of these prohibited drugs is amplified in the fingerprint samples by employing silver, gold as well as titanium oxide nanoparticles [19]. A nanosensor was developed in order to detect clonazepam drug from the blood and skeletal samples by using melamine altered gold nanoparticles [20]. Similarly, sensors have been developed for the better, rapid and inexpensive on-field test. Commercially available bioanalyzer, Agilent 2100 have the ability to increase the DNA samples even upto nanoliter scale within a very short span of time and thus, are used to quantify mitochondrial DNA samples in forensic research [21]. High quality DNA samples which are amplified by PCR can be extracted from different skeletal remains and body fluids with the help of magnetic nanoparticles, copper nanoparticles, and magnetic nanoparticles involving silica for the purpose of forensic analysis [22].

Role of nanotechnology in fingerprint analysis

Patent fingermarks can be seen with naked eyes and thus can be used directly for investigation process. However, latent fingermarks need to be developed in order to visualize them and for using in investigation process. Integration of nanotechnology with fingerprint analysis has a diverse range of implications, especially in enhancing clarity while developing the fingerprints. Clarity of the latent fingerprints can be enhanced by using gold nanoparticles bound with long hydrocarbon chains which form hydrophobic interactions with the sweat residues along with silver physical developer [23]. Identification of latent fingermarks on paper that has been wetted by water is a challenging task, as the amino acids dissolved in the sweat which are the main substrate for fingerprint enhancement are removed by water. Aqueous solution of silver nanoparticles stabilized by cationic surfactants act as appropriate silver physical developer (Ag-PD) in this case, as silver slowly deposits on the water-insoluble components of the sweat. However, this method suffers from glitches, such as lack of reproducibility, poor contrast, etc. This can be overcome by using Multimetal Deposition (MMD) method, where AuNPs bind to the fingermark residue through ionic interactions between the negatively charged gold colloids and the positively charged components of the fingermark. They further catalyze the precipitation of metallic silver from the Ag-PD solution (Figure 1). Another highly versatile and reproducible approach, where modified gold nanoparticles (AuNPs) are used to bind preferentially to paper through AuNPs–cellulose interactions and forms an invisible coating over the entire paper surface, while leaving the sebaceous ridges uncoated. The AuNPs are bound by bifunctional reagents, whose molecules contain active heads that have a high affinity to the paper, and active tails that can bind to the metallic nanoparticles. So second treatment with conventional Ag- PD solution will result in darkening of the gold coated region by the precipitation of black silver over them (**Figure 1**) [24]. Hybrid organicinorganic silica based particles are extremely helpful for latent fingermark analysis. Zinc sulfide/cadmium selenide nanoparticles are developed in order to enhance the fingerprints and these nanoparticles also have tendency to give fluorescence in the presence of UV light so that fingerprints under this system can be directly observed without the need of any additional method [25]. Hybrid cadmium sulfide quantum dots (QDs) nanocomposites assembled in a Porous Phosphate Heterostructures (PPH) functionalized with mercaptopropyl (PPH-SH) and Propionitrile (PPH-CN) tend to be fluorescent in nature that make them potentially useful for fingerprint analysis [26]. On the other hand, latent fingerprints present on patterned or multi-coloured background can be easily visualized by using hybrid Nano powders and gold nanoparticles functionalized by Poly (styrene-altmaleic anhydride)-b-Polystyrene (PSMA-b-PS) [27,28]. A mixture of the magnetically responsive diacetylene (DA) powder and magnetite nanoparticles, applied to a surface containing latent fingermarks, becomes immobilized along the ridge patterns of the fingerprints in presence of a magnetic field. This mixture composite immobilized on the latent fingerprint, on irradiation with UV light results in generation of blue-colored Polydiacetylenes (PDAs). Heat treatment of the blue-colored image promotes a blue-tored transition as well as fluorescence turn-on which enables efficient visual imaging of a latent fingerprint [29].

Outlook and Future Directions

Recently, nanotechnology had been utilized quite extensively in medical field for developing point-of care diagnostics for various diseases like cancer. Nanobiosensors have been explored for their clinical applications as an additional tool in diagnosis of a particular disease or an infectious micro-organism. Likewise, many materials and film assemblies with the aid of nanotechnology had constantly been developed to help in identifying the DNA fingerprints, especially in crime incidences. Enormous efforts have continuously been done for improving upon the sensitivity and specificity of the material in identifying the fingerprints which may be obtained in quite distorted situations. Forensic analysis certainly has a lot of room for improvement in terms of DNA analysis, which may be obtained from serum, saliva, blood or any other cell or tissue at the crime scene. Engineering new nanomaterials with the help of more sensitive and advanced techniques is quite challenging and is need of the hour for performing DNA analysis or identification of fingerprints.

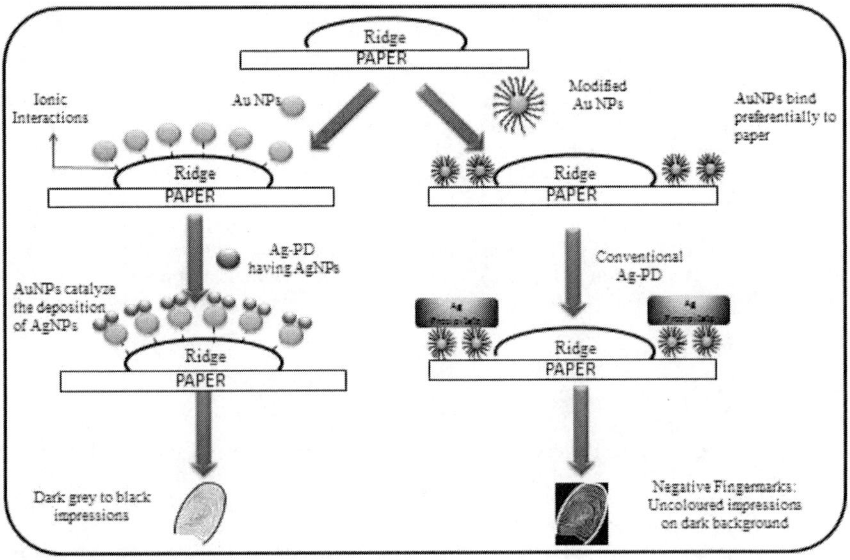

Figure 1: Identification of latent fingerprints on a wet paper using nanotechnology

ACKNOWLEDGEMENTS

Authors would like to acknowledge the research and development grant of University of Delhi, Delhi, India.

REFERENCES

1. Hazarika P, Russell DA (2012) Advances in fingerprint analysis. Angewandte Chemie International Edition 51: 3524-3531.
2. Daluz HM (2014) Fundamentals of Fingerprint Analysis. CRC Press.
3. (2015) Scientific American Editors Ask the Experts: The Human Body and Mind. Macmillan.
4. Maltoni D, Maio D, Jain A, Prabhakar S (2009) Handbook of fingerprint recognition. Springer Science & Business Media.

5. Leśniewski A (2016) Hybrid organic–inorganic silica based particles for latent fingermarks development: A review Synthetic Metals.
6. Spaulding J (2008) Careers in Fingerprint and Trace Analysis. The Rosen Publishing Group.
7. Hamilton SL (2008) Fingerprint Analysis: Hints from Print. ABDO Publishing Company.
8. Huynh C, Halamek J (2016) Trends in fingerprint analysis. TrAC Trends in Analytical Chemistry.
9. Muramoto S, Forbes TP, van Asten AC, Gillen G (2015) Test Sample for the Spatially Resolved Quantification of Illicit Drugs on Fingerprints Using Imaging Mass Spectrometry. Analytical chemistry 87: 5444-5450.
10. Goucher E, Kicman A, Smith N, Jickells S (2009) The detection and quantification of lorazepam and its 3-O-glucuronide in fingerprint deposits by LC-MS/MS. Journal of separation science 32: 2266-2272.
11. Rowell F, Hudson K, Seviour J (2009) Detection of drugs and their metabolites in dusted latent fingermarks by mass spectrometry. Analyst 134: 701-707.
12. Ifa DR, Jackson AU, Paglia G, Cooks RG (2009) Forensic applications of ambient ionization mass spectrometry. Analytical and bioanalytical chemistry 394: 1995-2008.
13. Wolstenholme R, Bradshaw R, Clench MR, Francese S (2009) Study of latent fingermarks by matrix-assisted laser desorption/ionisation mass spectrometry imaging of endogenous lipids. Rapid communications in mass spectrometry 23: 3031-3039.
14. Chen YF (2011) Forensic Applications of Nanotechnology. Journal of the Chinese Chemical Society 58: 828-835.
15. McCord B (2006) Nanotechnology and its Potential in Forensic DNA Analysis. Profiles in DNA 9: 7-9.
16. Javan GT (2015) Nanotechnology and Its Applications in Forensic and Criminal Cases.
17. Dongre NL Nanotechnology in Forensic Geosciences
18. Gohil S, Domb AJ, Kumar N (2011) Nanomaterials for regenerative medicine. Nanomedicine 6: 157-181.
19. Harvey JMH (2015) Nanoforensics: Forensic application of nanotechnology in illicit drug detection. J Forensic Res 6: 106.
20. Lodha A, Pandya A, Sutariya PG, Menon SK (2013) Melamine modified gold nanoprobe for "on-spot" colorimetric recognition of clonazepam from biological specimens. Analyst 138: 5411-5416.
21. Hallikeri VR, Bai M, Kumar AV (2012) Nanotechnology-The future armour of forensics: A short review. Journal of the Scientific Society 39: 10.
22. Lodha AS, Pandya A, Shukla RK (2016) Nanotechnology: An Applied and Robust Approach for Forensic Investigation. Forensic Res Criminol Int J 2: 00044.
23. Lang H, Hegner M, Gerber C (2007) Nanomechanical cantilever array sensors. In Springer Handbook of Nanotechnology 427-452.
24. Jaber N, Lesniewski A, Gabizon H, Shenawi S, Mandler D, et al. (2012) Visualization of latent fingermarks by nanotechnology: reversed development on paper-a remedy to the variation in sweat composition. Angewandte Chemie International Edition 51: 12224-122247.
25. Sametband M, Shweky I, Banin U, Mandler D, Almog J, et al. (2007) Application of nanoparticles for the enhancement of latent fingerprints. Chemical Communications 11: 1142-1144.
26. Algarra M, Jiménez-Jiménez J, Moreno-Tost R, Campos BB, da Silva JE, et al. (2011) CdS nanocomposites assembled in porous phosphate heterostructures for fingerprint detection. Optical Materials 33: 893-898.
27. Fernandes D, Krysmann MJ, Kelarakis A (2015) Carbon dot based nanopowders and their application for fingerprint recovery. Chemical Communications 51: 4902-4905.

28. Song K, Huang P, Yi C, Ning B, Hu S, et al. (2015) Photoacoustic and Colorimetric Visualization of Latent Fingerprints. ACS nano 9: 12344-12348.
29. Lee J, Lee CW, Kim JM (2016) A Magnetically Responsive Polydiacetylene Precursor for Latent Fingerprint Analysis. ACS applied materials & interfaces 8: 6245-6251.

CHAPTER 12

Metabolites Replace The Parent Drug in the Drug Arena. The Cases of Fonazepam and Nifoxipam

Maria Katselou[1] • Ioannis Papoutsis[1] • Panagiota Nikolaou[1] • Chara Spiliopoulou[1] • Sotiris Athanaselis[1]

[1].Department of Forensic Medicine and Toxicology, Faculty of MedicineNational and Kapodistrian University of AthensAthensGreece

ABSTRACT

Fonazepam (desmethylflunitrazepam) and nifoxipam (3-hydroxy-desmethylflunitrazepam) are benzodiazepine derivatives and active metabolites of flunitrazepam. They recently invaded the drug arena as substances of abuse and alerted the forensic community after being seized in powder and tablet forms in Europe between 2014 and 2016. A review of all the existing knowledge of fonazepam and nifoxipam is reported, concerning their chemistry, synthesis, pharmacology and toxicology, prevalence/use, biotransformation and their analysis in biological samples. To our knowledge, fonazepam and nifoxipam-related intoxications, lethal or not, have not been reported in the scientific literature. All the available information was gathered through a detailed search of PubMed and the World Wide Web.

Keywords: Fonazepam Nifoxipam Designer benzodiazepines NPS Active metabolites of flunitrazepam

INTRODUCTION

During recent years, a great number of new psychoactive substances (NPSs) or reappeared old ones have been introduced into the market of illicit drugs, and they have been reported to be used by drug addicts. According to the European Monitoring Centre for Drugs and Drug Addiction (EMCDDA), 101 NPSs were detected for the first time in Europe in 2014 and 100 NPSs in 2015, and these are the highest yearly numbers recorded ever [1]. These substances usually mimic the pharmacological effects of already known and abused drugs, possessing euphoric, stimulating or hallucinogenic effects, while avoiding detection and classification as illegal. Thus, being legally placed and generally available in Internet shops and online sales, they often become the causes of an increasing number of poisonings and fatal intoxications. They also represent an enormous challenge for clinical and forensic toxicologists, as well as policy makers in many countries [2, 3].

A relatively new phenomenon is the occurrence of designer benzodiazepines, like flubromazolam, meclonazepam, fonazepam, nifoxipam and pyrazolam, as they have been highlighted by the EMCDDA since 2011 [4, 5]. Some of them are common active metabolites of classic benzodiazepines (e.g., fonazepam), while others are prescribed medicinal products in some countries (e.g., phenazepam) [4].

Fonazepam and nifoxipam are the desmethyl- and the 3-hydroxy-desmethyl-derivatives of flunitrazepam (one of the 7-nitrobenzodiazepines), respectively, and they are two of its active metabolites. Flunitrazepam has been used as a hypnotic and pre-anesthetic drug for over 20 years, and it is also one of the most widely abused benzodiazepines [6, 7]. Thus, these two metabolites that are also used as new designer benzodiazepines are supposed to possess similar pharmacological and toxicological properties to the parent drug. Nifoxipam could also be a metabolite of fonazepam.

Fonazepam and nifoxipam have recently invaded the drug arena as NPSs and became known in the universal drug community during the last 2 years. A number of nifoxipam seizures were reported between December 2014 and July 2015 in Europe, while in the case of fonazepam, only two recent seizures of the drug have been reported in January 2016 in Germany and in March 2016 in Sweden. These two designer benzodiazepines are sold as research chemicals that are not intended for human or animal consumption [8, 9, 10].

The aim of this article is to review all the available information on chemistry and synthesis of fonazepam and nifoxipam, their pharmacology and toxicology, their biotransformation and their use as drugs of abuse, as well as the published methods for their determination in biological samples. All the reviewed information was gathered through a detailed search of PubMed and the World Wide Web using the keywords 'fonazepam', 'nifoxipam, 'desmethylflunitrazepam', '3-hydroxy-desmethylflunitrazepam', 'pharmacology', 'determination', 'analysis', 'toxicology', 'intoxications', 'fatalities', 'identification', 'biological fluids' and 'legal status'. Nevertheless, no data on their toxicology or extended information on their prevalence were found. Reported seizures around the world and their legal status are also presented.

CHEMISTRY

Fonazepam (desmethylflunitrazepam) and nifoxipam (3-hydroxy-desmethylflunitrazepam) are two of the metabolites of flunitrazepam that result from the in vivo demethylation of the parent compound followed by 3-hydroxylation of the benzodiazepine ring in the case of nifoxipam [9, 11, 12, 13]. Taking into consideration the structure of nifoxipam, we can also assume that it could be the 3-hydroxy metabolite of fonazepam. Fonazepam can also be a photodecomposition product of flunitrazepam [14].

The IUPAC name of fonazepam is 5-(2-fluorophenyl)-7-nitro-1,3-dihydro-1,4-benzodiazepin-2-one, while Ro-05-4435 or Ro 5-4435, *N*-desmethylflunitrazepam, desmethylflunitrazepam and norflunitrazepam are also used. Its CAS number is 2558-30-7. Fonazepam has the molecular formula $C_{15}H_{10}FN_3O_3$, a molecular weight of 299.26 g/mol, a boiling point of 470.74 °C at 760 mmHg and a melting point of 198.93 °C [9, 15, 16].

The IUPAC name of nifoxipam is 5-(2-fluorophenyl)-3-hydroxy-7-nitro-1,3-dihydro-2*H*-1,4-benzodiazepin-2-one. It is also named as 5-(2-fluorophenyl)-3-hydroxy-7-nitro-1*H*-benzo[e][1, 4]diazepin-2(3*H*)-one, DP370 and 3-hydroxydesmethylflunitrazepam. Its CAS number is 74723-10-7. Nifoxipam has the molecular formula $C_{15}H_{10}FN_3O_4$, a molecular weight of 315.26 g/mol and a log P value of 10.45. It is a crystalline solid that may be more soluble in water than its parent compound, due to the presence of the hydroxyl group in the 3 position of the benzodiazepine ring. In its powder form, it is also stable for 2 years [11, 12, 17, 18].

SYNTHESIS

The synthesis of fonazepam was achieved in 1963 in Hoffmann-La Roche laboratories by Sternbach et al. [19]. It was the product of the direct nitration of the corresponding unsubstituted benzodiazepinone with a mixture of concentrated sulfuric acid and potassium nitrate at 45–50 °C. Fonazepam is formatted in vivo via demethylation of flunitrazepam (the parent drug), during its biotransformation (Figure 1), and it was identified as its active metabolite in 1976. Nifoxipam is the product of 3-hydroxylation and demethylation of flunitrazepam [13, 20, 21].

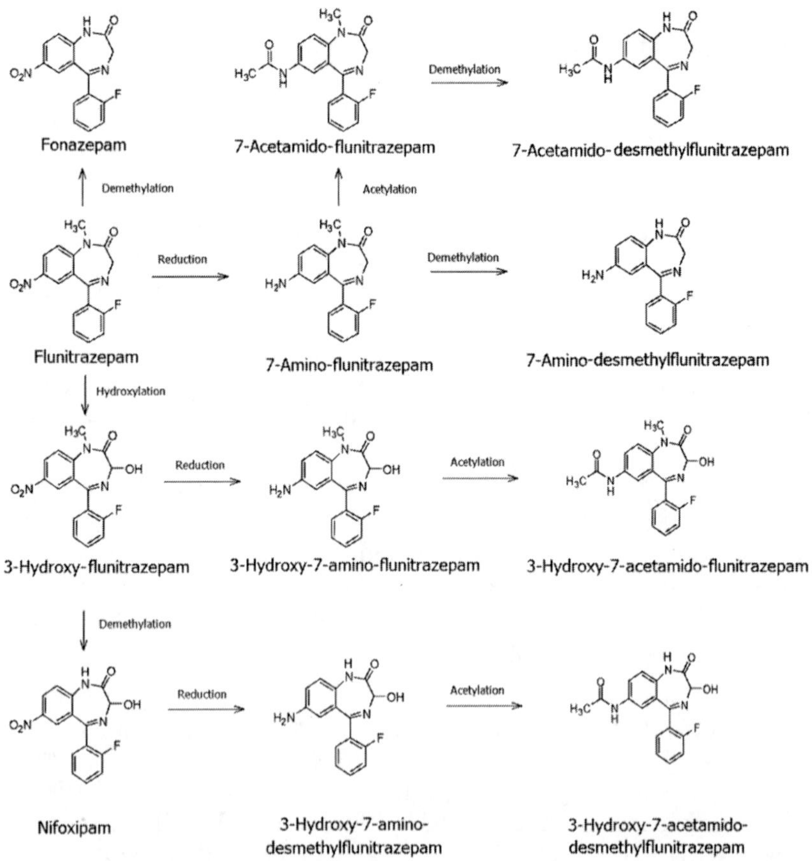

Figure 1. Phase I biotransformation pathways of flunitrazepam, fonazepam and nifoxipam in humans, as suggested by Wendt [20] and Meyer et al. [30]

Fonazepam and nifoxipam are normally used in tablet form and are administered in doses ranging from 0.5 to 3 mg [18, 22, 23]. They are also available in powder form [18]. The most common routes of administration are the oral and sublingual [22, 23, 24]. Users of these designer benzodiazepines self-reported the use of other benzodiazepines as well, like alprazolam, clonazepam, etizolam, diazepam, lorazepam, flubromazolam or meclonazepam [22].

BIOTRANSFORMATION

It is known that flunitrazepam undergoes biotransformation via *N*-demethylation, 3-hydroxylation and glucuronidation, and/or reduction of the nitro group to an amine with subsequent acetylation (Figure 1) [13, 20, 21]. Its formation is mediated by CYP2C19, CYP3A4 and CYP1A2 [25, 26, 27, 28, 29]. Over a 7-day period, an average of 84 % of an oral dose is eliminated in urine. At least 11 metabolites of flunitrazepam have been identified in urine. 7-aminoflunitrazepam (10 % of a dose), 3-hydroxyflunitrazepam (3.5 %), 7-acetamidonorflunitrazepam (2.6 %) and 3-hydroxy-7-acetamidoflunitrazepam (2.0 %) are the major ones, while less than 0.2 % of the initial dose is excreted unchanged [13, 20]. We can assume that fonazepam and nifoxipam undergo similar biotransformation pathways (Figure 1). There is no available information on the biotransformation of fonazepam. On the other hand, Meyer et al. [30] proved the metabolic pathway of nifoxipam based on its identified urinary metabolites. Thus, nifoxipam was mainly reduced to the respective 7-amino benzodiazepine and then acetylated (Figure 1). The acetylated metabolite was further conjugated with glucuronic acid. In that study, other metabolic steps, such as hydroxylation or sulfate conjugation were not observed, probably because the metabolic patterns and the metabolites detected depend on the ingested dose and the timespan between intake and sampling which were not known in detail. The suggested metabolic pathway of nifoxipam is presented in Figure 1 [30].

PHARMACOLOGY

Little is known about the pharmacology of fonazepam and nifoxipam. As it has already been mentioned, they are two active metabolites of flunitrazepam along with 7-aminoflunitrazepam (Figure 1) and, consequently, a similar pharmacological behavior is expected [13, 20].

Flunitrazepam is a low-dose sedative with therapeutic doses ranging from 0.5 to 2 mg and it is used in short-term treatment of moderate insomnia and as a premedication for minor surgical procedures [31, 32]. Fonazepam and nifoxipam, as benzodiazepine derivatives, probably possess similar pharmacological activity by binding the benzodiazepine receptors but they are not used therapeutically. More specifically, they bind with a subset of $GABA_A$ receptors, increasing the affinity of the neurotransmitter GABA for these receptors within the central nervous system, exactly as flunitrazepam does [31, 33, 34] and diminishing the transmission of several important signal substances such as dopamine, noradrenaline, serotonin and acetylcholine [32]. Research using neural artificial networks has predicted the binding affinity of fonazepam to benzodiazepine/$GABA_A$ receptors [35]. The drug was reported to have a binding activity (log IC_{50}) of 0.176 and a predicted value of 0.565 [35, 36]. Structure-activity relationship studies showed that position 7 of the benzodiazepine ring is the most important location on the molecule for increasing the receptor affinity, accounting for 30 % of the total connection weight. Increased lipophilicity and electronic charge by the presence of the nitro group have been directly related to increases in receptor affinity for both fonazepam and its parent drug, flunitrazepam. The fact that the latter possesses a higher binding activity as well as a predicted value (0.580 and 0.778, respectively) than its metabolite could be explained by the presence of the methyl group on the nitrogen of the benzodiazepine ring [35].

According to information gathered from users, the common oral dosage of nifoxipam is between 0.5 and 1 mg, reaching 2 mg in some cases. Its onset time of action after oral intake is 45–120 min and the duration of action is approximately 10–75 h [17, 37]. Users report that the intake of a 2 mg tablet of nifoxipam causes *"only a small relaxing feeling and body sedation"*, while a total dosage of 8 mg can cause benzodiazepine-like sedation and anti-anxiety effects along with slight euphoria and *"a little heavy"* feeling accompanied with nightmares [22]. Other nifoxipam users have experienced anxiety relief, greater hypnotic action than diazepam, as well as euphoria within 10–15 min after intake of 1 mg along with a drink [23]. In any case, euphoria and amnesia are expected effects after their use [17]. Anticonvulsant properties of fonazepam and nifoxipam are also expected, in part or entirely, due to binding to voltage-dependent sodium channels rather than benzodiazepine receptors [38].

TOXICOLOGY

To our knowledge, toxicology data on fonazepam and nifoxipam has not been published. It is assumed that as fonazepam and nifoxipam are active metabolites of flunitrazepam, they could result in respective toxicity and adverse effects, and that their abuse could have similar addictive potential as the parent drug.

Sedation is one of the most common physical effects of flunitrazepam and its active metabolites. These metabolites are moderately sedating and can potentially lead to a lethargic state. In some cases, users experience sleep deprivation that increases proportional to dosage and this sense becomes eventually enough to force a person into complete unconsciousness [17].

The predominant symptoms due to flunitrazepam overdose include ataxia, drowsiness, dizziness, hypotension, muscle relaxation, respiratory depression and coma that can be controlled successfully with supportive therapy [17, 18, 39]. Other adverse effects are loss of motor control, lack of coordination, slurred speech, confusion and gastrointestinal disturbances that last 12 h or more [40]. They also cause lack of coordination, which may lead to falls and injuries, and impair psychomotor functions, such as reaction time and driving skill with increased likelihood of road traffic accidents [17].

Chronic use of flunitrazepam can lead to physical dependence and appearance of withdrawal syndrome after discontinuation. Likewise, nifoxipam is extremely physically and psychologically addictive. Within a couple of days of continuous use, tolerance is developed along with sedative/hypnotic effects, while after cessation, it returns to baseline in 1–2 weeks. As with all benzodiazepines, withdrawal or rebound symptoms may occur after ceasing treatment abruptly following a few weeks or longer of steady dosing, while benzodiazepine discontinuation is difficult and potentially life-threatening for individuals that use it regularly without progressive reduction of the dosage. An increased risk of hypertension, seizures and death has been reported [17, 41].

It has to be mentioned here that nifoxipam presents cross-tolerance with all benzodiazepines, reducing in this way their pharmacological effects [17].

The use of flunitrazepam in combination with alcohol or opioids is always a particular concern since both of these central nervous system depressants potentiate each other's toxicity [17, 42, 43, 44]. The same is expected for its active metabolites. So, similar dangerous and potentially lethal combinations could include the simultaneous use of fonazepam or nifoxipam with other depressants, dissociatives and stimulants that usually result in fatal levels of respiratory depression, muscle relaxation, sedation, amnesia, and vomiting during unconsciousness [17].

Paradoxical effects, like increased seizures in epileptics, aggression, disinhibition, increased anxiety, irritability, violent behavior and suicidal behavior have been described during the therapeutic use of benzodiazepines although they are rare with an evidence rate below 1 % [45, 46]. These effects occur with greater frequency in recreational users, individuals with mental disorders, children and patients on high-dosage regimens [47, 48]. Similar paradoxical effects could be expected to appear to recreational users of fonazepam or nifoxipam.

SEIZURES

The first seizures of fonazepam were made in Europe in 2016. In January 2016, fonazepam was detected in a seized sample of 1 g of white/yellow powder that was sent by mail from an online chemical company based in China. The sample was identified as fonazepam by the Medical Center at the University of Freiburg, Institute of Forensic Medicine, Forensic Toxicology Department in Germany [10, 49]. Two months later, in March 2016, fonazepam was detected in 51 tablets (27 white, 15 blue and 9 grey tablets) seized by police in Linköping, Sweden. It was confirmed by gas chromatography–mass spectrometry (GC–MS) using a reference standard [10, 50].

The first seizure of nifoxipam was reported in April 2014 in Visby, also in Sweden. The drug was analytically confirmed, by the Swedish National Laboratory of Forensic Science, as the active ingredient of 20 brown tablets seized by the police using GC–MS, liquid chromatography–mass spectrometry (LC–MS) and nuclear magnetic resonance spectroscopy [12, 51]. In December 2014, four round, light-brown tablets marked as “Nifoxipam 1 MG” were sent from the UK to Finland where they were seized. The identification of its active ingredient, nifoxipam, was performed by using GC–MS and liquid chromatography–tandem mass spectrometry (LC–MS/MS) [12, 52]. A month later (January 2015), the Norwegian federal police seized 101 brown tablets found in a mail package sent from the UK to Lørenskog, Norway. Nifoxipam was identified by GC–MS [12, 53].

In April 2015, the Slovenian police seized a light-brown tablet of 0.21 g and the active ingredient was confirmed to be nifoxipam by high-performance liquid chromatography–time-of-flight-mass spectrometry (HPLC–TOF-MS). In May 2015, the Danish federal police seized 25 light-beige tablets of 0.1 g that were sent from the UK to Denmark. The tablets were seized at the Copenhagen International Post Office and were marked by the street and chemical name of their ingredient. The presence of nifoxipam was analytically confirmed by GC–MS and LC–QTOF-MS using a database [12, 54, 55]. Two more nifoxipam-related seizures were reported in France in April and July 2015. In the first case, 50 beige tablets sent from the UK were seized at the airport [12, 56], while the second seizure concerned a sample of 105 mg beige powder bought as flubromazepam (Lexomil®) [12, 57].

It seems that the problem of trafficking of these designer benzodiazepines concerns, for the time being, only Europe, while there is almost no information available in the USA, Australia or Japan despite the extensive use of other benzodiazepine derivatives for therapeutic or recreational purposes at these places [4, 58, 59].

DETERMINATION OF FONAZEPAM IN BIOLOGICAL SPECIMENS

Fonazepam and nifoxipam showed significant cross-reactivity, showing a high degree of detectability for the new designer benzodiazepines, by the most common commercially available immunochemical assays for screening purposes in urine (CEDIA, EMIT II Plus, HEIA, KIMS II) [60, 61]. For the confirmation of the results, several analytical methods, using known chromatographic techniques, have been developed for the determination of fonazepam as a metabolite of flunitrazepam in biological specimens. On the other hand, only three articles have been published that describe analytical methods for the detection and quantification of nifoxipam in biological specimens [30, 60, 61], and in one of these studies nifoxipam was determined as the parent drug along with its metabolites [30]. A summary of all the developed chromatographic methods for the determination of fonazepam and/or nifoxipam is shown in Table 1.

Liquid-liquid extraction and solid-phase extraction techniques have been used for the isolation of fonazepam and nifoxipam from biological specimens. Different chromatographic techniques have been used for the development of sensitive methodologies, such as GC combined with an electron capture detector [62, 63, 64, 65] or MS [66, 67, 68, 69], and LC combined with MS [70, 71], MS/MS [30, 61, 72, 73, 74, 75, 76], ultraviolet detection [77, 78], diode array detector [79] or photodiode array detection [71].

Table 1. Summary of analytical methods developed and used for the determination of fonazepam and nifoxipam in biological specimens

Reference	Biological fluid	Extraction method	Analytical technique	LOD/LOQ (ng/mL)	Recovery (%)	Concentration range (ng/mL)	Application
Fonazepam							
de Silva et al. [62]	Blood	LLE [alkalinization, benzene/dichloromethane (90:10, v/v)]	GC–ECD	n.r.	n.r.	n.r.	Application to clinical biological specimens
Faber et al. [63]	Serum	LLE (benzene)	GC–ECD	n.r.	98.6	10–300	n.r.
Drouet-Coassolo et al. [64]	Plasma	LLE (alkalinization, butyl acetate)	GC–ECD	0.5/1.5	n.r.	1–50	Pharmacokinetic study
Berthault et al. [77]	Serum	LLE [alkalinization, diethyl ether/chloroform (80:20, v/v)]	HPLC–UV (242 nm)	2.5/10	57	10–1000	n.r.
He and Parissis [79]	Plasma Urine	SPE (chloroform) (hydrolysis with *β*-glucuronidase in urine samples)	HPLC-DAD (gradient)	4.53/15.1 4.33/14.4	59.3 99.27	25–500 20–400	Application to forensic toxicology and pharmacokinetic studies in humans
Salamone et al. [60]	Urine	Hydrolysis with *β*-glucuronidase, LLE [alkalinization, dichloromethane/isopropanol (90:10, v/v)]	GC–MS	n.r.	n.r.	50–300	Pharmacokinetic study
Coller et al. [65]	Hepatic microsomal incubations	LLE [alkalinization, hexane/diethyl ether (50:50, v/v)]	RP HPLC–UV (210 nm)	n.r.	n.r.	0.2–10 *μ*M	Study of oxidative metabolism of flunitrazepam
Bogusz et al. [70]	Serum, blood, urine	SPE (methanol/0.5 M acetic acid)	LC–APCI-MS	1.0/n.r.	99	1–500	Application to real samples from subjects suspected of DUI of drugs
Nguyen and Nau [66]	Urine	SPE [dichloromethane/isopropanol/NH_3 (80:20:2, v/v/v)]	GC–MS	n.r.	n.r.	n.r.	n.r.
Borrey et al. [67]	Urine	SPE [hydrolysis with *β*-glucuronidase, elution with methanol, derivatization with pyridine/acetic anhydride (1:1, v/v)]	GC–MS	LOD = 18 (scan mode) LOD = 1.2 (SIM mode) LOQ < 50 (scan mode)	93.4–99.2	20–1000 (scan mode) 2–100 (SIM mode)	n.r.
El Mahjoub and Staub [78]	Urine	Online extraction with column switching First column: elution with 20 mM phosphate buffer (pH 7.2)/ACN (94:6, v/v) Second column: elution with 30 mM phosphate buffer (pH 7.5)/ACN (94:6, v/v)	HPLC–UV (245 nm)	n.r.	n.r.	n.r.	n.r.
Kollroser and Schober [72]	Plasma	SPE [dichloromethane/isopropanol/NH_3 (78:20:2, v/v/v)]	LC–APCI–MS/MS	2.0/5.0	93.7–103	5–100	Application to forensic and clinical toxicology
Jourdil et al. [73]	Plasma Urine	Automated SPE [dichloromethane/isopropanol/NH_3 (70:30:2, v/v/v)]	LC–ESI–MS/MS	0.04/0.5	81	n.r.	n.r.
Pirnay et al. [68] Pirnay et al. [69]	Rat plasma	LLE (dichloromethane/1,2-dichloroethane/heptane/isopropanol), derivatization with BSTFA with 1 % TMCS	GC–MS	n.r./0.125	81 (plasma)	0.125–12.5	Pharmacokinetic and toxicokinetic studies in rats.
Dussy et al. [71]	Whole blood Plasma	LLE (1-chlorobutane) or SPE [chloroform/isopropanol/NH_3 (69:29:2, v/v/v)]	HPLC–PDA-MS and PDA	n.r./3.0	n.r.	3–100	Application to forensic toxicology
Huang et al. [80]	Urine	SPE [dichloromethane/isopropanol/NH_3 (78:20:2, v/v/v)]	CE–MEKC	12/n.r.	n.r.	n.r.	n.r.
Forsman et al. [74]	Urine	Hydrolysis with *β*-glucuronidase, SPE [dichloromethane/isopropanol/NH_3 (80:20:2, v/v/v)]	LC–MS/MS	n.r.	n.r.	n.r.	n.r.
Verplaetse et al. [75]	Whole blood and urine	SPE [acetone/chloroform (1:1, v/v) and 2 % ammoniated ethyl acetate]. Hydrolysis with *β*-glucuronidase in urine samples	LC–ESI–MS/MS	0.050/2.0	n.r.	2–500	Application to forensic toxicology
Jeong et al. [76]	Urine	Centrifugation of urine samples, addition of internal standard, direct injection of 5 μL.	LC–MS/MS	3.0/10.0	n.r.	10–100	n.r.
Nifoxipam							
Salamone et al. [60]	Urine	Hydrolysis with *β*-glucuronidase, LLE [alkalinization, dichloromethane/isopropanol (90:10, v/v), derivatization with MTBSTFA, 1 % TBDMS]	GC–MS	n.r.	n.r.	50–300	Pharmacokinetic study
Meyer et al. [30]	Urine	Dilution in internal standard, centrifugation	Nano-LC–HR-MS/MS	n.r.	n.r.	n.r.	Identification of nifoxipam metabolites and application in drug testing analysis
Bergstrand et al. [61]	Urine	n.r.	LC–MS/MS	n.r.	n.r.	100–1000	Application to clinical and forensic toxicology

ACN acetonitrile, *APCI* atmospheric pressure chemical ionization, *BSTFA* *N,O*-bis (trimethylsilyl)trifluoroacetamide, *CE* capillary electrophoresis, *DUI* driving under the influence, *GC–ECD* gas chromatography–electron capture detector, *GC–MS* gas chromatography–mass spectrometry, *HPLC–DAD* high-performance liquid chromatography–diode array detection, *HR* high-resolution, *LC–ESI–MS/MS* liquid chromatography–electrospray ionization–tandem mass spectrometry, *LLE* liquid–liquid extraction, *LOD* limit of detection, *LOQ* limit of quantification, *MEKC* micellar electrokinetic chromatography, *MTBSTFA* (*N*-*tert*-butyldimethylsilyl)-*N*-methyltrifluoroacetamide, *n.r.* not reported, *PDA* photodiode array, *RP* reverse-phase, *SIM* selected ion monitoring, *SPE* solid-phase extraction, *TBDMS* tertiary butyldimethylsilyl trifluoroacetamide, *TMCS* trimethylchlorosilane, *UV* ultraviolet detection

El Mahjoub and Staub [78] used an on-line column switching HPLC method and evaluated two different extraction columns for the determination of fonazepam as a metabolite of flunitrazepam among other metabolites. The procedure was based on direct injection of benzodiazepines on the extraction column followed by the transfer of the compounds to the analytical column.

Another sensitive analytical technique, capillary electrophoresis combined with micellar electrokinetic chromatography, has been described by Huang et al. [80] for the determination of fonazepam along with the other active metabolite, 7-aminoflunitrazepam, and their parent drug, flunitrazepam, in urine after isolation via SPE.

A nano-LC–high-resolution-MS/MS chromatographic method was described for the identification of nifoxipam and its phase I and II metabolites together with other new designer benzodiazepines and their metabolites in urine. Urine samples needed no specific preparation and they were injected directly in the LC system for analysis [30].

No methods for the detection and determination of fonazepam and nifoxipam in seized materials, like tablets or powder, have been described in the literature.

LEGAL STATUS

Fonazepam and nifoxipam are newly introduced compounds in some European countries. Fonazepam is a controlled substance on Schedule IV of the Controlled Substances Act in the USA as a derivative of flunitrazepam [81]. Nifoxipam is illegal in the UK under the Psychoactive Substance Act, which came into effect on 26 May, 2016 [11]. It has also been controlled in Denmark since 18 February 2016, after amendment of the Executive Order on Euphoriant Substances. Thus, it may only be used for medical or scientific purposes [12].

It is not known to be specifically illegal within other countries around the world. However, people that possess fonazepam and nifoxipam or intend to sell or consume them may still be prosecuted, under certain circumstances, by “analogue” laws [82].

CONCLUSIONS

Fonazepam and nifoxipam, two designer benzodiazepines, appeared in Europe after 2014 as NPSs replacing the known controlled benzodiazepines in the drug arena. Although both drugs are active metabolites of flunitrazepam, extended information on their pharmacology and toxicology is not available. Due to their chemical structure, it is assumed that they possess similar pharmacological action and toxicological behavior with the parent drug, flunitrazepam, binding to the same $GABA_A$ receptors. These new designer benzodiazepines are available online labeled as a "research chemical" and "not intended for human or animal consumption". Since they have recently appeared in the drug arena to cover the legal loophole, the international drug enforcement agencies are expected to take actions and measures worldwide against their use or abuse. Increased public vigilance by hospitals, law enforcement and medical examiners is also needed. The final aim is to prevent the expansion of fonazepam and nifoxipam abuse, and the possible intoxications and deaths related to these new designer benzodiazepines.

COMPLIANCE WITH ETHICAL STANDARDS

Conflict of interest

There are no financial or other relations that could lead to a conflict of interest.

Ethical approval

This article does not contain any studies with human participants or animals performed by any of the authors.

REFERENCES

1. European monitoring centre for drugs and drug addiction (EMCDDA) (2016) EU drug market report. http://www.emcdda.europa.eu/system/files/publications/2373/TD0216072ENN.pdf. Accessed 17 Jul 2016
2. United Nations Office on Drugs and Crime (UNODC) (2012) World Drug Report 2012, Vienna. https://www.unodc.org/documents/data-and-analysis/WDR2012/WDR_2012_web_small.pdf. Accessed 17 Jul 2016
3. United Nations Office on Drugs and Crime (UNODC) (2013) World Drug Report 2013, Vienna. https://reliefweb.int/int/sites/reliefweb.int/files/resources/World%20Drug%20Report%202013.pdf. Accessed 17 Jul 2016
4. European Monitoring Centre for Drugs and Drug Addiction (EMCDDA) (2012) New drugs in Europe. http://www.emcdda.europa.eu/attachments.cfm/att_212366_EN_EMCDDA-Europol%202012%20Annual%Report_final.pdf. Accessed 28 May 2016
5. Moosmann B, King LA, Auwärter V (2015) Designer benzodiazepines: a new challenge. World Psychiatry 14:248
6. Mattila MAK, Larni HM (1980) Flunitrazepam: a review of its pharmacological properties and therapeutic use. Drugs 20:353–374

7. Woods JH, Winger G (1997) Abuse liability of flunitrazepam. J Clin Psychopharmacol 17:1S–57S
8. KFx (2015) Drug facts. Newer unregulated drugs. Research chemical briefings, pp 1–41. http://www.kfx.org.uk/drug_facts/drugs_images_and_pdfs/researchchemicals_4.2015.pdf#search='drug+facts.+newer+unregulated+drugs%2C+2015'. Accessed 28 Jul 2016
9. Wikipedia (2016) Desmethylflunitrazepam. http://en.wikipedia.org/wiki/Desmethylflunitrazepam. Accessed 9 Jun 2016
10. European Monitoring Centre for Drugs and Drug Addiction (EMCDDA) (2016) Formal notification of 5-(2-fluorophenyl)-1,3-dihydro-7-nitro-2*H*-1,4-benzodiazepin-2-one (fonazepam) by Sweden as a new psychoactive substance under the terms of Council Decision 2005/387/JHA. EU Early Warning System Formal Notification RCS ID EU-EWS-RCS-FN-2016-0029. https://ednd.emcdda.europa.eu/html.cfm/index7246EN.html?SUB_ID=608&detail. Accessed 28 Jul 2016
11. Wikipedia (2016) Nifoxipam. https://en.wikipedia.org/wiki/Nifoxipam. Accessed 12 Jul 2016
12. EDND (2016) Nifoxipam. https://ednd.emcdda.europa.eu/html.cfm/index7246EN.html?SUB_ID=479&detail. Accessed 11 Jul 2016
13. Baselt RC (2004) Disposition of toxic drugs and chemicals in man, 7th edn. Biomedical Publications, Foster City, pp 465–468
14. Busker RW, Beijersbergen van Henegowen GMJ, Kwee BMC, Winkens JHM (1987) Photobinding of flunitrazepam and its major photo-decomposition product N-desmethylflunitrazepam. Int J Pharm 36:113–120
15. Pubchem (2015) 2*H*-1,4-Benzodiazepin-2-one, 5-(2-fluorophenyl)-1,3-dihydro-7-nitro-. https://pubchem.ncbi.nlm.nih.gov/substance/250115239 #section=Depositor-Supplied-Synonyms. Accessed 12 Jun 2016
16. Bulgalupum (2016) Fonazepam. https://www.bulgalupum.com/products/Fonazepam. Accessed 9 Jun 2016
17. Psychonautwiki (2016) Nifoxipam. https://www.psychonautwiki.org/wiki/Nifoxipam. Accessed 9 Jul 2016
18. drugs-forum.com/forum/showwiki.php?title = Nifoxipam. Accessed 12 Jul 2016
19. Sternbach LH, Fryer RI, Keller O, Metlesics W, Sach G, Steiger N (1963) Quinazolines and 1,4-benzodiazepines. X. Nitro-substituted 5-phenyl-1,4-benzodiazepine derivatives. J Med Chem 6:261–265
20. Wendt G (1976) Schicksal des Hypnotikum Flunitrazepam in menschlichen Organismus. In: Hügin HG, Gemperle M (eds) Bisherigerfahrungen mit "Rohypnol" (flunitrazepam) in der Anästhesiologie und Intensivtherapie. Roche Editions, Basel, pp 27–38
21. Mandrioli R, Mercolini L, Raggi MA (2008) Benzodiazepine metabolism: an analytical perspective. Curr Drug Metab 9:827–844
22. Erowid (2016) Erowid experience vaults report ID: 105406. https://www.erowid.org/experiences/exp.php?ID=105406. Accessed 16 Jul 2016
23. Bluelight (2016) Nifoxipam/1 mg—first time—melting sedative warmth. http://bluelight. org/vb/archieve/index.php/t-720084.html. Accessed 12 Jul 2016
24. Drugs-form (2014) Six hours on meclonazepam and nifoxipam in the end (timeline). https://drugs-forum.com/forum/showthread.php?t=266079. Accessed 12 Jul 2016
25. Coller JK, Somogyi AA, Bochner F (1999) Flunitrazepam oxidative metabolism in human liver microsomes: involvement of CYP2C19 and CYP3A4. Xenobiotica 29:973–986
26. Hesse LM, Venkatakrishnan K, von Molthe LL, Shader RI, Greenblatt DJ (2001) CYP3A4 is the major CYP isoform mediating the in vitro hydroxylation and demethylation of flunitrazepam. Drug Metab Dispos 29:133–140

27. Kilicarslan T, Haining R, Rettie AE, Busto U, Tyndale RF, Sellers EM (2001) Flunitrazepam metabolism by cytochrome P450S 2C19 and 3A4. Drug Metab Dispos 29:460–465
28. Gafni I, Busto UE, Tyndale RF, Kaplan HL, Sellers EM (2003) The role of cytochrome P450 2C19 activity in flunitrazepam metabolism in vivo. J Clin Psychopharmacol 23:169–175
29. Hallifax D, Galetin A, Houston JB (2008) Prediction of metabolic clearance using fresh human hepatocytes: comparison with cryopreserved hepatocytes and hepatic microsomes for five benzodiazepines. Xenobiotica 38:353–367
30. Meyer MR, Bergstrand MP, Helander A, Beck O (2016) Identification of main human metabolites of the designer nitrobenzodiazepines clonazolam, meclonazepam, and nifoxipam by nano-liquid chromatography-high resolution mass spectrometry for drug testing analysis. Anal Bioanal Chem 408:3571–3591
31. Goodchild CS (1993) GABA receptors and benzodiazepines. Br J Anaesth 71:127–133
32. Robertson M, Raymon L (2001) Rohypnol and other benzodiazepines. In: LeBeau MA, Mozayani A (eds) Drug facilitated sexual assault. Academic Press, London, pp 89–105
33. Haefely W (1984) Benzodiazepine interactions with GABA receptors. Neurosci Lett 47:201–206
34. Xiang P, Shen M, Drummer OH (2015) Review: drug concentrations in hair and their relevance in drug facilitated crimes. J Forensic Leg Med 36:126–135
35. Maddalena DJ, Johnston GAR (1995) Prediction of receptor properties and binding affinity of ligands to benzodiazepine/$GABA_A$ receptors using artificial neural networks. J Med Chem 38:715–724
36. So SS, Karplus M (1996) Genetic neural networks for quantitative structure-activity relationships: improvements and application of benzodiazepine affinity for benzodiazepine/$GABA_A$ receptors. J Med Chem 39:5246–5256
37. http://drugs.tripsit.me/nifoxipam. Accessed 12 Jul 2016
38. McLean MJ, MacDonald RL (1988) Benzodiazepines, but not beta carbolines, limit high frequency repetitive firing of action potentials of spinal cold neurons in cell culture. J Pharmacol Exp Ther 244:789–795
39. Mignée C, Garnier R, Conso F, Efthymiou ML, Fournier E (1980) Acute overdose with flunitrazepam. Therapie 35:581–589
40. Bramness JG, Skurtveit S, Mørland J (2006) Flunitrazepam: psychomotor impairement, agitation and paradoxical reactions. Forensic Sci Int 159:83–91
41. Lann MA, Molina DK (2009) A fatal case of benzodiazepine withdrawal. Am J Forensic Med Pathol 30:177–179
42. Kales A, Scharf MB, Kales JD, Soldatos CR (1979) Rebound insomnia. A potential hazard following withdrawal of certain benzodiazepines. JAMA–J Am Med Assoc 241:1692–1695
43. Vermeeren A (2004) Residual effects of hypnotics: epidemiology and clinical implications. CNS Drugs 18:297–328
44. Gustavsen I, Bramness JG, Skurtveit S, Engeland A, Neutel I, Mørland J (2008) Road traffic accident risk related to prescriptions of the hypnotics zopiclone, zolpidem, flunitrazepam and nitrazepam. Sleep Med 9:812–822
45. Paton C (2002) Benzodiazepines and disinhibition: a review. Psychiatr Bull 26:460–462
46. Saïas T, Gallarda T (2008) Réactions d'agressivité sous benzodiazepines: une revue de la literature. Encephale 34:330–336
47. Bond JA (1998) Drug-induced behavioral disinhibition: incidence, mechanisms and therapeutic implications. CNS Drugs 9:41–57
48. Drummer OH (2002) Benzodiazepines—effects on human performance and behavior. Forensic Sci Rev 14:1–14

49. Europol (2016) Reporting form on new psychoactive drug. 5-(2-fluorophenyl)-1,3-dihydro-7-nitro-2*H*-1,4-benzodiazepin-2-one (fonazepam), Germany. 13.01.2016. Hague
50. Europol (2016) Reporting form on new psychoactive drug. 5-(2-fluorophenyl)-1,3-dihydro-7-nitro-2*H*-1,4-benzodiazepin-2-one (fonazepam), Sweden. 19.04.2016. Hague
51. Europol (2015) Reporting form on new psychoactive drug. 5-(2-fluorophenyl)-3-hydroxy-1,3-dihydro-7-nitro-2*H*-1,4-benzodiazepin-2-one (nifoxipam), Sweden. 07.01.2015. Hague
52. Europol (2015) Reporting form on new psychoactive drug. 5-(2-fluorophenyl)-3-hydroxy-1,3-dihydro-7-nitro-2*H*-1,4-benzodiazepin-2-one (nifoxipam), Finland. 18.02.2015. Hague
53. Europol (2015) Reporting form on new psychoactive drug. 5-(2-fluorophenyl)-3-hydroxy-1,3-dihydro-7-nitro-2*H*-1,4-benzodiazepin-2-one (nifoxipam), Norway. 28.10.2015. Hague
54. Europol (2015) Reporting form on new psychoactive drug. 5-(2-fluorophenyl)-3-hydroxy-1,3-dihydro-7-nitro-2*H*-1,4-benzodiazepin-2-one (nifoxipam). Slovenia. 29.07.2015. Hague
55. Europol (2015) Reporting form on new psychoactive drug. 5-(2-fluorophenyl)-3-hydroxy-1,3-dihydro-7-nitro-2*H*-1,4-benzodiazepin-2-one (nifoxipam), Denmark. 30.06.2015. Hague
56. Europol (2015) Reporting form on new psychoactive drug. 5-(2-fluorophenyl)-3-hydroxy-1,3-dihydro-7-nitro-2*H*-1,4-benzodiazepin-2-one (nifoxipam), France. 19.11.2015. Hague
57. Europol (2015) Reporting form on new psychoactive drug. 5-(2-fluorophenyl)-3-hydroxy-1,3-dihydro-7-nitro-2*H*-1,4-benzodiazepin-2-one (nifoxipam), France. 26.10.2015. Hague
58. Lader M (1980) The present status of benzodiazepines and psychiatry and medicine. Drug Res 30:910–913
59. Ashton H (2002) Benzodiazepines: how they work and how to withdraw, "The Ashton Manual". http://www.benzo.org.uk/manual/bzcha03.htm. Accessed 30 Jul 2016
60. Salamone SJ, Honasoge S, Brenner C, McNally AJ, Passarelli J, Goc-Szkutnicka K (1997) Flunitrazepam excretion patterns using the Abuscreen OnTrak and OnLine immunoassays: comparison with GC-MS. J Anal Toxicol 21:341–345
61. Bergstrand MP, Helander A, Hansson T, Beck O (2016) Detectability of designer benzodiazepines in CEDIA, EMIT II Plus, HEIA, and KIMS II immunoassays. Drug Test Anal.
62. de Silva JAF, Bekersky I, Puglisi CV, Brooks MA, Weinfeld RE (1976) Determination of 1,4-benzodiazepines and -diazepin-2-ones in blood by electron-capture gas-liquid chromatography. Anal Chem 48:10–19
63. Faber DB, Kok RM, Rempt-Van Dijk EM (1977) Quantitative gas chromatographic analysis of flunitrazepam in human serum with electron-capture detection. J Chromatogr 133:319–326
64. Drouet-Coassolo C, Iliadis A, Coassolo P, Antoni M, Cano JP (1990) Pharmacokinetics of flunitrazepam following single dose oral administration in liver disease patients compared with healthy volunteers. Fundam Clin Pharmacol 4:643–651
65. Coller JK, Somogyi AA, Bochner F (1998) Quantification of flunitrazepam's oxidative metabolites, 3-hydroxyflunitrazepam and desmethylflunitrazepam, in hepatic microsomal incubations by high-performance liquid chromatography. J Chromatogr B 719:87–92
66. Nguyen H, Nau DR (2000) Rapid method for the solid phase extraction and GC–MS analysis of flunitrazepam and its major metabolites in urine. J Anal Toxicol 24:37–45
67. Borrey D, Mayer E, Lambert W, Van Peteghem C, De Leenheer AP (2001) Simultaneous determination of fifteen low-dosed benzodiazepines in human urine

by solid-phase extraction and gas chromatography–mass spectrometry. J Chromatogr B 765:187–197
68. Pirnay S, Bouchonnet S, Hervé F, Libong D, Milan N, d'Athis P, Baud F, Ricordel I (2004) Development and validation of a gas chromatography-mass spectrometry method for the simultaneous determination of buprenorphine, flunitrazepam and their metabolites in rat plasma: application to the pharmacokinetic study. J Chromatogr B 807:335–342
69. Pirnay S, Megarbane B, Declèves X, Risède P, Borron SW, Bouchonnet S, Bérangère P, Debray M, Milan N, Duarte T, Ricordel I, Baud FJ (2008) Buprenorphine alters desmethylflunitrazepam disposition and flunitrazepam toxicity in rats. Toxicol Sci 106:64–73
70. Bogusz MJ, Maier R-D, Krüger K-D, Früchtnicht W (1998) Determination of flunitrazepam and its metabolites in blood by high-performance liquid chromatography–atmospheric pressure chemical ionization mass spectrometry. J Chromatogr B 713:361–3369
71. Dussy FE, Hamberg C, Briellmann TA (2006) Quantification of benzodiazepines in whole blood and serum. Int J Legal Med 120:323–330
72. Kollroser M, Schober C (2002) Simultaneous analysis of flunitrazepam and its major metabolites in human plasma by high performance liquid chromatography tandem mass spectrometry. J Pharm Biomed Anal 25:1173–1182
73. Joudril N, Bessard J, Vincent F, Eysseric H, Bessard G (2003) Automated solid-phase extraction and liquid-chromatography–electrospray ionization–mass spectrometry for the determination of flunitrazepam and its metabolites in human urine and plasma. J Chromatogr B 788:207–219
74. Forsman M, Nyström I, Roman M, Berglund L, Ahlner J, Kronstrand R (2009) Urinary detection times and extraction patterns of flunitrazepam and its metabolites after a single oral dose. J Anal Toxicol 33:491–501
75. Verplaetse R, Cuypers E, Tytgat J (2012) The evaluation of the applicability of a high pH mobile phase in ultrahigh performance liquid chromatography tandem mass spectrometry analysis of benzodiazepines and benzodiazepine-like hypnotics in urine and blood. J Chromatogr A 1249:147–154
76. Jeong Y-D, Kim MK, Suh SI, In MK, Paeng K-J (2015) Rapid determination of benzodiazepines, zolpidem and their metabolites in urine using direct injection liquid chromatography–tandem mass spectrometry. Forensic Sci Int 257:84–92
77. Berthault F, Kintz P, Mangin R (1996) Simultaneous high-performance liquid chromatographic analysis of flunitrazepam and four metabolites in serum. J Chromatogr B 685:383–387
78. El Mahjoub A, Staub C (2001) High-performance liquid chromatography determination of flunitrazepam and its metabolites in plasma by use of column switching technique: comparison of two extraction columns. J Chromatogr B 754:271–283
79. He W, Parissis N (1997) Simultaneous determination of flunitrazepam and its metabolites in plasma and urine by HPLC/DAD after solid phase extraction. J Pharm Biomed Anal 16:707–715
80. Huang C-W, Jen H-P, Wang R-D, Hsieh Y-Z (2006) Sweeping technique combined with micellar electrokinetic chromatography for the simultaneous determination of flunitrazepam and its major metabolites. J Chromatogr A 1110:240–244
81. Wikipedia (2016) Controlled Drugs and Substances Act. http://en.wikipedia.org/wiki/Controlled_Drugs_and_Subtances_Act#Schedule _IV. Accessed 9 Jun 2016
82. Legislation.gv.uk. (2016) Psychoactive Substances Act. http://www.legislation.gov.uk/ukpga/2016/2/contents/enacted. Accessed 9 Jul 2016

CHAPTER 13

Facial Index Based 2D Facial Composite Process for Forensic Investigation in Sri Lanka

P. B. Jayasekara[1], L. Sivaneasharajah[1], M. A. S. Perera[1], J. Perera[2], D. D. Karunaratne[1], K. D. Sandaruwan[1], R. N. Rajapakse[1]

[1]*University of Colombo School of Computing, Colombo, Sri Lanka*

[2]*Department of Forensic Medicine and Toxicology, Faculty of Medicine, University of Colombo, Colombo, Sri Lanka*

ABSTRACT

The "facial composite" is one of the major fields in the forensic science that helps the criminal investigators to carry out their investigation process. The survey conducted by United States Law Enforcement Agencies confirms that 80% of the law enforcement agencies use computer automated composite systems whereas Sri Lanka is still far behind in the process of facial composite with lot of inefficiencies in the current manual process. Hence this research introduces a novel approach for the manual facial composite process, while eliminating the inefficiencies of the manual procedure in Sri Lanka. In order to overcome this situation, this study introduces an automated image processing based software solution with 2D facial feature templates targeting the Sri Lankan population. Thus, this was the first ever approach that creates the 2D facial feature templates by incorporating both medically defined indexes and relevant aesthetic aspects. Hence, this research study is comprised of two separate analyses on anthropometric indices and facial feature shapes which were carried out targeting the local population. Subsequently, several evaluation techniques were utilized to evaluate this methodology where we obtained an overall success rate as 70.19%. The ultimate goal of this research study is to provide a system to the law enforcement agencies in order to carry out an efficient and effective facial composite process which can lead to increase the success rate of suspect identification.

Keywords: Facial Composite, Facial Index, Facial Feature Templates, Sri Lanka

1. INTRODUCTION

In the occurrence of crime incidents, the law enforcement agencies and police officers often depend on eyewitness testimony in order to establish the complete view of the crime situation. Eyewitness testimony is an account given by a victim unfolding the occurrence of a criminal activity [1] . The most prominent usage of the eyewitness testimony is when the investigators are impotent to find a pathway to carry out the interrogation as no threat of physical evidence is found at the scene.

In the event of criminal identification, the way Sri Lanka is analyzing the physical evidences is more similar to a certain extent with international context but still suspect identification through facial composite is far behind than we could ever imagine.

A facial composite is a graphical representation of a human face according to eyewitnesses' description. In order to overcome the current situation in Sri Lanka, an automated image processing based software solution has been introduced which includes 2D facial feature templates created using the anthropometric indices and shapes.

The failure rate of suspect identification through facial composite sketch has been declared as 92.86% in the year of 2014 [2] . The reason may be the failures in manual hand drawing procedure carried out by Sri Lankan forensic sketch artists [2] . When considering the current situation in Sri Lanka, still there are only two forensic artists who have not been given the opportunity of a professional training for drawing composite images [2] . Apart from this, still there's no secondary layer to carry out the job currently done by Sri Lankan forensic artists. This raises questions especially in terms of continuity in a case where forensic artists leave their profession.

However there is several recognized composite sketch software available in the international market such as FACES 4.0, E-FIT which is currently used by CIA, FBI and the US Military [3] . Even though, the data set which has been used in these available software has a vast difference with that of the Sri Lankan facial features. Thus it is not practical to use this software to create composite images of Sri Lankan suspects.

Since Sri Lanka is at its infancy level in facial composition, our attempt is to pinpoint the gap in order to find out the optimum way to automate this traditional procedure by incorporating the information technologies (image processing) and statistical analysis techniques and its related aspects. Since this research proposes a template based composite system, as a first step the need is to construct 2D facial feature templates by analyzing the Sri Lankan dataset. In early studies it was proven that the ability of assessing facial feature appearance has been achieved by anthropometric proportion indices and this has been used for a number of forensic and clinical practices [4] . Additionally, the facial features' shapes have also been identified to figure out the most frequently occurring shapes of Sri Lankan population. Eventually both these indexes and shapes were incorporated in the 2D facial feature template creation.

Therefore, to create these templates, in the first place a preliminary study was done with 140 undergraduate students (both male and female) aged 20 - 25 to find out the anthropometric proportion indices measurements of eyes, nose, upper vermilion, lower vermilion and these facial feature shapes of Sri Lankans.

Furthermore, this research also consolidates a detailed rule set based decision support system to filter the most possible facial feature templates according to eyewitness description. Thus the incorporation of Information Technology into the Sri Lankan facial composite process will be a guide to a more accurate and efficient suspect identification procedure with the help of more advanced facial composite process. Hence we have taken the very first step to provide Information Technology based solution for facial composite process in Sri Lanka.

2. RELATED WORKS

The "Facial composite" process has a very long history where it evolved from manual composition to computer based facial composition solutions. The entire facial composite evolution can be categorized into four [5] generations. In the beginning of 1911 [5] , the first generation came up with the manual facial composite process where the sketch artist converted the eyewitness description into a hand drawn facial image based on their own impression [5] . This manual process is still prevailing in certain countries as the only technique in compositing the culprits face image. The second generation of the facial composite techniques was known as "Mechanical Systems" [5] such as Identikit and Photo fit which used the idea of facial composite by manually synthesizing the individual facial feature templates to get the ultimate image. "Identiti Kit" was a mechanical system for a facial resemblance which made up on 568 facial features drawn on transparent acetate sheets [5] .

The "Photo fit" was developed in three versions in order to cover the population of male Caucasian, female Caucasian and male Afro Asian [6] . The overall design of the photo fit was a book format which contains thin paper strips along with an index list.

The third generation evolved with the software solutions where the individual facial feature templates are composited with the help of the computer technology [5] . E-FIT is unique and innovative computer based facial composite software which emerged in this third generation to carry out the holistic composite which ultimately creates near photo realistic and color image of the suspect based on eyewitness testimony [7] [8] .

Finally, the fourth generation of facial composite system has been evolved with the concept of genetic algorithm [7] where the eyewitness needs to select the whole face rather than combining the individual facial feature [5] . "EvoFit" was one such system which was initiated in mid stage of 1990's. A 25% to 60% of world leading composites have been constructed using EvoFit which seems more accurate than the previous systems used by law enforcement authorities [9] . However EvoFit differs from other solutions as EvoFit carried out a genetic algorithm in synthesizing facial proportional aspects where the need was to integrate genetic recombination and mutation operators to create the best likeness face [9] - [11] .

From 1990, feature based selection method was used for most of the composite software solutions. The products such as FACES, Mac-a-Mug Pro are commonly used software by the United States [12] . Similarly as EvoFit, FACES was also used by thousands of law enforcement agencies including FBI and CIA [12] . Generally it consists of features like eyes, mouth, nose, eyebrows, lips, mouth lines, tattoos and scars.

The entire aforementioned computer based software systems such as EFit, EvoFit and Faces increase the efficiency of traditional manual methods. Since almost all these international composite systems were developed targeting the facial features of foreign people, these are not feasible to utilize these software systems in Sri Lanka [2] .

Facial Feature Analysis

Anthropometric proportion indices have been used in a considerable number of facial analysis studies conducted for various facial feature types and different races in several fields such as medicine, forensic, plastic surgery. Previous researches related to numerical measurements of facial feature analysis have done similar calculations with the use of medically accepted facial indexes using anthropometric soft tissue landmarks [13] .

According to the literature, different formulas have been stated for Facial Index, Nasal Index, and Eye Index. A similar anthropometric calculations was applied in the upper and lower vermilion index calculations as there were no such formulas were defined for these two indexes.

As stated in previous studies facial index [14] , eye index [6] nasal index [6] [13] [15] [16] can be defined as follows;

$$\text{Facial index} = \frac{\text{Facial Height} \times 100}{\text{Upper Facial Width}} \tag{1}$$

$$\text{Eye index} = \frac{\text{Eye Width} \times 100}{\text{Eye Height}} \tag{2}$$

$$\text{Nasal Index} = \frac{\text{Nose width} \times 100}{\text{Nose height}}. \tag{3}$$

Research conducted in [17] has identified that in certain faces upper vermilion values differ from lower vermilion values based on the person's facial type and shape whereas in certain faces the height of the upper vermilion equals to the height of the lower vermilion [18] . Considering these facts it is essential to have two separate formulas for the calculation of upper vermilion index and lower vermilion index.

$$\text{Upper vermilion Index} = \frac{\text{Mouth Width} \times 100}{\text{Upper Vermilion Height}} \tag{4}$$

$$\text{Lower vermilion Index} = \frac{\text{Mouth Width} \times 100}{\text{Lower Vermilion Height}} \tag{5}$$

3. METHODOLOGY

Mainly, the proposed solution consists of two steps which need to be carried out to create an ultimate composite image. They are Pre requirement stage and Building composite image.

3.1. Pre Requirement Stage

As the pre requirement stage of our proposed process, the 2D facial feature templates were created based on the analysis of two relevant parameters known as facial feature indices and facial feature shapes. Henceforth the index measurements of facial features were taken to obtain the size whereas the available shapes of main facial features of Sri Lankan people were considered to obtain the shapes in the process of template creation.

3.1.1. Facial Feature Template Creation

The dataset which consists of 140 frontal standard facial photographs of undergraduate students comprising both male and female between 20 to 25 age categories, were captured in order to create the facial feature templates. The same standard environment setup was used to provide a similar condition (frontal view, neutral expression, daylight condition, same distance from camera to personal object) for the entire process of capturing photographs.

This research study has focused only on main facial components and subsequently the 2D facial feature templates have been created for eyes, nose, mouth and face. These templates were created using the results obtained from the two sub studies carried out for index analysis and facial shape analysis of Sri Lankans. Since the weight of a person can affect the width and height of the facial features, the collected data set was categorized based on the weight categories using BMI (Body Mass Index) values, as BMI can be used to screen for different weight categories.

3.1.2. Index Measurement of the Facial Features

In the creation of facial feature templates, it was required to carry out an in-depth analysis of facial features. Even though there have been initial studies conducted with regard to facial indices in Sri Lanka [19] - [21] , the index analysis hasn't been conducted for Sri Lankans for all the indexes required in this study. According to the literature, numerical analysis of the linear measurements and proportional indices were carried out for main facial features of the targeted subjects through anthropometric soft tissue landmarks as shown in Figure 1 [17] .

The facial feature detection library, Luxand Face SDK was used for the facial feature detection from facial photographs as it is comparatively more reliable and accurate over manual measuring techniques and other related tools. Using such libraries would overcome inefficiencies related to the manual measuring techniques such as subjectivity and high time consumption. Subsequently the distance between the feature points and the indices were calculated via the Java based tool stated above. Initially, the facial feature index formulas which are mentioned in the literature review were used to measure the index values of each facial feature. Subsequently they were classified further into different categories as shown in Tables 1-5 in order to obtain a more accurate categorization. Once after classifying each facial feature, the width and height of these facial features were measured in millimeters and the average value of these measurements were calculated to decide the most possible width and height to represent the whole data set for each category.

Table 1. Average eye index, average width and average height along with the eye classifications for both male and female.

Eye classification	Eye index range		Eye index mean		Eye width mean		Eye height mean	
	Male	Female	Male	Female	Male	Female	Male	Female
Narrow	<274.99	<249.937	255.95	228.94	7.46 ±0.44	7.26 ± 0.66	2.88 ± 0.22	3.10 ± 0.22
Medium	275 - 325.1	250 - 293.999	296.91	272.59	7.44 ± 0.51	7.25 ± 0.49	2.51 ± 0.28	2.69 ± 0.22
Wide	>325.29	>294	341.37	326.32	7.44 ± 0.68	7.56 ± 0.33	2.13 ± 0.21	2.31 ± 0.14

Table 2. Average nasal index, average width and average height along with the Nasal classifications for both male and female.

Nasal classification	Nasal index range		Nasal index mean		Nasal width mean		Nasal height mean	
	Male	Female	Male	Female	Male	Female	Male	Female
Narrow	<86	<82.999	83.54	78.04	9.38 ± 0.90	8.44 ± 0.58	11.40 ± 0.88	10.80 ± 0.90
Medium	86.1 - 99.2	83 - 91.999	94.84	88.24	9.54 ± 0.71	8.65 ± 0.59	10.39 ± 0.84	9.90 ± 0.67
Wide	>99.3	>92	102.69	94.49	9.80 ± 0.68	9.23 ± 0.64	9.62 ± 0.89	9.64 ± 0.58

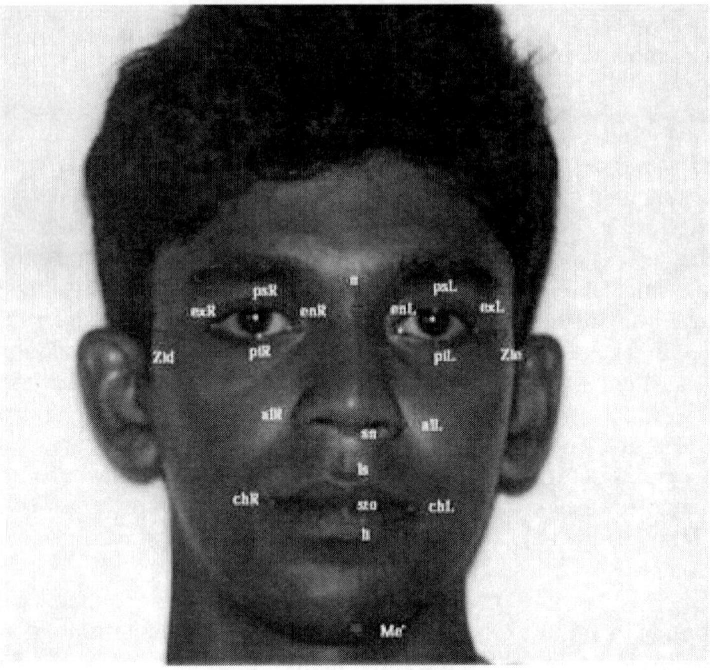

Figure 1. Anthropometric soft tissue landmarks [3] Nasion (n), subnasale (Sn), Alare (alL/R), endocanthion (enL/R), exocanthion (exL/R), palpebraleinferius (piL/R), palpebralesuperius (psL/R), cheilion (ch), labialesuperius (ls), labialeinferius (li), stomion (sto), right zigion (Zid), Menton (Me).

Table 3. Average Upper Vermilion (U.V) index, average width and average height along with the upper vermilion classifications for both male and female.

U.V classification	U.V index range		U.V Index mean		U.V width mean		U.V height mean	
	Male	Female	Male	Female	Male	Female	Male	Female
Narrow	<594.1	<808.999	514.13	1487.95	12.06 ± 0.95	11.75 ± 1.05	2.33 ± 0.31	1.87 ± 0.32
Medium	594.2 - 789.8	809 - 1191.999	686.94	897.92	12.67 ± 1.04	11.59 ± 0.65	2.07 ± 1.53	1.30 ± 0.1
Wide	>789.9	>1192	879.49	630.97	12.63 ± 1.08	11.81 ± 0.98	1.49 ± 0.34	0.79 ± 0

Table 4. Average Lower Vermilion (L.V) index, average width and average height along with the Lower vermilion classifications for both male and female.

L.V classification	L.V index range		L.V index mean		L.V width mean		L.V height mean	
	Male	Female	Male	Female	Male	Female	Male	Female
Narrow	<443	<444.999	430.37	396.71	12.35 ± 1	3.05 ± 0.29	2.87 ± 0.34	2.94 ± 0.31
Medium	443.1 - 553.1	445 - 535.9	541.11	484.06	47.54 ± 1.03	11.25 ± 2.44	2.36 ± 0.26	2.98 ± 2.04
Wide	>553.2	>536	648.46	586.96	13.37 ± 0.84	12.07 ± 2.83	1.95 ± 0.29	2.00 ± 0.20

Table 5. Average face index, average width and average height along with the face classifications for both male and female.

Face classification	Face index range		Face index mean		Face width mean		Face height mean	
	Male	Female	Male	Female	Male	Female	Male	Female
Narrow	<79.9	<77.9	76.49 ± 2.82	75.39 ± 1.64	178.04 ± 12.80	168.59 ± 10.21	227.34 ± 15.35	216.34 ± 14.60
Medium	80 - 88.9	78 - 82.9	84.29 ± 2.49	79.76 ± 1.44	174.64 ± 8.85	166.18 ± 8.89	242.27 ± 17.32	220.36 ± 13.75
Wide	> 89	>83	91.99 ± 2.42	84.93 ± 1.97	174.69 ± 9.59	168.42 ± 11.22	243.40 ± 28.24	233.86 ± 12.59

4. RESULTS

The results of the eye, nasal, upper vermilion, lower vermilion and face index analysis of each facial feature are presented in Tables 1-5 respectively. As the main outcome of this analysis, average values of index, height and width were calculated. The below mentioned mean values of width and height were used to construct the 2D facial feature templates.

4.1. Commonly Available Shapes of Facial Features

Since there are no other studies conducted on facial feature shapes for Sri Lankan context, shape categorization has been conducted based on aesthetic aspects with the help of the domain experts' knowledge. Therefore, two domain experts in the aesthetic field were consulted to classify the facial feature shapes.

The same 140 frontal photographs were used to detect the different shapes of Sri Lankan people's facial features. All the 140 photographs were inspected by the domain experts in aesthetic field Mr. Winnie Hettigoda, the Senior lecturer, of multimedia art Faculty of Visual Art in University of the

Visual and Performing Arts and the forensic artists at Criminal Records Division in Sri Lanka categorized the different shapes of facial features.

During the classification process the two domain experts were able to identify 6, 10, 8 and 7 types of shapes for face, eye, nasal and mouth respectively for males. Similarly 6, 8, 8 and 7 types of shapes for face, eye, nasal and mouth were identified respectively for females. Figure 2(a) and Figure 2(b) represent the identified shapes for each facial feature along with their frequency percentage for male and female correspondingly.

4.2. Building Composite Image

A facial feature template based composite system was implemented to computerize the manual hand drawn facial composite process to enhance the efficiency of the current manual process. The created 2D facial feature templates were then added to the system afterwards the facial composite process will be take place at this stage with the help of the eyewitness description. Subsequently an interface has been implemented to derive the facial feature templates and each type of templates will be shown in different tabs in composite software system from which the eyewitness can select the appropriate facial feature templates to produce the ultimate suspect's face.

5. EVALUATION

Face pool comparison and resemblance techniques were carried out as the evaluation methods for this study. The evaluation was conducted only for male category as the facial feature templates were constructed only targeting male indices and shape categories. As shown in Figure 3, the face pool technique was conducted with 75 responded where they were asked to participate in the online survey. Table 6 illustrates the results of 06 test cases which were obtained from the face pool comparison conducted via the aforementioned online survey.

As the second evaluation method the resemblance evaluation technique was conducted via online survey with 51 respondents. As shown in Figure 4 the respondents were asked to rate the resemblance rate of the each feature in the composite image with the original image. The options suggested for the rating was Very Similar, Similar, Average, Slightly Similar and Not at all which range from 5 to 1 were in order to analyze the evaluation results. Table 7 illustrates the results of 04 test cases which were obtained from the resemblance evaluation which conducted via the aforementioned online survey.

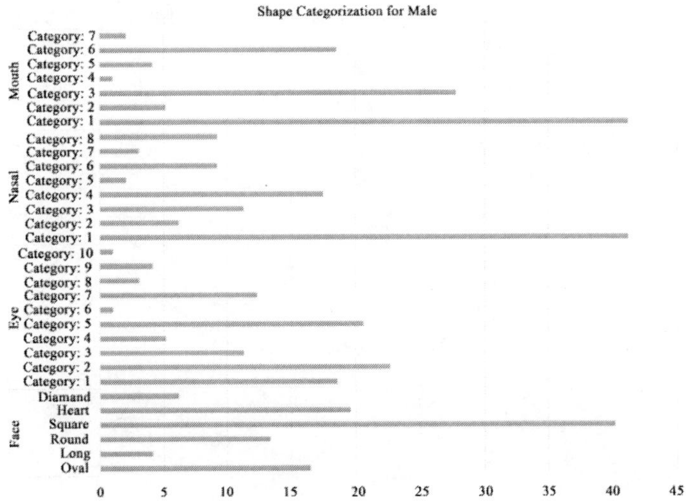

(a)

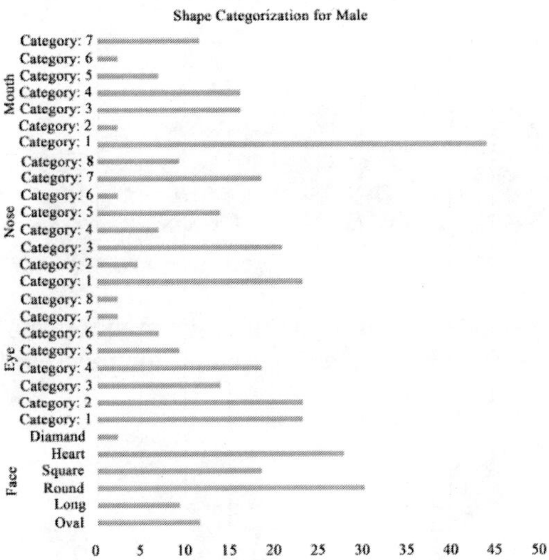

(a)

Figure 2. (a) Facial feature shape categorization for Male (b) Facial feature shape categorization for female.

Table 6. Results of the face pool comparison and the highlighted cell depicts the correct image along with its percentage.

Image No	Test Case 1	Test Case 2	Test Case 3	Test Case 4	Test Case 5	Test Case 6
Image 1	13%	0%	**72.7%**	6.1%	0%	0%
Image 2	13.3%	12.1%	9.1%	6.1%	18.2%	**81.8%**
Image 3	4%	**69.75%**	12.1%	12.1%	18.2%	0%
Image 4	**78.7%**	12.1%	6.1%	**72.7%**	18.2%	6.1%
Image 5	2.7%	6.1%	0%	3%	**45.5%**	12.1%

Table 7. Results from the resemblance rating evaluation.

Features	Case 1	Case 2	Case 3	Case 4	Case 5	Case 6
Face	Similar 47.4%	Similar 31.6%	Similar 42.1%	Average 42.1%	Similar 36.8%	Similar 37.3%
Eye	Similar 42.1%	Very Similar 36.8%	Similar 47.4%	Similar 47.4%	Average 36.2%	Similar 47.4%
Nose	Similar 36.8%	Similar 21.1%	Similar 42.1%	Similar 47.4%	Slightly similar 31.6%	Similar 35.3%
Mouth	Average 47.4%	Similar 31.6%	Average 47.4%	Average 47.4%	Slightly similar 42.1%	Average 33.3%

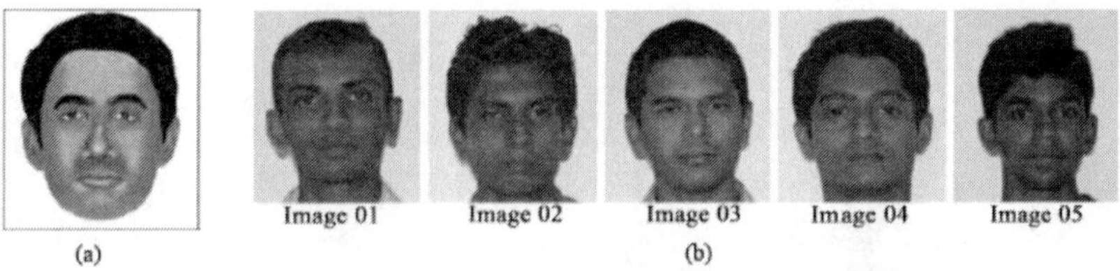

Figure 3. (a) The composite image generated created for the Test Case 01 face pool evaluation; (b) The image list created for the Test Case 01 face pool evaluation.

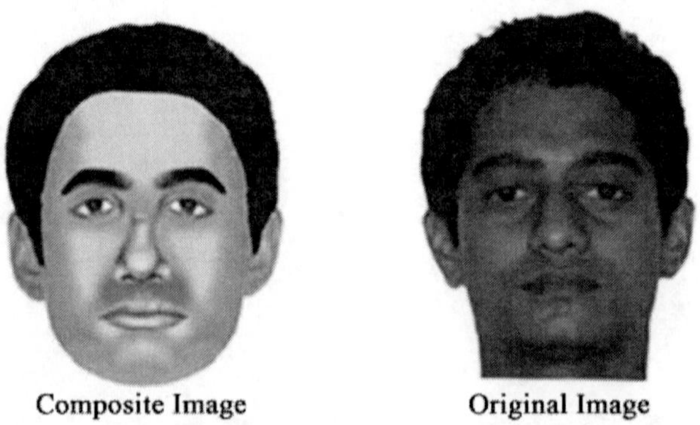

Figure 4. The composite image and the original image of the provided Test Case 1.

6. DISCUSSION

The overall success rate which obtained from this research study was 70.19% for the 6 test cases whereas the success rate of the prevailing international software gives 60% [22] . Since the accuracy level of the proposed approach could reach to the success rate of available international software, it is obvious that this solution can be used in the identification of suspects through facial composite in Sri Lanka.

According to the evaluation phase, test cases 1, 2, 3, 4 and 6 have taken the higher accuracy rate than the success rate gain from international software solutions. In the proposed solution, the facial feature templates were constructed using the medically defined indexes which were ideal for Sri Lankan people and the aesthetic aspects were also considered when placing the facial templates for the ultimate composite image. These well accepted measurements and aesthetic aspects were put together in constructing facial composite image via this process. Therefore this has become one major reason behind the highest accuracy rate achieved from the proposed process.

However case study 5 has obtained the lowest result as 45.5% and it clearly depicts that this attained success rate is lower than the average success rate which was derived from the proposed process. Moreover the results obtained from the resemblance evaluation shows face, eye features of composite image of case 5 were presented high equivalence rate whereas the features that presented a low equivalence were nose and mouth. Hence it is obvious that the nose and mouth features were the major causes which mislead the audience to select the wrong person for case study 5. And the root cause for this mismatching nose and mouth templates were selected to construct this composite image because only limited nose and mouth shapes were identified from the process as our data set was 140 and it was not enough to represent the whole population. But still this drawback can be eliminated, if the same procedure was applied for large dataset and as a result more nose and mouth shapes can be identified through the shape classification method as mentioned in section "Commonly available shapes of facial features".

The present study has been able to establish an efficient facial composite process for Sri Lankans which can achieve a considerable amount of high success rate. Therefore the findings and the results of this study will definitely helpful in the forensic science and will eliminate the drawbacks in suspect identification via facial composite.

7. CONCLUSIONS

In the field of forensic science, the facial composite plays a major role in the process of tracking down the criminal when there is no physical evidence found from the crime incident. Henceforth the facial composite image becomes one of the major clues in the investigation procedure.

The study introduces an automated image processing based composite software solution which includes 2D facial feature templates created for the first time targeting the local population to eliminate the inefficiencies in the current manual facial composite procedure in Sri Lanka.

Since these 2D Facial Features templates have been created targeting Sri Lankan context, the analysis of facial feature variations of Sri Lankans were required before creating these facial feature templates. Therefore two major sub researches have been carried and the results obtained from these sub studies were incorporated to construct the 2D facial features templates needed to carry out the proposed facial composite process.

The study conducted was evaluated with face pool and the resemblance techniques and obtained 70.19% accuracy indicating that this proposed process has a significant impact on dramatically increasing the success rate of the facial composite comparative to the prevailing success rate. Hence this computerized system will significantly improve the existing manual process and benefit professionals involved in the criminal justice process mainly police and crime investigators and ensure justice for the victimized person.

ACKNOWLEDGEMENTS

A special acknowledgement goes to the advisor Mr. Winnie Hettigoda, senior lecturer of University of the Visual & Performing Arts and Senior Superintendent of Police Mr. Priyantha Jayakody, Director of Criminal Record Division in Sri Lanka who helped us in gaining the relevant domain knowledge. Furthermore, authors are thankful to Mr. Nadaun Liyanage for the immense assistance given for the success of the automated composite system.

CITE THIS PAPER

P. B. Jayasekara, L. Sivaneasharajah, M. A. S. Perera, J. Perera, D. D. Karunaratne, K. D. Sandaruwan, R. N. Rajapakse (2016) Facial Index Based

2D Facial Composite Process for Forensic Investigation in Sri Lanka. *Forensic Medicine and Anatomy Research*,**04**,7-16. doi: 10.4236/fmar.2016.41002

REFERENCES

1. McLeod, S. (2009) Eyewitness Testimony. http://www. simplypsychology.org/eyewitness-testimony.html
2. Artist, F. (2015) Interviewee, Manual Process of Suspect Identification through Facial Composite in Sri Lanka.
3. Frowd, C., McQuiston-Surrett, D., Anandaciva, S., Ireland, C. and Hancock, P. (2007) An Evaluation of US Systems for Facial Composite Production. Ergonomics, 50, 1987-1998.
4. Rajapakse, R., Madugalla, A., Amarasinghe, I., Padmathilake, V., Dharmaratne, A., Sandaruwan, D. and Vidanapathirana, M. (2012) Facial Muscle Anatomy Based Approach for Forensic Facial Reconstruction in Sri Lanka. 2012 International Conference on Advances in ICT for Emerging Regions (ICTer), Colombo, 27-35.
5. Lindsay, R.C.L., Ross, D.F., Read, J.D. and Toglia, M.P. (2007) The Handbook of Eyewitness Psychology. Vol. II: Memory for People, Lawrence Erlbaum Associates Publishers, Mahwah, 59-83. [Citation Time(s):7]
6. Ellis, H.D., Davies, G.M. and Shepherd, J.W. (1978) A Critical Examination of the Photofit* System for Recalling Faces. Ergonomics, 21, 297-307.
7. Gibson, S.J., Maylin, M.I.S., Solomon, C.J. and George, B. (2008) EFIT-V—Interactive Evolutionary Strategy for the Construction of Photo-Realistic Facial Composites. GECCO Genetic and Evolutionary Computation Conference, New York, 12-16 July 2008, 1485-1490.
8. Miran International (2011) Vision Metric. http://www.miranintl.com.pk/VisionMetric.html
9. University of Stirling (2014) EvoFIT: Applying Psychology to the Identification of Criminals.http://impact.ref.ac.uk/casestudies2/refservice.svc/GetCaseStudyPDF/44434
10. Hancock, P. and Frowd, C. (2015) EvoFIT: Applying Our Understanding of Face Perception to the Identification of Criminals. http://www.bps.org.uk/impact/evofit-applying-our-understanding-face-perception-identification-criminals
11. Frowd, C.D., Hancock, P. and Bruce, V. (2011) Catching More Offenders with Evofit Facial Composites: Lab Research and Police Field Trials. Global Journal of Human Social Science, 11, 35-46.
12. Key, L. (2014) New Methods to Improve Recognition Performance of 3D Face Composite Software.
13. Othman, S., Ahmad, R. and Asi, S. (2014) Three-Dimensional Quantitative Evaluation of Facial Morphology in Adults with Unilateral Cleft Lip and Palate, and Patients without Clefts. The British Journal of Oral & Maxillofacial Surgery, 52, 208-213.
14. Morosini, I.d.A.C., Peron, A.P.L.M., Correia, K.R. and Moresca, R. (2012) Study of Face Pleasantness Using Facial Analysis in Standardized Frontal Photographs. Dental Press Journal of Orthodontics, 17, 24-34.
15. Esomonu, U., Ude, R., Lukpata, P. and Nandi, E. (2013) Anthropometric Study of the Nasal Index of Bekwara, Ethnic Group of Cross River State, Nigeria. International Research Journal of Applied and Basic Sciences, 5, 1262-1265.
16. Edler, R., Agarwal, P., Wertheim, D. and Greenhill, D. (2006) The Use of Anthropometric Proportion Indices in the Measurement of Facial Attractiveness. European Journal of Orthodontics, 28, 274-281.

17. Sekharan, P.C. (2012) Lip Forensics for Criminal Identification. International Conference on Forensic Research & Technology, Chicago-Northshore.
18. Daenecke, S., Bianchini, E.M. and da Silva, A.P. (2006) Anthropometrical Measurements of the Height of the Upper Lip and Length of the Philtrum. Pró-Fono Revista de Atualização Científica, 18, 249-258.
19. R.N.Rajapakse, K.D.Sandaruwan, D.D.Karunaratne, M.Vidanapathirana (2014) Anthropometric classification of nasal indexes for a Sri Lankan undergraduate student population. International Conference on Advances in ICT for Emerging Regions (ICTer), Colombo, 10-13 December 2014, 261.
20. Chandimal, K.M., Yasawardene, S.G. and Adikari, G. (2012) Facial (Prosopic) Index in Purana Inhabitants at the Foot of Sigiriya, Sri Lanka. Proceedings of the Annual Research Symposium, Faculty of Graduate Studies, University of Kelaniya, 27.
21. Madugalla, A.K., Rajapakse, R.N., Amarasinghe, I.U., Padmathilake, V.H., Dharmaratne, A.T. and Vidanapathirana, M. (2013) Face ID: A 3D Computer Graphic Application for Forensic Medicine: A Novel Semi-Automated Muscle Based Digital Sculpting Initiative for Forensic Facial Reconstruction in Sri Lanka. International Conference on Computer Medical Applications (ICCMA), Sousse, 20-22 January 2013, 1-6.
22. Pitchford, M., Skelton, F., Petkovic, A. and Frowd, C.D. (2012) Catching Even More Offenders with EvoFIT Facial Composites. Third International Conference on Emerging Security Technologies (EST), Lisbon, 5-7 September 2012, 20-26.

CHAPTER 14

Factors Affecting Species Identifications of Blow Fly Pupae Based upon Chemical Profiles and Multivariate Statistics

William Kranz [1], Clinton Carroll [1], Darren A. Dixon [1], John V. Goodpaster [1,2,*] and Christine J. Picard [2,3,*]

[1]*Department of Chemistry and Chemical Biology, Indiana University Purdue University Indianapolis, Indianapolis, IN 46202, USA*

[2]*Forensic and Investigative Sciences Program, Indiana University Purdue University Indianapolis, Indianapolis, IN 46202, USA*

[3]*Department of Biology, Indiana University Purdue University Indianapolis, Indianapolis, IN 46202, USA*

ABSTRACT

Alternative methods for the identification of species of blow fly pupae have been developed over the years that consist of the analyses of chemical profiles. However, the effect of biotic and abiotic factors that could influence the predictive manner for the tests have not been evaluated. The lipids of blowfly pupae (*Cochliomyia macellaria*, *Lucilia cuprina*, *Lucilia sericata*, and *Phormia regina*) were extracted in pentane, derivatized, and analyzed by total-vaporization solid phase microextraction gas chromatography-mass spectrometry (TV-SPME GC-MS). Peak areas for 26 compounds were analyzed. Here we evaluated one biotic factor (colonization) on four species of blow flies to determine how well a model produced from lipid profiles of colonized flies predicted the species of flies of offspring of wild-caught flies and found very good species identification following 10 generations of inbreeding. When we evaluated four abiotic factors in our fly rearing protocols (temperature, humidity, pupation substrate, and diet), we found that the ability to assign the chemical profile to the correct species was greatly reduced.

Keywords: *Phormia regina*; *Lucilia sericata*; *Lucilia cuprina*; *Cochliomyia macellaria*; forensic entomology; TV-SPME; blow fly pupae; lipid analysis

1. INTRODUCTION

Insects make up a large proportion of the environment; therefore, it is not surprising that insects would be present in crime scenes. Insects play an important role in decomposition, therefore they become forensically useful in estimating the postmortem interval [1]. Insects recovered from the body can be used as a "clock" to determine the minimum postmortem interval (PMI_{MIN}), or the time since the body has been available to the insects, as most species in North America will only lay eggs or larvae on a resource if it is dead [2,3].

The most common fly that will be encountered in casework is the blowfly (Diptera: Calliphoridae). The species of the collected specimens is determined, and some measure of their age—usually the length of the larvae or the life cycle stage of the insects' development—is used to extrapolate PMI_{MIN}, referencing published developmental data (e.g., [4,5,6,7,8,9]). To estimate the PMI_{MIN}, investigators should attempt to collect the oldest insect specimens present at the crime scene, and in some cases, these are pupae. Due to their lack of morphological characters, it is often difficult, if not impossible, to identify the species. Therefore, pupae are typically placed in an environmental chamber at known temperature and allowed to complete their development. Once they are adults, the species can be determined using published morphological keys [10]. Two shortcomings of this technique are the additional time needed for the pupae to complete development, as well as the assumption that the pupae are still viable. Any damaged or dead pupae will never be identified by this method, and few crime laboratories have the resources available to rear flies. An alternative is to extract the DNA from the pupae and determine the species typically by sequencing a mitochondrial DNA locus [11,12,13,14,15,16,17], however, expertise is necessary for the analysis of the DNA sequence, and a thorough representation of all possible species is required [18].

Further complicating the use of blow fly pupae for forensic postmortem interval estimations is the process of estimating the age of the pupae. Various methods have been proposed, which include an analysis of the differential expression of genes [19,20]. However, this requires a well-preserved, high-quality sample with intact mRNA. Other proposed methods include the analysis of the internal morphology of the pupa [21,22,23], however, this requires a great deal of expertise, and has only been done on a single species to date.

Considering these issues, experts in the field of forensic entomology have suggested chemical analyses to complement, or substitute for, traditional morphological examinations [2] and/or DNA-based methods. Such chemical analyses would be easily conducted on instruments common to all forensic laboratories, alleviating the burden on smaller labs. They would also be more objective, subject to empirical scrutiny, and able to report on their own propensity for error.

Several possible methods of analysis have been proposed in recent years, but the idea that has shown the most promise relies upon the analysis of the organic compounds that can be extracted from the insect cuticle [2]. Associated with the cuticle are a multitude of chemicals—hydrocarbons, free fatty acids, alcohols, aldehydes, wax esters, and fatty acid methyl esters [24]—which serve a variety of purposes with respect to the insect's physiology and life cycle.

All insects have cuticular hydrocarbons, those hydrocarbons can often be individualized to species—or, even further, to the sex of a species, to a part of the life cycle, or to a caste [25]. The evolution of such compounds, which include long-chain hydrocarbons and hydrophobic moieties, appears to have originated from the need, by insects, to retain water within their bodies and prevent desiccation [26]. This knowledge is supported by the observation that species of Drosophila found in arid environments have a complement of hydrocarbons whose chains are longer than those in more temperate climates [27,28]. In addition to this water retention function, the cuticular compounds serve to protect the insect in several other ways—notably, by providing a barrier to infiltration by insecticides and toxins [29], as well as fungi and bacteria [24]. The hydrocarbons of the cuticle can also convey information via chemical signaling. When one insect encounters another, the cuticular hydrocarbon profile may signal any or all of the following: (i) that the individual is a member of the same species [25]; (ii) of the same colony; (iii) male or female [30]; (iv) caste membership [31]; (v) whether the individual is a close family relation, or a dominant member of the colony attempting to pass along an order, or an inferior member standing by to take a command [32]; and (vi) whether the other insect is fertile and ready to mate, or has already copulated [33]. Flies are able to detect cuticular hydrocarbons via the olfactory organs of the maxillary palps and antennae, or by taste, via the organs of the proboscis [28].

Likewise, a constellation of fatty acids and sterols have been isolated from flies, as described by Golebiowski et al. [24,33,34]. Free fatty acids ranging between hexanoic acid (6:0) and hexacosanoic acid (26:0) have been described, although many of these are at trace levels in comparison to the dominant palmitic acid, palmitoleic acid, oleic acid, linoleic acid, and stearic acid contributions . Like the hydrocarbons, these lipids are known to provide resistance to desiccation, fungal infection, and bacterial infection. A number of sterols have also been reported, including cholesterol, campesterol, and sitosterol, although cholesterol appears to be the most consistently dominant [35].

As there appear to be qualitative and quantitative differences in the occurrence of these compounds among the various species, the chemical profiles may be useful as a tool for identification and discrimination. Solvent extraction has been the traditional method for the analysis of insect cuticular compounds [24,26,36,37,38]. The process is straightforward: the insect is placed into a nonpolar solvent, such as methylene chloride, ether, pentane, or hexane, during which time the biological compounds leach into the liquid. Those compounds may then be concentrated and analyzed by gas chromatography-mass spectrometry (GC-MS) or by high performance liquid chromatography (HPLC).

Previous work has been done demonstrating the discriminating power of chemical profiles for blow fly larvae and pupae [26,39,40,41,42]. Headspace solid phase microextraction (SPME) experiments of a single species (*Calliphora vicina*) utilized a polydimethylsiloxane/divinyl benzene/carboxen (PDMS/DVB/CAR) SPME fiber to collect hydrocarbons, fatty acids, and sterols from larvae, and a PDMS/CAR SPME fiber to do the same for pupae [39]. The presence and absence of the compounds were noted and the qualitative data was analyzed by agglomerative hierarchical clustering (AHC). The results in this case indicated that the chemical profiles of larvae and pupae were dissimilar, as were the profiles of pupae of differing ages [39].

The purpose of this project was to interrogate the chemical profiles that could be obtained from pupae by solvent extraction followed by total vaporization solid-phase microextraction (TV-SPME) [43]. The advantage of this method over headspace or immersion SPME is the elimination of the need to determine the optimal organic solvent, and that extracts do not need to be filtered reducing buildup and potential contamination of the GC column. Furthermore, TV-SPME can add additional selectivity and sensitivity based on fiber choice, and requires no changes in instrumentation.

Four species of blowfly were investigated: *Cochliomyia macellaria*, *Lucilia cuprina*, *Lucilia sericata*, and *Phormia regina*. The objectives were to: (i) determine whether the chemical profiles could be used to effectively discriminate between different species; (ii) to determine the extent to which genetic variation affected the chemical profile and the integrity of these determinations; and, (iii) to examine the effects of various abiotic variables on the chemicals extracted from the flies. These scenarios were used to produce a model of blow fly species identifications.

2. MATERIALS AND METHODS

2.1. Rearing of Fly Colonies and Collection of Specimens

Colonies of *Phormia regina*, *Cochliomyia macellaria*, *Lucilia sericata*, and *Lucilia cuprina* were reared through ten successive generations (G). At the G1 and G10 generations, 19 × 100 maggots were transferred into a Dixie cup containing 50 g chicken liver, which were in turn placed inside 19 separate glass mason jars half-filled with a pupation substrate. The mason jars were kept in an incubator maintained at 25 °C temperature and 60% humidity. Seven to ten days were allotted for the maggots to develop through their larval stages. At 1 d following the observation of pupation, all of the pupae were harvested from three randomly-selected jars; these pupae were stored separately in a freezer at −80 °C and classified as Timepoint 1. Samples were collected each additional 24 h following for a total of six time points (18 jars). This data was not used to estimate age, but rather to determine the consistency of the chemical profiles over time to accurately identify the species. The maggots in the 19th jar served as a control, the time of adult emergence was recorded, and the number of adult flies was compared against the number of puparia and unhatched pupae to ensure proper development.

To gauge the chemical effects wrought by changes in temperature, humidity, pupation substrate, and diet, the same experimental design as above was carried out using pupae of G10+ *Phormia regina* (Table 1). During this set of experiments, pupae were likewise collected over six timepoints.

Table 1. Experimental design for the evaluation of abiotic factors of temperature, humidity, feeding, and pupation substrates on lipid profiles for the determination of age of *Phormia regina* pupae. L:D: Light: Dark; CL: chicken liver; LFB: low fat beef; HDB: high fat beef; SD: sawdust; Ver: vermiculite.

Abiotic Factor	Temperature (°C)	Humidity (%)	Feeding Substrate	Pupation Substrate
Temperature	15	55	CL	SD
	20	55	CL	SD
	25	55	CL	SD
	30	55	CL	SD
Humidity	25	40	CL	SD
	25	70	CL	SD
	25	85	CL	SD
Larval Feeding Substrate	25	55	LFB	SD
	25	55	HFB	SD
	25	55	Rat	SD
Pupation Substrate	25	55	CL	SD
	25	55	CL	Sand
	25	55	CL	Ver

2.2. Instrumental Analysis

A 6890 gas chromatograph coupled to a 5975 mass spectrometer (Agilent, Santa Clara, CA, USA) served as the principal instrument, with autosampler functionality provided by an MPS2 (Gerstel, Mülheim an der Ruhr, Germany). The column was a J&W DB-5ms (30 m × 0.25 mm × 0.25 µm). All GC-MS analyses utilized H_2 carrier gas with a flow rate of 2.5 mL/min operated in splitless mode, with a scan range of m/z 40–550.

For each timepoint, three haphazardly selected pupae from each of the three jars were thawed and extracted in 1 mL pentane. After 4 day, the extracted solution was siylated and submitted to GC-MS analysis via the TV-SPME method described [43]. This method provided excellent sensitivity an order of magnitude greater than what could be achieved by liquid injection, allowing for the analysis of liquid extracts without a reconstitution step.

2.3. Statistical Analysis

Following instrumental analysis, a qualitative assessment of the chromatograms was undertaken, retention times converted to retention indices (RI), and 26 compounds of interest were identified across all four species (see Results). Compounds were identified based upon their retention times and mass spectra. Authentic standards of the most common free fatty acids, squalene and cholesterol, were also analyzed alongside unknown samples. For compounds for which we did not have a standard, a computerized search of the National Institute of Standards and Technology (NIST) mass spectral database was conducted. In these cases, only forward match scores greater than 800/1000 were considered probative. The mass spectra for these compounds were compiled into an Automated Mass Spectral Deconvolution and Identification System (AMDIS) library, and automated peak integration proceeded using AMDIS mass spectral deconvolution software. The peak areas were normalized to the square root of the sum of squares for all compounds of interest in the chromatogram.

The normalized dataset was investigated by principal component analysis (PCA) and discriminant analysis (DA) to illuminate the underlying trends among the variables and timepoints. Therefore, a complex system of 26 variables (i.e., 26 compounds of interest) can be efficiently described by a system of three PCs, which can themselves be plotted on a Cartesian coordinate plane, thereby aiding in the visualization of the data.

DA is another technique whose purpose is to visualize groupings in the data, as well as to predict and ascribe group membership for new samples. DA strives to maximize the discrimination among different groups. The group memberships are pre-defined, but the integrity of these memberships can be assessed by a leave-one-out cross-validation, which is reported in a table known as a confusion matrix.

3. RESULTS AND DISCUSSION

3.1. Determination of Species

The relative abundance of the 26 compounds of interest for the G1 generations are presented in a heat map in Table 2. Wherever possible, the identity of the compound has been reported. However, due to similar retention times and indistinguishable mass spectra, the exact identities of some of the hydrocarbons remain elusive. These have been compared against the retention times of an n-alkane mix and differentiated based upon their retention indices. Similarly, the compounds labeled “14:1 FFA” and “18:1 FFA” are mono-unsaturated 14:1 and 18:1 free fatty acids, but the precise location of the double bond is unknown.

Table 2. Heat map depicting the log of the normalized peak areas for the G1 generation of *C. macellaria* ($n = 53$), *L. cuprina* ($n = 54$), *L. sericata* ($n = 54$), and *P. regina* ($n = 36$) across all time points. Twenty-six compounds were selected for statistical analysis. The color scale is logarithmic, varying from black (10^{-7}) to red (10^{0}). "TCAIE" (Compound E) is 2,2,4-trimethyl-3-carboxyisopropyl- pentanoic acid isobutyl ester. FFA: free fatty acid; ND: not determined.

Identifier	Compound	*C. macellaria*	*L. cuprina*	*L. sericata*	*P. regina*	Identifier	Compound	*C. macellaria*	*L. cuprina*	*L. sericata*	*P. regina*
A	Tetradecane	−3.80	−2.79	−2.32	−5.12	N	Linolenic Acid	−1.82	−1.84	−2.16	−2.09
B	FFA	ND	−2.73	−3.05	ND	O	Linoleic Acid	−0.60	−0.47	−0.42	−0.53
C	Alkane A	ND	ND	−2.94	ND	P	Oleic Acid	−0.32	−0.24	−0.24	−0.33
D	Pentadecane ($C_{15}H_{32}$)	−4.94	−2.74	−2.13	ND	Q	Stearic Acid	−1.83	−2.01	−2.18	−1.71
E	TCAIE	−2.07	−1.79	−2.04	−2.88	R	18:1 FFA	−2.43	−3.31	−3.44	−2.50
F	Lauric Acid	−2.32	−2.34	−2.48	−2.48	S	Arachidonic Acid	−1.23	−1.51	−1.62	−1.72
G	Isopropyl Myristate	−3.14	−2.72	−3.31	−3.46	T	Alkane B	−3.31	−1.83	−2.36	−2.65
H	14:1 FFA	−3.21	−3.31	−3.14	−3.24	U	Alkane C	−5.86	−3.45	−3.99	ND
I	Myristic Acid	−1.16	−1.28	−1.61	−1.22	V	Alkane D	−3.24	ND	ND	ND
J	Pentadecanoic Acid	−2.58	−2.30	−3.13	−2.97	W	Alkane E	−2.31	−1.57	−2.85	−3.23
K	Palmitelaidic Acid	−1.31	−0.99	−1.47	−1.42	X	Alkane F	−3.26	−5.26	−4.63	ND
L	Palmitoleic Acid	−0.27	−0.35	−0.34	−0.25	Y	Alkane G	−5.02	−3.00	−6.27	ND
M	Palmitic Acid	−0.22	−0.25	−0.29	−0.24	Z	Cholesterol	−2.56	−2.98	−3.52	−2.88

There is a high degree of qualitative similarity between the four species, particularly with respect to the expression of fatty acids. For example, palmitoleic acid, palmitic acid, linoleic acid, and oleic acid are observed at relatively high levels in all samples for all species. Substantially more variability is evident in the hydrocarbon compound class, with certain alkanes commonly seen in some species, yet wholly absent in other species. For example:

1. alkane A (retention index (RI) = 1465) was only present in *L. sericata*.
2. pentadecane, alkane C (RI = 2800), alkane F (RI = 2934), and alkane G (RI = 3084) were present in all species except *P. regina*.
3. alkane D (RI = 2867) was only present in *C. macellaria.*

Less obvious in Table 2, but still just as important to the statistical analysis, are the more nuanced quantitative differences in the detection of these compounds among the different species (i.e., differences in the relative amounts of each of the major compounds). It is often customary to reduce the dimensionality of a dataset first via PCA prior to attempting DA. This is typically done to sidestep one of the inherent limitations of DA, where each grouping must possess more observations (i.e., pupa exemplars) than variables (i.e., compounds of interest, or principal components thereof). A preliminary round of PCA therefore serves as a convenient way to reduce the number of variables whilst still capturing a sizeable amount of the variance in the data. For this data set, PCA of the G1 generation (data not shown) tended to over-emphasize the differences between *Lucilia cuprina* and *Lucilia sericata* at the expense of the differences between *Cochliomyia macellaria* and *Phormia regina*, and therefore obfuscated more than it aided.

Improved discrimination of species was achieved in this study by performing DA on the dataset directly without the intermediate PCA step. DA results for the G1 generation were prepared by this method and an observations plot is shown in Figure 1. This plot illustrates the two species that are readily differentiable (*L. cuprina* and *L. sericata*) and two species that are more apt to be confused (*C. macellaria* and *P. regina*).

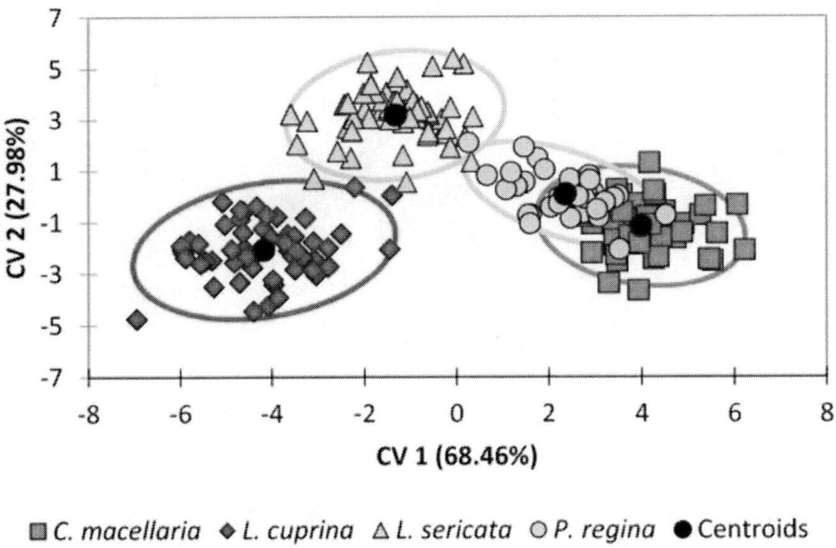

Figure 1. Discriminant analysis of the original variables for all four species, G1 generation, at all timepoints. The ellipses correspond to the 95% confidence limits of each species based upon its centroid and distribution of observations in the canonical variate (CV) space.

A variables plot is also provided in Figure 2. This type of plot serves as a “roadmap” for interpreting the DA results in Figure 1. Each ray in the variables plot represents one of the 26 compounds of interest. Rays that form an acute angle represent compounds whose abundances are highly correlated. For example, rays I, R, and S, correspond to myristic acid, octadecenoic acid (18:1), and arachidonic acid, which tend to rise and fall together. Similarly, rays T, U, W, and Y, represent long-chain alkanes (waxes) that are also correlated. Rays that form obtuse angles are highly anti-correlated. For example, linoleic acid (ray O) and arachidonic acid (ray S) are negatively correlated. Rays at right angles reflect compounds with little or no correlation.

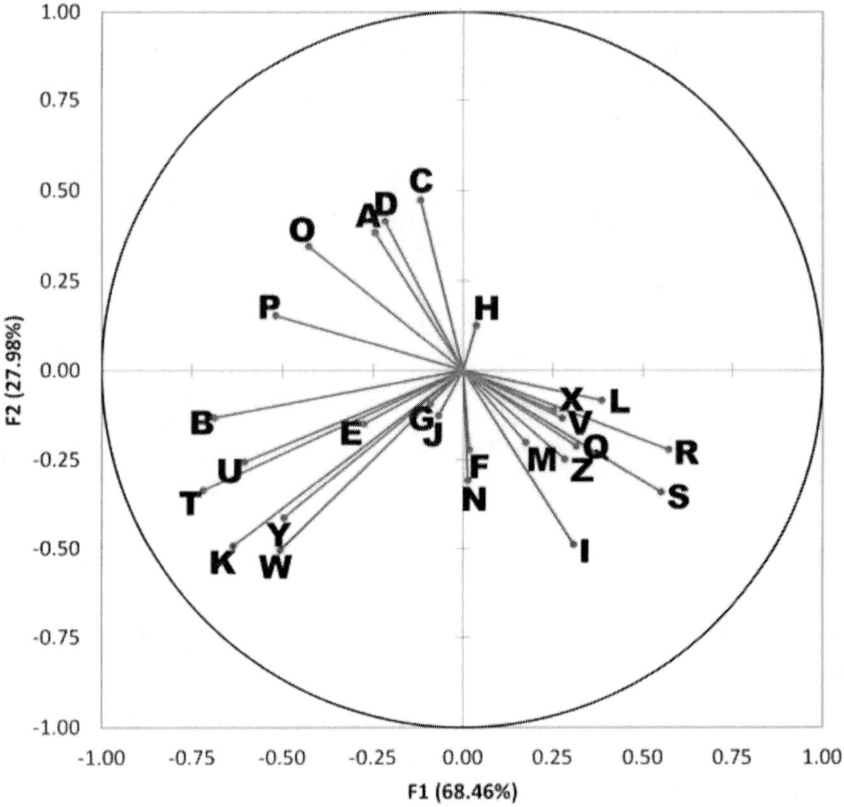

Figure 2. Variables plot for discriminant analysis (DA) results in Figure 1. Identities for the rays projected from the origin correspond to the lettered compounds in Table 2.

The quadrants of the variables plot (Figure 2) are also directly comparable to the quadrants of the corresponding DA chart (Figure 1). The length of each ray indicates the magnitude of the association. Long rays indicate variables that are important in describing the variation in sample profiles, whereas short rays that fall close to the origin are compounds that describe little to none of the variation. The distribution of the rays in Figure 2 is quite instructive. For example, rays B, K, T, U, W, and Y project into the lower left quadrant and tend to be found at higher levels in *L. cuprina*. Rays I, R, and S project into the lower right quadrant and tend to be found at higher levels in *P. regina* and *C. macellaria*. Lastly, rays A, C, D, O, and P project into the top left quadrant and tend to be found at higher levels in *L. sericata*.

Thus, the following observations are manifest:

- *Lucilia cuprina* is distinguishable by high concentrations of palmitoleic acid and several long-chain alkanes (alkane B, alkane C, alkane E, and alkane G).
- *Phormia regina* and *Cochliomyia macellaria* are distinguishable by high concentrations of myristic acid, 18:1 FFA, and arachidonic acid.
- *Lucilia sericata* is distinguishable by high concentrations of several alkanes, linoleic acid, and oleic acid.

Lastly, when the DA model for the G1 generation was tested using leave-one-out cross-validation (Table 3), classification accuracies ranged from 89% (*P. regina*) to 96% (*L. sericata*) with an overall classification accuracy of 93%. Whereas Figure 1 allows for a qualitative assessment of the discrimination of the four species, classification accuracy is a quantitative way to assess the level of confusion between classes. Therefore, it is worth noting that of the four species, *P. regina* and *C. macellaria* had the lowest accuracies. For *P. regina*, 75% (three out of four) of the misclassifications were due to *P. regina* observations erroneously classified as belonging to the *C. macellaria* species. Similarly, 75% (three out of four) of the misclassifications were due to *C. macellaria* observations erroneously classified as belonging to the *P. regina* species. Achieving an overall classification accuracy of greater than 90% is an excellent result and this is sound evidence for the theory that taxonomic information can be gleaned from the lipid profiles.

Table 3. Leave-one-out confusion matrix for the G1 generation chemical profiles.

From/To	*C. macellaria*	*L. cuprina*	*L. sericata*	*P. regina*	Total	% Correct
C. macellaria	49	2	0	3	53	92.5
L. cuprina	0	62	3	0	54	94.4
L. sericata	0	3	52	0	54	96.3
P. regina	3	0	1	32	36	88.9

3.2. Effect of Genetic Diversity

A key issue this project sought to address was the effect of genetic diversity on a specimen's chemical profile. Specifically, do G1 pupae yield a markedly different suite of compounds than pupae belonging to the G10 generation, whose genetic diversity has been homogenized over successive generations of inbreeding?

When the results from the G10 generations were analyzed by DA (see Figure 3 and Figure 4 and Table 4), the results were like those for the G1 generation. Note that the orientation of the species within the DA plot is rotated ~180° relative to Figure 2. This is not as significant as the relative positions of the classes and the overall separation between species, which is more evident in this data set (Figure 3). The correlations between variables and the observations are also very similar (Figure 4), although rotated relative to Figure 2 as mentioned above. One exception was compound R (18:1 or octadecenoic acid), which was no longer strongly correlated with any of the other compounds.

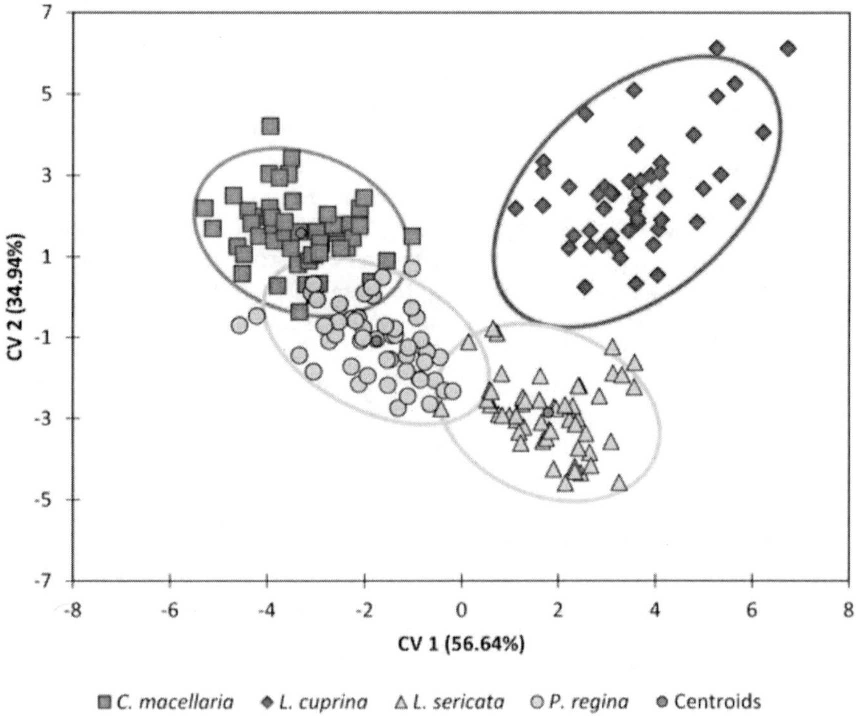

Figure 3. Discriminant analysis of all four species, G10 generation, at all timepoints.

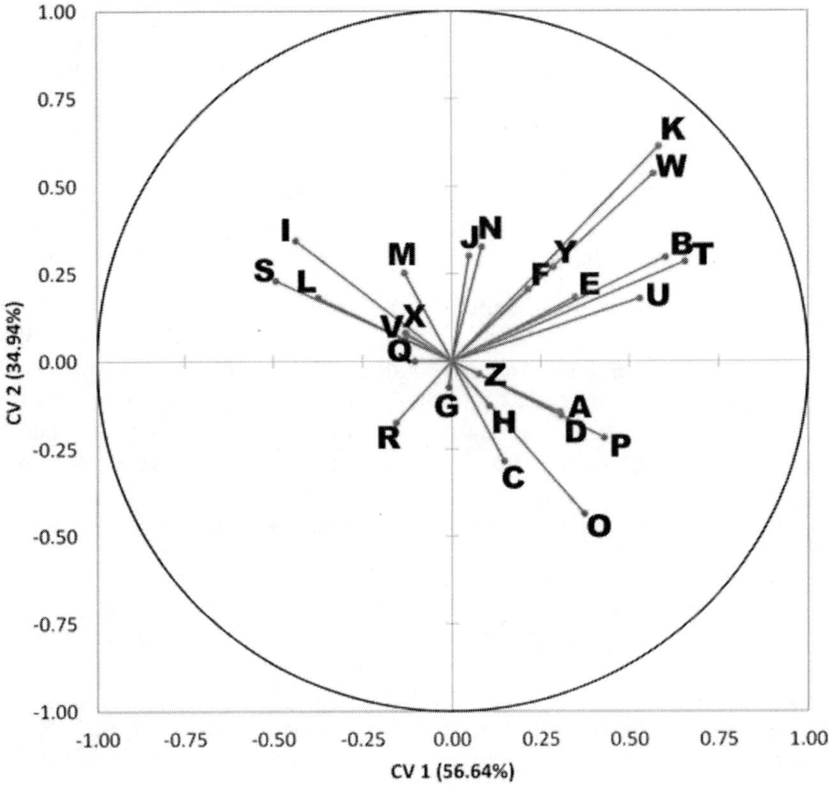

Figure 4. Variables plot for the discriminant analysis results in Figure 3. Identities for the projected rays from the origin correspond to the lettered compounds in Table 2.

Table 4. Leave-one-out confusion matrix for the G10 generation chemical profiles.

From/To	*C. macellaria*	*L. cuprina*	*L. sericata*	*P. regina*	Total	% Correct
C. macellaria	52	0	0	2	54	96.3
L. cuprina	0	48	1	0	49	98.0
L. sericata	0	0	52	2	54	96.3
P. regina	2	0	1	51	54	94.4

Lastly, when the DA model for the G10 generation was tested using leave-one-out cross-validation (Table 4), classification accuracies ranged from 94% (*P. regina*) to 98% (*L. cuprina*) with an overall classification accuracy of 96%. In similar fashion to the G1 generation, most of the misclassifications for *P. regina*, 67% (four out of seven) were due to *P. regina* observations that were erroneously classified as belonging to the *C. macellaria* species, and vice versa. Similarly, all misclassifications for *C. macellaria* (two out of two) were due to *C. macellaria* observations that were erroneously classified as belonging to the *P. regina* species. The overall performance of the DA model is better for the G10 generation (96%) than it was for the G1 generation (93%). This small increase in accuracy may be explained by reducing the intra-species variation through repeated inbreeding.

DA was also used to attempt to discriminate samples hailing from the G1 and G10 generation of each species (i.e., *C. macellaria* G1 samples were compared to *C. macellaria* G10 samples, *L. cuprina* G1 samples were compared to *L. cuprina* G10 samples, etc.). When DA is applied to only two classes (as in a series of pair-wise comparisons), a Receiver Operating Characteristic (ROC) curve can be constructed. ROC curves depict the binary classification accuracy of the system, and the area under a ROC curve varies from 0.5 (no discrimination) to 1.0 (perfect discrimination).

The classification accuracy and ROC results for the G1 and G10 generations of all four species are shown in Table 5. The species whose G1 and G10 generations were the most distinguishable was *P. regina*, with a classification accuracy of 82.2% and a ROC area under the curve (AUC) of 0.88. Therefore, the effects of genetic homogenization were the most significant for this species. The species whose G1 and G10 generations were the least distinguishable was *L. sericata*, with a classification accuracy of 58.9% and ROC AUC of 0.65. Therefore, the effects of genetic homogenization were the least significant for this species.

Table 5. Classification accuracy and area under the curve (AUC) for the G1 and G10 pupae of each species.

		G10 Generation			
	From/to	*C. macellaria*	*L. cuprina*	*L. sericata*	*P. regina*
G1 Generation	*C. macellaria*	68.9% (0.832)			
	L. cuprina		73.5% (0.839)		
	L. sericata			58.9% (0.648)	
	P. regina				82.2% (0.883)

DA was also performed by treating samples of G1 pupae as unknowns and classifying them per the DA model constructed from G10 pupae. This models a situation where a laboratory would have a database of lipid profiles taken from flies that could reproduce and inbreed for 10 generations. That same laboratory would then take case samples, analyze them immediately, and classify them per this database. If genetic differences are irrelevant, then the G1 "casework" pupae would be expected to be grouped per their species with near 100% accuracy, as was seen for the G10 "database" generation provided in Table 4. Poor classification accuracy, conversely, would indicate that the model is indeed betrayed by genetic differences. The results are shown in Table 6.

Table 6. Leave-one-out confusion matrix for the G1 generation chemical profiles.

		G10 Generation				Total	% Correct
	From/To	*C. macellaria*	*L. cuprina*	*L. sericata*	*P. regina*		
G1 Generation	*C. macellaria*	47	0	0	6	53	88.7
	L. cuprina	2	49	2	1	54	90.7
	L. sericata	0	0	52	2	54	93.0
	P. regina	7	0	2	27	36	75.0

Although the confusion matrix shows lower classification accuracies across the board (when compared to Table 4), the accuracies for *L. cuprina* and *L. sericata* are excellent (>90%) and the accuracy for *C. macellaria* is very good (89%), despite the mismatch between the training set (G10 samples) and the test set (G1 samples). It is also interesting to see that the species whose G1 and G10 generations were reliably differentiated in Table 5 (i.e., *P. regina*), and therefore, exhibited the most significant chemical changes, had low classification accuracies (75%) in this example. Conversely, a species whose G1 and G10 generations were not differentiated in Table 5 (e.g., *L. sericata*), and therefore, exhibited the least significant chemical changes, had high classification accuracy (93%). It should be noted that no *L. sericata* were classified as *L. cuprina*, and 2 of 54 (7%) of *L. cuprina* were classified as *L. sericata*. This is important as both species belong to the same genus and share a close evolutionary history.

Based on these results, it appears that G10 pupae may be a suitable model for G1 pupae, if the effect of genetic homogenization is not significant. For species such as *P. regina*, genetic homogenization does indeed alter the chemical profile in a statistically-significant way. Future researchers seeking to extrapolate useful information about G1 and wild-type populations based on data gained from advanced generations are cautioned to bear this in mind.

3.3. Effects of Abiotic Variables (P. regina)

DA was carried out and confusion matrices were generated for the data from the studies of temperature, humidity, substrate, and diet. Although overall classification accuracies of 60% or more were achieved for all four experiments, in most cases the actual differences observed among the classes were marginal. For example, almost no difference in the lipid profile was manifest among specimens allowed to pupate in sawdust, in sand, or in vermiculite (data not shown). Classification accuracies were therefore poor, with a third of samples misclassified in the cases of pupae reared in sawdust and sand, and half of all samples in the case of vermiculite. In these cases, only the data in the experiment (for example, the pupation substrate) were used, therefore, it appeared that it did not matter what the pupation substrate was, there were no distinct classes formed.

Results for pupae reared under different humidity conditions were almost as poor (data not shown), with discrimination only achieved for the sample set reared under normal 55% humidity conditions. This is perhaps a surprising result, as the cuticular hydrocarbons of insects are known to play a part in resistance to desiccation, and might be expected to vary in response to adverse environmental conditions such as low humidity. It should be noted that although discrimination could not be achieved in these cases, this is, in fact, beneficial. In the context of insect specimens recovered from a crime scene, it is to the advantage of an investigator if the chemical profiles do not vary significantly with temperature, humidity, and other uncontrollable factors. A reliable method for age and species determination that is not undermined or biased by biotic and abiotic variables is the gold standard for any analyst. Therefore, the lack of discrimination is, in this case, the desired result.

Temperature appeared to have more of an impact on the lipid profiles, although its influence was still relatively small, as attested by the poor classification accuracies of 41.7% and 62.3% for pupae reared at 15 °C and 20 °C, respectively (data not shown). More substantial chemical differences were observed at 25 °C and 30 °C. In general, lauric acid, myristic acid, pentadecanoic acid, palmitic acid, and alkane G increased with temperature.

Diet was the single-largest contributing factor to differences in the lipid profile (Figure 5). Classification accuracies (Table 7) were 100% for pupae reared on chicken liver and 87.5% for pupae reared on feeder rats. The statistical model was unsuccessful at distinguishing between pupae raised on high-fat ground beef and pupae fed on low-fat ground beef. Neglecting that minor distinction, pupae raised on ground beef were still correctly categorized 100.00% of the time. The following trends are clear:

- Pupae reared on chicken liver have lipid profiles with higher quantities of palimitic acid, palimitoleic acid, unknown alkane B, and cholesterol.
- Pupae reared on ground beef have lipid profiles with higher quantities of lauric acid, myristic acid, palmitelaidic acid, linolenic acid, and stearic acid.
- All other compounds appear in greatest abundance in feeder rats.

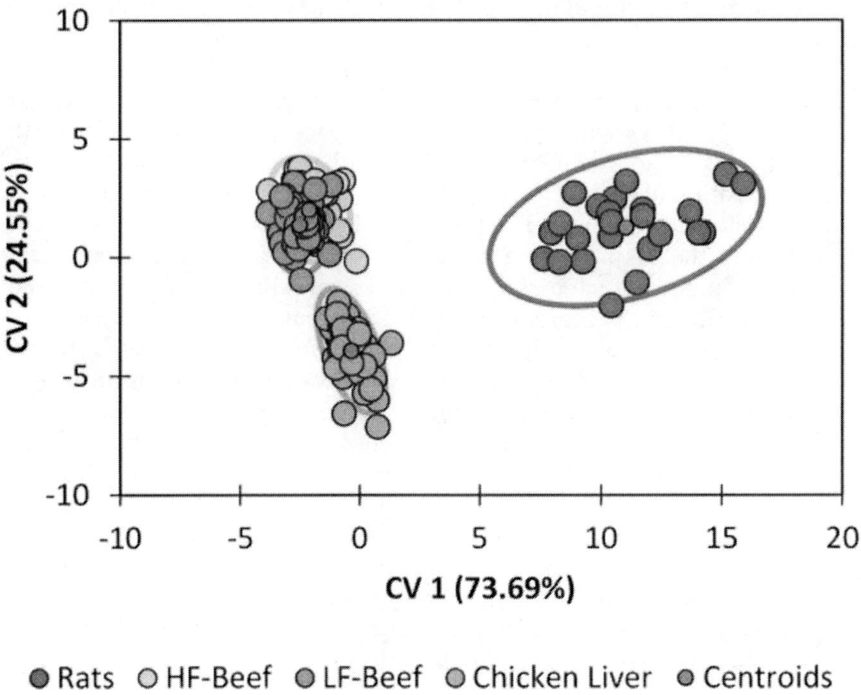

Figure 5. Discriminant analysis of *P. regina* pupae reared on various diets.

Table 7. Leave-one-out confusion matrix for *P. regina* pupae reared on larval diets.

From/To	Rats	HF-Beef	LF-Beef	Chicken Liver	Total	% Correct
Rats	21	2	0	1	24	87.5
HF-Beef	0	36	17	0	53	67.9
LF-Beef	0	13	41	0	54	75.9
Chicken Liver	0	0	0	53	53	100

The most meaningful takeaway from this series of experiments is that, which the exception of diet, changes in any given abiotic factors did result in reliably differentiable classes for *P. regina*. However, each factor had a larger capacity to obfuscate accurate species determinations when comparing to a population of *P. regina* reared under normal conditions. When each chemical profile under every single differing abiotic condition was validated using discriminant analysis with the four species (G10) as the training set, most species classifications were less than 50% correct (Table 8). However, when comparing the control conditions (these conditions are identical to the G10 conditions) with expected classification accuracies of >82% (see Table 5), only 37.7% classification accuracy was recovered. This is indicative that conditions changed between the G10 rearing conditions and that of the abiotic factor experiments conducted sometime later. Nevertheless, it should be noted that classification accuracy using discriminant analysis is only as good as the similarity between the training and validation data sets. In this case, the data sets differ, thus obscuring an accurate species identification. Our data indicate that diet has the greatest impact (with classification accuracies less than 2%), whereas temperature and pupation substrate appear to be most like the control profiles (Table 8). This is not surprising; given that pupae produce lipids in their cuticles for the protection against desiccation and predation, temperature and pupation substrate (our substrates were all dry) should not have much of an effect. However, humidity (different lipids/lipid ratios would be needed to protect the pupa), and diet (the lipids would be obtained in different ratios) would significantly influence the profiles.

Table 8. Leave-one-out confusion matrix for *P. regina* larvae reared under different abiotic conditions and to which species they assign. The control abiotic conditions were 25 °C, 55% r.h., chicken liver and sawdust—the same conditions as the G10 generation. Only the abiotic condition in column one changed, other conditions remained as in the control condition.

From/To	*C. macellaria*	*L. cuprina*	*L. sericata*	*P. regina*	Total	% Correct
Rats	0	3	21	0	24	0
HF-Beef	2	50	0	1	53	1.9
LF-Beef	2	51	0	1	54	1.9
15 °C	0	8	10	6	24	25.0
20 °C	4	8	18	23	53	43.4
30 °C	20	5	12	16	53	30.2
40% r.h.	0	9	21	12	42	28.6
70% r.h.	3	26	19	4	52	7.7
85% r.h.	0	15	25	13	53	24.5
Sand	2	8	18	26	54	48.1
Vermiculite	0	10	11	23	44	52.3
CONTROL	0	6	27	20	53	37.7

4. CONCLUSIONS

This study aimed to better understand how biotic (reduced genetic variation) and abiotic factors (temperature, humidity, pupation substrate, and diet) influence the uses of chemical analyses for species identification of forensically important pupae. Considerable chemical differences were documented among *Cochliomyia macellaria*, *Lucilia cuprina*, *Lucilia sericata*, and *Phormia regina* as a function of species and genetics. Classification accuracies of 89–94% were obtained for the G1 generation, and classification accuracies of 93%–98% were obtained for the G10 generation. The compounds most correlated and anti-correlated with species were identified. Data from the G10 generations were shown to be a suitable model for G1 pupae, although performance on *P. regina* suffered, with classification accuracy falling to 75%. However, genetic homogenization was found to have an impact on the observed chemical profiles, which should serve as a cautionary note for future researchers hoping to extract data with real world relevance from colonies at advanced generations. Temperature and pupation substrate were found to have minimal impact on the lipid profiles, whereas small chemical changes were observed with changes in humidity, and significant changes were observed with variations in diet.

If comparing the classification accuracies of the different abiotic conditions relative to the control, it is immediately clear that diet plays the most important role in the composition of lipid profiles in developing pupae (Table 8). Therefore, in "real-world" scenarios where pupae are recovered from a scene but their diet is unknown, we suspect that using chemical profiles to determine species is not ideal, and thus more research is necessary. Although we used a single classification method here (discriminant analysis), other methods such as (Soft Independent Modeling of Class Analogy (SIMCA), Neural networks or Bayesian classifiers, etc.) could change the classification accuracies presented here. However, the effect of diet is a biochemical process and not one that could be easily corrected for via statistical methods. It is important that additional data be collected to establish a greater understanding of the changes in chemical profiles over time under different conditions to establish a reliable database for species identification purposes.

ACKNOWLEDGMENTS

We would like to acknowledge Hillary Veron (IUPUI) for her tireless efforts in colony maintenance, and with aiding with sample collection. We would like to acknowledge Jeffrey D. Wells (Florida International University) for providing entomological specimens for the purposes of establishing colonies. Supported was provided by the National Institute of Justice grant 2013-DN-BX-K019. Views represented here do not reflect views or opinions of the National Institute of Justice.

AUTHOR CONTRIBUTIONS

Christine J. Picard and John V. Goodpaster conceived and designed the experiments. William Kranz, Clinton Carroll and Darren A. Dixon performed the experiments. John V. Goodpaster and William Kranz analyzed the data. Clinton Carroll and Darren A. Dixon aided in sample collection. William Kranz, John V. Goodpaster and Christine J. Picard wrote the paper.

CONFLICTS OF INTEREST

The authors declare no conflict of interest.

REFERENCES

1. Byrd, J.H.; Castner, J.L. Forensic Entomology: The Utility of Arthropods in Legal Investigations; CRC Press: Boca Raton, FL, USA, 2010.
2. Greenberg, B. Flies as forensic indicators. J. Med. Entomol. 1991, 28, 565–577.
3. Greenberg, B.; Kunich, J.C. Entomology and the Law. Flies as Forensic Indicators; Cambridge University Press: Cambridge, UK, 2002.
4. Byrd, J.H.; Allen, J.C. The development of the black blow fly, Phormia regina (Meigen). Forensic Sci. Int. 2001, 120, 79–88.
5. Byrd, J.H.; Butler, J.F. Effects of temperature on Cochliomyia macellaria (Diptera: Calliphoridae) development. J. Med. Entomol. 1996, 33, 901–905.
6. Byrd, J.H.; Butler, J.F. Effects of temperature on Chrysomya rufifacies (Diptera: Calliphoridae) development. J. Med. Entomol. 1997, 34, 353–358.
7. Boatright, S.A.; Tomberlin, J.K. Effects of Temperature and Tissue Type on the Development of Cochliomyia macellaria (Diptera: Calliphoridae). J. Med. Entomol. 2010, 47, 917–923.
8. Kamal, A.S. Comparative study of thirteen species if sarcosaprophagous Calliphoridae and Sarcophagidae (Diptera). Annu. Rev. Entomol. Soci. 1958, 51, 261–270.
9. Anderson, G.S. Minimum and maximum development rates of some forensically important Calliphoridae (Diptera). J. Forensic Sci. 2000, 45, 824–832.
10. Whitworth, T. Keys to the genera and species of blow flies (Diptera: Calliphoridae) of America North of Mexico. Proc. Entomol. Soc. Wash. 2006, 108, 689–725.
11. Sperling, F.A.; Anderson, G.S.; Hickey, D.A. A DNA-based approach to the identification of insect species used for postmortem interval estimation. J. Forensic Sci. 1994, 39, 418–427.
12. DeBry, R.W.; Timm, A.E.; Dahlem, G.A.; Stamper, T. mtDNA-based identification of Lucilia cuprina (Wiedemann) and Lucilia sericata (Meigen) (Diptera: Calliphoridae) in the continental United States. Forensic Sci. Int. 2010, 202, 10210–10219.
13. Boehme, P.; Amendt, J.; Disney, R.H.L.; Zehner, R. Molecular identification of carrion-breeding scuttle flies (Diptera: Phoridae) using COI barcodes. Int. J. Leg. Med. 2010, 124, 577–581.
14. Caine, L.M.; Real, F.C.; Salona-Bordas, M.I.; De Pancorbo, M.M.; Lima, G.; Magalhaes, T.; Pinheiro, F. DNA typing of diptera collected from human corpses in Portugal. Forensic Sci. Int. 2009, 184, e21–e23.
15. Guo, Y.D.; Cai, J.F.; Wang, X.H.; Lan, L.M.; Liu, Q.L.; Li, X.A.; Chang, Y.F.; Ming, Z.; Wang, X.; Wen, J.F. Identification of forensically important sarcophagid flies (Diptera: Sarcophagidae) based on COI gene in China. Rom. J. Leg. Med.2010, 18, 217–224.
16. Malewski, T.; Draber-Monka, A.; Pomorski, J.; Los, M.; Bogdanowicz, W. Identification of forensically important blowfly species (Diptera: Calliphoridae) by high-resolution melting PCR analysis. Int. J. Leg. Med. 2010, 124, 277–285.
17. Nelson, L.A.; Wallman, J.F.; Dowton, M. Identification of forensically important Chrysomya (Diptera: Calliphoridae) species using the second ribosomal internal transcribed spacer (ITS2). Forensic Sci. Int. 2008, 177, 238–247.
18. Wells, J.D.; Wall, R.; Stevens, J.R. Phylogenetic analysis of forensically important Lucilia flies based on cytochrome oxidase I sequence: A cautionary tale for forensic species determination. Int. J. Leg. Med. 2007, 121, 229–233.
19. Tarone, A.M.; Foran, D.R. Gene expression during blow fly development: Improving the precision of age estimates in forensic entomology. J. Forensic Sci. 2011, 56, S112–S122.
20. Zehner, R.; Amendt, J.; Boehme, P. Gene expression analysis as a tool for age estimation of blowfly pupae. Forensic Sci. Int. Genet. Suppl. Ser. 2009, 2, 292–293.

21. Davies, K.; Harvey, M.L. Internal morphological analysis for age estimation of blow fly pupae (Diptera: Calliphoridae) in postmortem interval estimation. J. Forensic Sci. 2013, 58, 79–84.
22. Brown, K.; Harvey, M. Optical coherence tomography: Age estimation of Calliphora vicina pupae in vivo? Forensic Sci. Int. 2014, 242, 157–161.
23. Brown, K.; Thorne, A.; Harvey, M. Calliphora vicina (Diptera: Calliphoridae) pupae: A timeline of external morphological development and a new age and PMI estimation tool. Int. J. Leg. Med. 2015, 129, 835–850.
24. Golebiowski, M. Comparison of free fatty acids composition of cuticular lipids of Calliphora vicina Larvae and Pupae. Lipids 2012, 47, 1001–1009.
25. Howard, R.W. Cuticular Hydrocarbons and Chemical Communication. In Insect Lipids: Chemistry, Biochemistry, and Biology; Stanley, D.W., Nelson, D.R., Eds.; University of Nebraska Press: Lincoln, NE, USA, 1993; pp. 179–226.
26. Zhu, G.H.; Ye, G.Y.; Hu, C.; Xu, X.H.; Li, K. Development changes of cuticular hydrocarbons in Chrysomya rufifacies larvae: Potential for determining larval age. Med. Vet. Entomol. 2006, 20, 438–444.
27. Toolson, E.C.; Kupersimbron, R. Laboratory evolution of epicuticular hydrocarbon composition and cuticular permeability in Drosophila pseudoobscura—Effects on sexual dimorphism and thermal acclimation ability. Evolution1989, 43, 468–473.
28. Ferveur, J.F. Cuticular hydrocarbons: Their evolution and roles in Drosophila pheromonal communication. Behav. Genet. 2005, 35, 279–295.
29. Gilby, A.R. Cuticle and Insecticides. In Biology of the Integument; Bereiter-Hahn, J., Matoltsy, A.G., Richards, K.S., Eds.; Springer: Berlin/Heidelberg, Germany, 1984; pp. 694–702.
30. Ginzel, M.D.; Blomquist, G.J.; Millar, J.G.; Hanks, L.M. Role of contact pheromones in mate recognition in Xylotrechus colonus. J. Chem. Ecol. 2003, 29, 533–545.
31. Wagner, D.; Brown, M.J.; Broun, P.; Cuevas, W.; Moses, L.E.; Chao, D.L.; Gordon, D.M. Task-related differences in the cuticular hydrocarbon composition of harvester ants, Pogonomyrmex barbatus. J. Chem. Ecol. 1998, 24, 2021–2037.
32. Monnin, T.; Malosse, C.; Peeters, C. Solid-phase microextraction and cuticular hydrocarbon differences related to reproductive activity in queenless ant Dinoponera quadriceps. J. Chem. Ecol. 1998, 24, 473–490.
33. Farine, J.P.; Ferveur, J.F.; Everaerts, C. Volatile drosophila cuticular pheromones are affected by social but not sexual experience. PLoS ONE 2012, 7, e40396.
34. Golebiowski, M.; Urbanek, A.; Oleszczak, A.; Dawgul, M.; Kamysz, W.; Boguś, M.I.; Stepnowski, P. The antifungal activity of fatty acids of all stages of Sarcophaga carnaria L. (Diptera: Sarcophagidae). Microbiol. Res. 2014, 169, 279–286.
35. Golebiowski, M.; Cerkowniak, M.; Bogus, M.I.; Wloka, E.; Przybysz, E.; Stepnowski, P. Developmental Changes in the sterol composition and the glycerol content of cuticular and internal lipids of three species of flies. Chem. Biodivers. 2013, 10, 1521–1530.
36. Ingleby, F.C.; Hosken, D.J.; Flowers, K.; Hawkes, M.F.; Lane, S.M.; Rapkin, J.; Dworkin, I.; Hunt, J. Genotype-by-environment interactions for cuticular hydrocarbon expression in Drosophila simulans. J. Evol. Biol. 2013, 26, 94–107.
37. Huberman, L.; Gollop, N.; Mumcuoglu, K.Y.; Breuer, E.; Bhusare, S.R.; Shai, Y. Antibacterial substances of low molecular weight isolated from the blowfly, Lucilia sericata. Med. Vet. Entomol. 2007, 21, 127–131.
38. Fedina, T.Y.; Kuo, T.H.; Dreisewerd, K.; Dierick, H.A.; Yew, J.Y.; Pletcher, S.D. Dietary effects on cuticular hydrocarbons and sexual attractiveness in Drosophila. PLoS ONE 2012, 7, e49799.
39. Frederickx, C.; Dekeirsschieter, J.; Brostaux, Y.; Wathelet, J.P.; Verheggen, F.J.; Haubruge, E. Volatile organic compounds released by blowfly larvae and pupae: New perspectives in forensic entomology. Forensic Sci. Int. 2012.
40. Moore, H.E.; Adam, C.D.; Drijfhout, F.P. Potential use of hydrocarbons for aging Lucilia sericata blowfly larvae to establish postmortem interval. J. Forensic Sci. 2013, 58, 404–412.

41. Goodrich, B.S. Cuticular lipids of adults and puparia of the Australian sheep blowfly Lucilia cuprina (Wied.). J. Lipid Res. 1970, 11, 1–6.
42. Zhu, G.H.; Xu, X.H.; Yu, X.J.; Zhang, Y.; Wang, J.F. Puparial case hydrocarbons of Chrysomya megacephala as an indicator of the postmortem interval. Forensic Sci. Int. 2007, 169, 1–5.
43. Rainey, C.L.; Bors, D.E.; Goodpaster, J.V. Design and optimization of a total vaporization technique coupled to solid-phase microextraction. Anal. Chem. 2014, 86, 11319–11325.

CHAPTER 15

New Psychoactive Substance A-PVP in a Traffic Accident Case

Sebastian Rojek[1] • Karol Kula[1] • Martyna Macio´w-Gła˛b[1] • Małgorzata Kłys[1]

[1].Department of Forensic MedicineJagiellonian University Medical CollegeKrakówPoland

ABSTRACT

The problems of new psychoactive substances (NPSs), especially related to drivers, constitute an open research area. In this case report, we present a traffic accident case, in which two passengers of five individuals died instantly, while the other three persons survived the accident with minor injuries only. From the blood samples of the driver and the passengers, α-pyrrolidinovalerophenone (α-PVP), an NPS belonging to the category of cathinone derivatives, was disclosed. Therefore, we established a detailed procedure for analysis of α-PVP in blood samples by liquid chromatography–tandem mass spectrometry. After careful validation tests of this method, α-PVP concentration in blood samples from the surviving driver and passengers, and from the two deceased, were measured. The concentrations varied from 20 to 650 ng/mL. Access to detailed information originating from the court files and from explanations provided by the driver and eye witnesses revealed extremely valuable illustrative details addressing the symptoms and pharmacological effects of α-PVP on the human organism, thus contributing to enriching the body of knowledge of α-PVP abuse.

Keywords: Traffic accident New psychoactive substances (NPSs) α-pyrrolidinovalerophenone (α-PVP) Psychological symptoms DRUID

INTRODUCTION

The issues associated with driving under the influence of alcohol and narcotic substances are subject to continuous interest not only for legal practitioners, but also for the police, prosecution, and medico-legal teams. Among initiatives implemented in this field, there is the European Union DRUID program carried out by 19 EU countries in the years 2006–2011, aiming at evaluating the phenomena. The final report claimed that although the gravest danger in traffic in EU member countries was posed by alcohol consumption, especially when combined with medications, considerable problems also arose by other psychoactive substances, such as amphetamines and cannabis that are the most popular [1].

Practice shows that preliminary testing of a driver by police road patrol to detect the presence of alcohol and classic narcotic substances often yields a negative result; the driver may be under the influence of other psychoactive agent(s), the presence of which is not detected by most of testers. This may cause the driver to develop a sense of impunity and thus leads to increased chance of traffic accidents. Inasmuch as the issue of alcohol and classic narcotic substances in relation to traffic has been legally regulated in keeping with legislative directives of particular European countries, the problem of new psychoactive substances (NPSs) as related to drivers constitutes an open research area. Although research papers addressing this subject have been published [2, 3, 4, 5, 6], the problem continues to be poorly investigated.

In this case report, we present a traffic accident case in which two passengers were killed, and the driver and two passengers survived; from the blood samples of all individuals, variable concentrations of α-pyrrolidinovalerophenone (α-PVP), an NPS belonging to the category of cathinone derivatives, were detected. Access to detailed information originating from the court files and derived from explanations provided by the driver and the eye witnesses revealed extremely valuable illustrative material addressing the effect of α-PVP on the human organism, thus contributing to enriching the body of knowledge of the NPS.

CASE HISTORY

According to the court files, in a small town in Poland, a group of young people was driving home from a nearby discotheque in early morning hours. Approximately at 6 a.m. the driver, exercising no special caution, drove onto the curb of a safety island, lost control of the car, and drove into a ditch, where the car overturned and hit a tree. In consequence of the accident, two passengers, a 17-year-old female and a 28-year-old male, were killed instantly; one of the remaining passengers, a 19-year-old female, suffered a minor spine injury, while another passenger, a 36-year-old male, and the 21-year-old driver succumbed to minor injuries. Immediately upon arrival of the police, the surviving individuals involved in the accident, the driver and the passenger sitting in front were tested for state of intoxication by means of breath analyzers. The wounded female passenger was tested upon arrival at a hospital ~1 h later, while the deceased individuals were tested by means of postmortem blood analysis. After the accident, blood samples were collected and analyzed for psychoactive substances, with the exception of the injured 19-year-old female, whose blood was tested for ethyl alcohol upon admission. As follows from the court files, all people involved in the accident except the driver were under the influence of alcohol, which was later confirmed by tests.

The sensations felt by the driver following his taking the narcotic substance were described on his allegation. The psychoactive substance was insufflated through the nose by him and his friends. They had a form of crystals called FETA, KATAR, or KRYSTYNA. The persons attending the party had taken the first fix of the psychoactive substance ~6 h prior to the accident. The driver informed that the fix of the substance in an amount of ~0.4 g "was good" for about 6 h. One person had prepared the so-called toot, which was then divided and insufflated through the nose. The doses were not calculated, but rather applied using a "hit-and-miss" method. Having insufflated the first fix of the substance, the driver felt agitated, started to sweat profusely, and experienced palpitations. He exploded with energy in spite of the fact that he was tired, having previously performed physical labor in the forest. He had not taken any alcohol. Together with his friends, he took another fix 4 h later, i.e., ~2 h prior to the accident. Afterwards, he was agitated, no longer sleepy, and did not experience palpitations or sweating; he was only thirsty, but generally felt well and full of inner energy. Immediately before the accident, however, he recounted: "...suddenly I saw spots before my eyes...for a moment I felt as if I had lost consciousness, when I snapped out, there was a safety island in front of me...then I jerked the steering wheel to the right and to the left.... I put the car in neutral...I pressed the gas and clutch...I did not use brakes...I could not control the car...". Then everything went very fast. The car overturned. The passengers sitting in the backseat occupying side seats, without their seat-belts fastened, died instantly, the remaining individuals sustained minor injuries.

It was obvious that the driver and the fault that he committed would be burdened by the highest degree of guilt for causing the accident;

nevertheless, according to the information available, the other individuals present in the car and involved in the accident might contribute to its happening indirectly. At the discotheque, through various forms of pressure, the driver was forced by the other participants to drive, because he had not drunk alcohol. However, he did not have a driving license. During the drive, the passengers were agitated, and they shouted to pressure the driver into going faster.

Addressing a question to experts whether the presence of α-PVP at the level determined could have affected the psychomotor abilities of the driver leading to the miserable accident, the public prosecutor's office expected a justified and positive answer.

It should be added that the two passengers sitting in the backseat died at the place of the event. According to the available data, the cause of death of both individuals was severe multiorgan damage; moreover, the 17-year-old female sustained a spinal fracture, and the 28-year-old male had a fracture of the skull base.

MATERIALS AND METHODS

Biological materials

Sample of peripheral blood was collected from the driver and from two living passengers after accident. Autopsy blood was also taken from two dead victims during autopsy carried out within 24 h after death. The samples were kept frozen (−20 °C) until analyses. Blank blood samples for development and validation of analytical method were taken from unpoisoned subjects and from the Regional Centre of Blood Donation and Blood Treatment in Kraków.

Standards and chemicals

Standard solutions of 1-(4-methoxyphenyl) piperazine (4-MeOPP), 1-(*m*-trifluoromethylphenyl) piperazine (TFMPP), 1-benzylpiperazine (BZP), 1-methyl-4-(phenylmethyl)-piperazine (MPMP), 2,5-dimethoxy-4-methy lamphetamine (DOM), 2-methylmethcathinone (2-MMC), 3,4-dimethy lmethcatinone (3,4-DMMC), 3,4-methylenedioxyamphetamine (MDA), 3,4-methylenedioxyethylamphetamine (MDEA), 3,4-methylenedioxymethyl-amphetamine (MDMA), 3,4-methylenedioxypyrovalerone (MDPV), 3,4-methylenedioxy-α-pyrrolidinobutiophenone (MDPBP), 3-methylmethcathinone (3-MMC), 4-bromo-2,5-dimethoxyamphetamine (DOB), 4-bromo-2,5-dimethoxyphenethylamine (2C-B), 4-

chloromethcathinone, 4-fluoroamphetamine, 4-hydroxymethamphetamine, 4-methoxyamphetamine (PMA), 4-methoxymethamphetamine (PMMA), 4′-methoxy-α-pyrrolidinopropiophenone (MeOPPP), 4-methylethcathinone (4-MEC), 4-methylmethcathinone (4-MMC, mephedrone), 5-(2-aminopropyl) benzofuran (5-APB), 5-(2-ethylaminopropyl) benzofuran (5-EAPB), 5-(2-methylaminopropyl) benzofuran (5-MAPB), 6-(2-aminopropyl) benzofuran (6-APB), amphetamine, brephedrone, buphedrone, butylone, caffeine, cathinone, ephedrine, ephedrone, ethylphenidate, flephedrone, methamphetamine, methcathinone, methyl- benzylpiperazine, methylone, methylphenidate, *N,N*-dimethylamylamine, naphyrone, *N*-ethylbuphedrone (NEB), *N*-ethylcathinone, *N*-methyl-1,3-benzodioxolylbutanamine (MBDB), pentedrone, phentermine, phenylpropanolamine (norepherdrine), pseudoephedrine, α-PVP, and α-PVP-d_8 used as internal standard (IS), were purchased from LGC Standards (Łomianki, Poland); ammonium formate, ammonium carbonate, formic acid, and acetic acid (all ≥99.9 % for LC-MS) from Sigma (Poznań, Poland); acetonitrile, methanol (all ≥99.9 % for LC–MS) and solid-phase extraction (SPE) columns LiChrolut RP-18 containing 500 mg C18-RP bonded silica from Merck (Warszawa, Poland).

Extraction procedure

Each 1-mL sample was put into a clean 15-mL tube and diluted with 0.01 M carbonate buffer to pH 9.3 (1:5, v/v). α-PVP-d_8 at a concentration of 0.1 mg/L was added to each sample. Then, the samples were vortexed and centrifuged for 7.5 min at 4400×*g*.

The SPE cartridge was firstly equilibrated with 1 mL each of methanol, distilled water, and the above carbonate buffer pH 9.3, and then the supernatant of each blood sample was applied on the cartridge and slowly passed through it. In the next step, matrix impurities were cleaned with 2 mL of the carbonate buffer pH 9.3, and then vacuum was applied to each cartridge for 30 min to remove residual moisture. The analytes were eluted with 2 mL of the 1 M acetic acid in methanol (1:9, v/v). The eluate was evaporated to dryness under a gentle nitrogen stream at 40 °C, and the residue was dissolved in 0.1 mL of a mixture of mobile phases (95 % phase A + 5 % phase B) and injected into a liquid chromatography–tandem mass spectrometry (LC–MS-MS) system.

The calibrators for quantitative analysis were prepared by spiking α-PVP (0.010, 0.025, 0.050, 0.250, and 0.500 mg/L) into the drug-free blood prior to extraction.

Instrumentation

Liquid chromatography

An Agilent Technologies 1200 liquid chromatograph (Santa Clara, CA, USA) equipped with a binary pump (G1312 A) and an autosampler (G1329 A) was used. A 0.01 mL injection loop was used in the gradient mode. The chromatographic separation was performed with a Poroshell 120 EC-C18 column (100 × 3.0 mm id, particle size 2.7 μm, Agilent Technologies) and a Poroshell 120 EC-C18 guard column (5 × 3.0 mm id, Agilent Technologies). The phase A was water, which contained 0.2 % formic acid and 0.002 M of ammonium formate, and the phase B was acetonitrile with 0.2 % formic acid and 0.002 M of ammonium formate.

The gradient for samples concerning α-PVP and IS was programmed as follows: 95 %A/5 %B at a flow rate of 0.5 mL/min, followed by a linear change to 5 %A/95 %B at a flow rate of 1 mL/min in 10 min and held for 5 min.

Tandem mass spectrometry

A 6410 triple quadrupole mass spectrometer (Agilent Technologies) equipped with an electrospray ionization (ESI) source, operated under the positive mode was used. The operational parameters of the ESI source were as follows: vaporizing temperature, 350 ºC; pressure of the nebulising gas, 40 psi; flow of the drying gas, 9 L/min; capillary potential, 3.5 kV; fragmentor voltages, 84 V for α-PVP and 99 V for α-PVP-d_8. The mass spectrometer was operated in the multiple reaction monitoring (MRM) mode with the transitions at m/z 232.1/91.1 (quantification) and 232.1/77.1 (qualification) for α-PVP and at m/z 240.3/91.1 and 240.3/77.1 for α-PVP-d_8, respectively. The optimized collision-induced dissociation (CID) energies were 25 eV for transitions at m/z 232.1/91.1 (α-PVP) and at m/z 240.3/91.1 (α-PVP-d_8), and 53 eV for transitions at m/z 232.1/77.1 (α-PVP) and at m/z 240.3/77.1 (α-PVP-d_8).

Validation of the LC–ESI-MS-MS method

Selectivity

To evaluate peak-purity and selectivity, blank blood samples (neither analyte nor IS added) were analyzed with each batch to check for impurity peaks that might interfere with detection of α-PVP or IS. To assess possible interferences of α-PVP, quality control (QC) samples were spiked individually to contain 10 mg/L each of 4-MeOPP, TFMPP, BZP, MPMP, DOM, 2-MMC, 3,4-DMMC, MDA, MDEA, MDMA, MDPV, MDPBP, 3-MMC, DOB, 2C-B, 4-chloromethcathinone, 4-fluoroamphetamine, 4-hydroxymethamphetamine, PMA, PMMA, MeOPPP, 4-MEC, 4-MMC, 5-APB, 5-EAPB, 5-MAPB, 6-APB, amphetamine, brephedrone, buphedrone, butylone, caffeine, cathinone, ephedrine, ephedrone, ethylphenidate, flephedrone, methamphetamine, methcathinone, methylbenzylpiperazine, methylone, methylphenidate, *N,N*-dimethylamylamine, naphyrone, NEB, *N*-ethylcathinone, MBDB, pentedrone, phentermine, phenylpropanolamine (norepherdrine), and pseudoephedrine.

Linearity and limits of quantification and detection

Calibration curves were constructed after the analysis of drug-free blood sample containing known amounts of α-PVP. To prepare these standards, blood samples were spiked with α-PVP at the following concentrations: 0.010, 0.025, 0.050, 0.250, and 0.500 mg/L. Each level was prepared in triplicate. The samples were extracted according to the procedure described before. Calibration curves were constructed by plotting the peak-area ratios (y) of the analyte (α-PVP)/(α-PVP-d_8) against α-PVP concentration (x).

Validation samples were prepared in triplicate at the following concentrations: 0.020, 0.100, 0.500 mg/L of α-PVP to assess the accuracy of the method. Negative QC samples were analyzed after each linearity sample to evaluate potential carry-over. The limit of detection (LOD) of the method was determined by analyzing validation samples ($n = 5$) to determine if acceptance criteria were met for the analyte. The LOD was defined as the lowest concentration at which the analyte ion signal-to-noise ratio (determined by peak area) was ≥10. The limit of quantification (LOQ) was defined as the lowest concentration that met LOD criteria and also had a standard deviation within ±20 %.

Accuracy and precision

Inter- and intra-assay accuracy and precision data for α-PVP were determined with the low (0.020 mg/L), medium (0.100 mg/L), and high (0.500 mg/L) QC samples. Intra-assay data were assessed by comparing data from within one run ($n = 5$) and inter-assay data were determined with five separate runs ($n = 15$). Data were evaluated using one-way analysis of variance with day as the grouping variable. Accuracy, expressed as a percentage, was calculated by dividing the mean calculated concentration by the spiked concentrations, and multiplying by 100. Precision, expressed as percent relative standard deviation (%RSD), was determined by calculating the percent ratio of the standard deviation divided by the calculated mean concentration and multiplied by 100.

Matrix effect, extraction efficiency, and process efficiency

Extraction efficiency, matrix effect and process efficiency were evaluated via three sets of samples as described by Matuszewski [7]. Extraction efficiencies for α-PVP and IS were measured at each QC concentration. Blank blood was fortified with α-PVP and IS solution before and after SPE. Percent extraction efficiency from blood was expressed as mean analyte area of QC samples ($n = 5$) fortified with references standard solution before extraction divided by mean area of samples ($n = 5$) with references standard solution added after SPE. The matrix effect (relative) was calculated by comparing analyte peak areas obtained from six different blank blood samples fortified with α-PVP and IS solutions after SPE with peak areas of references standards solution at the same nominal concentrations prepared in a 95:5 (v/v) mixture of mobile phases A and B (neat samples). Matrix suppression or enhancement (absolute matrix effect) was calculated as follows: (mean peak area of fortified blood after SPE/mean peak area of the neat sample) × 100 – 100. The process efficiency shows the overall effect of SPE extraction efficiency and matrix effect on the quantification of analytes of interest. This was determined by comparing the mean analyte peak area of five samples fortified before SPE with the mean peak area of five neat samples prepared in the mobile phase.

Stability

Processed sample stability was investigated by reinjecting low, medium, and high QC extracted samples stored at 4 °C for 24, 48, and 72 h on the autosampler ($n = 3$) and at room temperature for 72 h and by calculating the results against the original calibration curve as a percent differences.

RESULTS

In the present study, we experienced five people who were involved in a traffic accident; among them, two were killed and three were injured, but survived the accident. It was disclosed that all of them had taken the NPS α-PVP. To accurately quantify α-PVP in their blood samples, we have established a detailed procedure for its analysis by LC–ESI-MS-MS together with the validation of the method. First of all, we tested the selectivity of this method using more than 50 of psychoactive compounds, but none of them interfered with the peck of α-PVP or IS (Figure 1). The linearity of the calibration curve was confirmed in the range of 0.010–0.500 mg/L for α-PVP in blood samples; the regression equation and correlation coefficient r were : $y = 0.0062x - 0.047$, and $r = 0.9994$, respectively. The LOD and LOQ were 0.005 and 0.010 mg/L, respectively. Accuracy and precision data were generally satisfactory (Table 1).

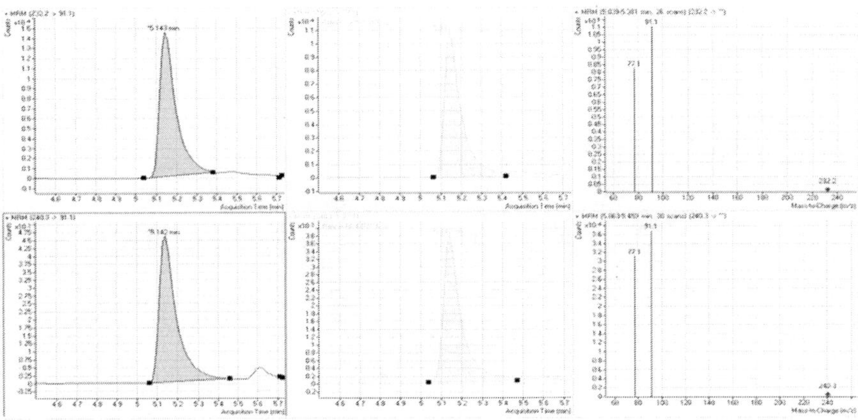

Figurc 1. Multiple reaction monitoring chromatograms and product ion mass spectra of α-PVP (*top*) and α-PVP-d_8 (*bottom*) extracted from an autopsy blood sample of passenger 3 recorded by liquid chromatography–electrospray ionization-tandem mass spectrometry

Table 1. Accuracy and precision data of the developed method

Concentration of α-PVP (mg/L)	Intra-assay (n = 5)		Inter-assay (n = 15)	
	Accuracy (%)	Precision (%RSD)	Accuracy (%)	Precision (%RSD)
0.020	16.7	2.9	10.2	3.5
0.100	−0.8	3.0	−6.5	7.1
0.500	1.2	1.0	−0.9	1.8

RSD relative standard deviation

Because we used stable isotopic α-PVP as IS (isotope dilution method), it seems not necessary to present matrix effects and efficiencies. However, the data seem to be of interest for readers. The results showed that the method suffered from appreciable suppressive matrix effects (Table 2).

Table 2. Matrix effects, extraction efficiencies, and process efficiencies of the developed method

Compound	Blood concentration (mg/L)	Matrix effect (%)		Coefficient of variation (%RSD)	Extraction efficiency (%)	Process efficiency (%)
		Relative	Absolute suppression/enhancement			
α-PVP	0.020	0.24–0.36	−75.8 to −63.3	6.7	17.5–44.9	6.3–15.1
	0.100	0.25–0.35	−74.7 to −64.8	5.6	26.4–84.6	8.1–21.4
	0.500	0.31–0.35	−68.8 to −64.8	2.2	57.4–77.1	19.4–25.9
α-PVP-d_8	0.100	0.33–0.45	−67.5 to −54.7	7.7	31.0–87.0	11.8–28.3

The stability of α-PVP at 4 °C and room temperature showed that this compound is relatively stable (Table 3).

Table 3. Stability of α-PVP

Compound	Blood concentration (mg/L)	Percent difference			
		4 °C			Room temperature
		24 h	48 h	72 h	72 h
α-PVP	0.020	−0.16	−7.0	11.9	6.8
	0.100	−2.1	−11.5	11.6	11.5
	0.500	−0.48	−2.6	14.0	12.5
α-PVP-d_8	0.100	−3.9	−8.3	11.5	5.7

Table 4. Toxicology results of the participants in the accident

Individual	Blood collection time	Concentration of α-PVP (ng/mL)	Time	Post-accident alcohol blood level (‰)
Living subjects				
Driver, male—21 years	9:30 a.m.	230	6:45 a.m.	0.0
	10:30 a.m.	360		
Passenger 1, male—36 years, front seat	10:10 a.m.	30	6:50 a.m.	1.2
	01:30 p.m.	20		
Passenger 2, female—19 years, back seat, center	–[a]	–[a]	7:10 a.m.	0.4 (medical records)
Autopsy subjects				
Passenger 3, female—17 years back seat, side	–	650	–	1.9
Passenger 4, male—28 years back seat, side	–	290	–	2.6

[a] Not sampled

In blood samples collected from the surviving and deceased participants in the accident, the analysis demonstrated α-PVP in a diversified range of concentration values. The results of toxicological tests performed in individuals involved in the accident are presented in Table 4.

The blood samples were collected from the driver 3.5 and 4.5 h after the accident, arriving at an increasing sequence of α-PVP concentration values of 230 and 360 ng/mL, respectively. As it follows from data originating from the court records, the male insufflated the psychoactive twice in the period preceding the accident, i.e., ~6 and 2 h previously, which explains the results as the effect of interposition of the elimination and absorption phases of both doses taken at different times. The demonstrated concentrations of α-PVP in the blood of the driver should be considered high as compared with the literature data available [4, 8]. It may be also assumed that at the time of the accident, the blood α-PVP level of the driver might have been higher than the level demonstrated in the tests. Passenger 1, a survivor of the accident, had taken α-PVP ~7.5 and 13.5 h before blood collection, and after that period, the substance continued to be present in his blood at 30 and 20 ng/mL, respectively. On the other hand, in the blood of the deceased individuals involved in the accident, α-PVP concentration values were high (650 and 290 ng/mL), because they reflected the levels of the drug after 6 and 2 h after insufflating α-PVP doses. The numerical values were different, which most likely resulted from the difference in the size of doses apportioned to both abusers using the "hit-and-miss" method and in their body parameters.

DISCUSSION

In the analyzed case, chemical and toxicological tests demonstrated a psychoactive substance α-PVP in the blood samples collected from the driver and perpetrators of the accident. The alleged substance belongs to the group of NPSs, being in the category of synthetic cathinone derivatives. Until 2015, α-PVP was not included in the list of narcotic substances and psychotropic agents of the legal regulations in force [9]. The cathinone α-PVP belongs to the group of pyrrolidinophenones. The group also includes two structural α-PVP analogues, namely pyrovalerone and MDPV, which are included in the list of psychotropic substances in the above regulation.

Although α-PVP was synthesized in the 1960s, at the present time it has gained some popularity in the narcotic market. Among various types of cathinones, derivatives of α-PVP have become popular as NPSs due to the presence of the lipophilic pyrrolidinyl group that increases the ability to cross the blood-brain barrier to obtain a better psychedelic effect [10, 11].

The mechanism underlying α-PVP activities is poorly understood, and for this reason, while commenting on the biological activities of the compound, the likeness of its activities to those of other compounds is also taken into consideration, namely pyrovalerone and especially MDPV, which show close similarity in their chemical structures and allow assuming that they may be similar to their biological activities. Generally, all three above-mentioned compounds belong to selective reuptake inhibitors of neurotransmitters dopamine, noradrenaline, and serotonin. In view of their mechanism of activity, pyrrolidinophenones are considered central nervous system stimulants. As a rule, the basic anticipated symptoms of people often taking the above compounds include enhanced energy, euphoria, empathy, openness, and increased libido. Nevertheless, soon afterwards, there may appear such negative side effects as tachycardia, increased arterial pressure, confusion, agitation, insomnia, aggression, and hallucinations [4].

The employed α-PVP doses have not been precisely determined. However, based on the presently analyzed case and data derived from Internet forums frequented by young people, there may be a fairly broad range of 100–500 mg.

The available literature described that as compared to cocaine and amphetamine, which belong to the group of classic central nervous system stimulants, both α-PVP and its structurally close derivative MDPV are much more potent selective reuptake inhibitors of dopamine and noradrenaline. Moreover, in studies performed in mice, a significant increase in locomotor activity by the compounds was demonstrated, being higher than that by cocaine and amphetamine and resulting from an increased dopamine level [12]. Psychophysical tests employed in assessment of ability to operate motor vehicles and performed in individuals under the influence of MDPV demonstrated considerable functional impairment of psychomotor abilities in 84 % of testees, and in 7 %, there was moderate or complete inability to operate motor vehicles. The determined blood MDPV concentration values in the tested individuals were below 500 ng/mL [12]. In view of a lack of data on toxic α-PVP levels, interpretation of toxicology results in the analyzed case is most assuredly complex and ambiguous.

Knoy et al. [4] described a case of a driver under the influence of α-PVP. The driver was stopped by the police due to bravado driving; following the intervention, the policemen noticed symptoms that might suggest his being under the influence of a narcotic substance, namely verbal confusion, disorientation, excitation, tachycardia, dilated pupils, and tremors. The symptoms observed in the course of the above police intervention are similar to those previously noted in cases of other representatives of the pyrrolidinophenone group. During subsequent procedures involving the above driver, his blood sample collected ~2 h after the police intervention demonstrated α-PVP at a concentration of 63 ng/mL, being thus several times lower than the values noted in the presently analyzed driver case (α-PVP 230 and 360 ng/mL).

Analyzing possible behavior patterns of individuals under the influence of α-PVP, one may also quote other cases that well illustrate the patterns, although they are not related to traffic. Nagai et al. [8] provide a description of a male who exhibited sudden violent behavior at a place where several individuals lived together. The roommates tried to restrain the excited man, who collapsed after approximately 2 h of struggling, and after 30 min of routine resuscitation efforts, he died. His restraint was thought to induce death from asphyxia due to chest compression. During autopsy, intra-neck muscle hemorrhage, submental and high neck petechiae, and facial congestion were observed. Toxicological analysis of a postmortem blood sample demonstrated 411 ng/mL of α-PVP.

The present authors experienced a case of a death of a 19-year-old man with an underlying mechanism similar to that described above [13]. Because of his aggressive and dangerous behavior, the man was brought to a mental hospital by the police. In the hospital, following a short period of calming down, he developed progressive motor agitation and started to destroy a patient room, and it was necessary to restrain him using immobilization belts. In spite of intense efforts at resuscitation, the patient died of acute circulatory and respiratory failure associated with psychomotor excitation. A postmortem blood sample showed α-PVP at the level of 850 ng/mL.

In both above cases, the mechanism of death should be considered within the category of complications of pathological motor excitation. The mechanism is most commonly encountered in individuals who are incapable of controlling their behavior, being under the influence of psychoactive substances or exhibiting alcohol delirium. Pathological excitation causes the affected individual to thrash around purposelessly, chaotically employing all muscles at the same time, which markedly intensifies oxygen consumption.

Nagai et al. [8] drew attention to the danger of employing sustained restraint of excited persons with hallucinations resulting from α-PVP or other stimulants, suggesting the use of sedation and adrenergic blockade. Thus, as it follows from the practice of the present authors [13] and published literature data [8], as compared to methamphetamine and MDMA, which can be lethal, α-PVP and related drugs appear to be less lethal, but seem more hallucinogenic and agitating.

The metabolism of α-PVP may be fairly slow. Such a thesis results from observations of its concentration values demonstrated for example in the blood of the participants in the traffic accident. As shown in Table 4, in the surviving passenger 1, the compound (20 ng/mL) was detected, even 13.5 h after his having taken the last dose.

Analyzing the available court records pertaining to the present case, including the driver's medical and psychological examination results, there is no doubt that at the time of the accident, he might have lost normal psychomotor ability resulting in increased energy and euphoria, and might have been agitated and unable to control his car fully; he also might have had a brief episode of unconsciousness. Moreover, the behaviors of the passengers who were also under the influence of the drug might have influenced the decreased alertness of the driver while driving a car. The very well-documented description of the behavior of the driver accused of causing a traffic accident, being under the influence of α-PVP and being based mainly on the explanations that he provided is in accordance with the classic features of the substance effects observed in other individuals and described in the literature. Additionally, attention was focused on the mental aspect of the analyzed case, which cannot be unrelated to the tragic accident.

CONCLUSIONS

The presented case report points to a unique opportunity to understand the psychological effects of α-PVP in a real driving situation, emphasizing—along with another report [4]—the importance of police observation on the road, especially when alcohol and/or classic narcotic substances have been ruled out at the site.

ACKNOWLEDGMENTS

This study was funded by Jagiellonian University Medical College (Grant Number: K/ZDS/004604).

COMPLIANCE WITH ETHICAL STANDARDS

Conflict of interest

The authors declare that they have no conflict of interest.

Ethical approval

Informed consent was obtained from all individual participants included in the study. All procedures performed in studies involving human participants were in accordance with the ethical standards of the institutional and/or national research committee and with the 1964 Helsinki declaration and its later amendments or comparable ethical standards (in case humans are involved).

REFERENCES

1. DRUID (2012) 6th Framework programme. Final report: work performer, main results and recommendations. http://www.druid-project.eu. Accessed August 2012
2. Heis T, Lyckegaard A, Simonsen KW, Steentoft A, Bernhoft IM (2013) Risk of severe driver injury by driving with psychoactive substances. Accid Anal Prev 59:346–356
3. Musshoff F, Madea B, Kernbach-Wighton G, Bicker W, Kneisel S, Hutter M, Auwärter V (2014) Driving under influence of synthetic cannabinoids ("Spice"): a case series. Int J Legal Med 128:59–64
4. Knoy JL, Peterson BL, Couper FJ (2014) Suspected impaired driving case involving α-pyrrolidinovalerophenone, methylone and ethylone. J Anal Toxicol 38:615–617
5. Maas A, Wippich C, Madea B, Hess C (2015) Driving under the influence of synthetic phenethylamines: a case series. Int J Legal Med 129:997–1003
6. Karinen R, Tuv SS, Øiestad EL, Vindenes V (2015) Concentrations of APINACA, 5F-APINACA, UR-144 and its degradant product in blood samples from six impaired drivers compared to previous reported concentrations of other synthetic cannabinoids. Forensic Sci Int 246:98–103
7. Matuszewski BK (2006) Standard line slopes as a measure of relative matrix effect in quantitative HPLC-MS bioanalysis. J Chromatogr B 830:293–300
8. Nagai H, Saka K, Nakajima M, Maeda H, Kuroda R, Igarashi A, Tsujimura-Ito T, Nara A, Komori M, Yoshida K (2014) Sudden death after sustained restraint following self-administration of the designer drug α-pyrrolidinovalerophenone. Int J Cardiol 172:263–265
9. The law on preventing drug addiction dated 29 July 2005 (2012) Ustawa o przeciwdziałaniu narkomanii z dnia 29 lipca 2005 r. (tekst jednolity: Dz. U. z 2012 r. poz. 124 z poź. zm.)
10. Zaitsu K, Katagi M, Tatsuno M, Tsuchihashi H, Ishii A (2014) Recently abused synthetic cathinones, α-pyrrolidinophenone derivatives: a review of their pharmacology, acute toxicity and metabolism. Forensic Toxicol 32:1–8
11. Wurita A, Hasegawa K, Minakata K, Gonmori K, Nozawa H, Yamagishi I, Susuki O, Watanabe K (2014) Postmortem distribution of α-pyrrolidinobutiophenone in body fluids and solid tissues of a human cadaver. Leg Med 16:241–246
12. Kaizaki A, Tanaka S, Numazawa S (2014) New recreational drug 1-phenyl-2-(1-pyrrolidinyl)-1-pentanone (alpha-PVP) activates central nervous system via dopaminergic neuron. J Toxicol Sci 39:1–6
13. Konopka T (2014) Archives of Department of Forensic Medicine Jagiellonian University Medical College in Krakow (S-626)

CHAPTER 16

Methods and Tools of Digital Triage in Forensic Context: Survey and Future Directions

Vacius Jusas [1,*], Darius Birvinskas [2] and Elvar Gahramanov [1]

[1]*Software Engineering Department, Kaunas University of Technology, Studentu St. 50, LT-51368 Kaunas, Lithuania*

[2]*Computer Department, Kaunas University of Technology, Studentu St. 50, LT-51368 Kaunas, Lithuania*

ABSTRACT

Digital triage is the first investigative step of the forensic examination. The digital triage comes in two forms, live triage and post-mortem triage. The primary goal of the live triage is a rapid extraction of an intelligence from the potential sources. The live triage raises legitimate concerns. The post-mortem triage is conducted in the laboratory and its main goal is ranking of the seized devices for the possible existence of the relevant evidence. The digital triage has the potential to quickly identify items that are likely to contain the evidential data. Therefore, it is a solution to the problem of case backlogs. However, existing methods and tools of the digital triage have limitations, especially, in the forensic context. Nevertheless, we have no better solution for the time being. In this paper, we critically review published research works and the proposed solutions for digital triage. The review is divided into four sections as follows: live triage, post-mortem triage, mobile device triage, and triage tools. We conclude that many challenges are awaiting for the developers in creating methods and tools of digital triage in order to keep pace with the development of new technologies.

Keywords: digital forensic; digital triage; live triage; post-mortem triage; triage tools

1. INTRODUCTION

The volume of data for forensic investigation keeps constantly growing. This is a result of the continuing technology development when scale and bounds of the Internet rapidly change and social networks come to everyday use. The storage capacity expands to new areas when smart phones become part of the Internet devices and cloud storage services are offered. The digital forensic process is very time consuming, because it requires the examination of all available data volumes collected from the cybercrime scene. The digital forensic process commences with the collection, duplication, and authentication of every piece of digital media prior to examination. Moreover, every action taken has to adhere to the legitimacy rules so that the obtained digital evidence could be presented in the court. However, life is very dynamic, and the situations, in which some information about a possible cybercrime has to be obtained as promptly as possible without adhering to the rules of long legal scrutiny, arise. Of course, the information obtained in a such way cannot be directly used in the court; however, a quick access to such knowledge can speed up the future process of digital forensics and, is some situations, can even save somebody's life. Therefore, such actions are justifiable.

A process that takes place prior to the standard forensic methodology is called digital triage. It can provide valuable intelligence without subjecting digital evidence to a full examination. This quick intelligence can be used in the field to guide the search and seizure, and in the laboratory to determine if a media is worth to be examined.

The term "triage" comes from the field of medicine, in which it refers to the situations when because of having limited resources, the injured people are ranked according to the necessity to receive treatment. Such ranking ensures the achievement of the least damage to patients when resources are limited [1].

Rogers et al. [2], the authors of the first field triage model in computer forensics, define triage as a process of ranking objects in terms of importance or priority. Casey et al. [3] define triage in digital forensics as part of forensic examination process. The forensic examination is described as three-tier strategy consisting of three levels: (i) survey/triage forensic inspection, (ii) preliminary forensic examination, and (iii) in-depth forensic examination. The first stage, in which many potential sources of digital evidence for specific information are reviewed, is alternatively referred to as survey or triage. The same idea that triage is part of forensic examination, is supported in later works [4,5,6,7,8]. Casey [4] underlines that triage is effective for prioritizing, but it is not a substitute for a more thorough review. Casey [5] argues that triage is a technical process, which can be performed outside a laboratory by professionals with basic training and limited oversight. Categorizing digital triage as a technical process makes it more clear that the information has not undergone rigorous quality assessment and its legitimacy has not been evaluated.

There are many other definitions of triage, which slightly differ depending on the attributed qualities [7,8,9,10,11,12]. The diversity of triage definitions reflects the variety of the views and indicates the immaturity of the field. However, it is not the main problem. The focus should be devoted to the decision whether digital triage is a forensic process. As Cantrell et al. [13] state, "Digital triage is not a forensic process by definition". It is not clear to which definition Cantrell et al. [13] refer. It is possible to suppose that it is the definition by Rogers et al. [2]. However, other definitions exist, and this statement is not true for all the cases [7,11,14]. Koopman and James [11], and Roussev et al. [7] use the term "digital forensic triage". If digital triage is not the forensic process, then the term "forensic" cannot be used together with the term "digital triage", because it misleads. Hong et al. [14] introduce a triage model that is adapted to the requirements of the legal Korean system. Consequently, the proposed triage model adheres to the rules of the forensic process. Moreover, Hong et al. [14] suggest establishing a triage model individually for the legal system of a specific country.

To summarize the diversity of views on digital triage, we stress the following features:

1. Digital triage is a technical process to provide information for the forensic examination, but does not involve the evaluation of digital evidence
2. The goal of digital triage is to rapidly review many potential sources of digital evidence for specific information and prioritize the digital media to make the subsequent analysis easier
3. The term "forensic" cannot be used together with the term "digital triage" if the process of digital triage does not adhere to the rules of the forensic process specific to the country

Digital triage comes in two forms: live and post-mortem. The post-mortem form of triage, which is conducted on the digital image, is not always recognized as triage. We suppose that both forms of digital triage are equally important. Live triage raises many concerns, because it is conducted on the live system, and the destruction of the likely evidence is possible. However, live digital triage has several advantages:

1. It enables a rapid extraction of intelligence that can be used for suspect interrogation
2. Some data can be lost if the computer is shut down

The primary concern inherent to both forms of digital triage is that the evidential data can remain unnoticed [15]. Pollitt [16] argues that the process of digital triage in the context of forensics is an admission of failure. However, he recognizes that for now a better approach does not exist.

Moreover, the term "triage" becomes the common word to indicate the initial and rapid step in the different areas of the forensic investigation. For example, it is used in the retail industry [17], in the internet of things [18], in the fraud of identity and travel documents [19].

We review the research works related to digital triage. We divide the review into four sections as follows: live triage, post-mortem triage, mobile device triage, and triage tools. The largest section is on the triage tools. Such abundance of research works highlights the practical need for triage tools.

In the next section, we review the models and methods of live triage.

2. MODELS AND METHODS OF LIVE TRIAGE

Rogers et al. [2] introduce the model for the field triage process in computer forensics and name it the Cyber Forensic Field Triage Process Model (CFFTPM). The CFFTPM has six phases: planning, triage, usage/user profiles, chronology/timeline, internet activity, and case specific evidence. Each phase has several sub-tasks and considerations that vary according to the specifics of the case and operating system under investigation. The CFFTPM originates from child pornography cases. Nevertheless, it is general enough to be applicable to other possible cases; however, the model cannot be considered as the ultimate solution for every case. It is also important to note that the proposed model does not preclude transporting the system to a laboratory environment for a more thorough investigation.

Cantrell et al. [13] discuss a proposed model for digital triage. The proposed model is a linear framework, except the preservation phase that is an investigative principle preserved throughout all the phases. The first phase is planning and readiness that occurs before the investigation onsite. The next phase is live forensic that is included as an optional step, depending on the need and expertise, and it must occur prior to the following phases because the volatile memory can be lost very quickly. The middle three phases: computer profile phase, crime potential phase, and presentation phase are intended to be an automated process, coded as a computer program or script using the existing tools. The last phase, triage examination phase, is optional depending on the need. The triage examination should be an automated process that is guided by the examiner using predefined templates specific to each case.

Hong et al. [14] propose a theoretical framework for implementing a triage model. The requirement for the triage model is to consider the limiting factors of the onsite search and seizure. The framework consists of three phases: assessment, triage model, and reassessment. The proposed framework is based on the assumption that reassessments are performed periodically according to the changes in search and the conditions of the onsite seizure. To establish a triage model, a questionnaire that consists of 48 questions, which are provided in the paper, was prepared; it was answered by 58 respondents in total. The paper presents a large discussion of the results. After assessing the results of the questionnaire, a new triage model is proposed. The triage process is divided into four steps: planning, execution, categorization, and decision. The properly collected information mostly depends on the execution step. The execution step prioritizes the file types for the search according to three types of crime: personal general crime, personal high-tech crime, and corporate general crime. Next, the file search is conducted in the following order: timeline of interest; filename- or contents-based keywords search; and file/directory path-based search. Another important procedure in the execution step is the detection of suspicious files. The proposed triage model can be applied only to personal computers and it is tailored to the Korean legal system requirements for the privacy protection.

Overill et al. [20] propose an attractive idea to introduce triage template pipelines into the investigative process for the most popular types of digital crimes, enabling digital evidence to be examined according to a number of prioritised criteria. Each specific digital crime has its own template of prioritised devices and the data based on the cost-effectiveness criteria of front-loading probative value and back-loading resource utilisation. The authors declare that about 80% of all digital crimes in Hong Kong are accounted for just five types of crime. However, they do not enumerate these types of crime. The authors state that "the work this far has addressed the set of five digital crime templates", however, the examples of templates for only two digital crimes are provided. To be more precise, they are the Distributed Denial of Service (DDoS) template diagram and the Peer-to-Peer (P2P) template diagram. Moreover, the construction of these example templates is not discussed in detail. An advantage of the triage template pipeline approach over the triage tools is that the evidential recovery process can be terminated as soon as it becomes apparent that the probative value criterion has been fulfilled. Therefore, the triage time can be shorter in some cases. The essence of the proposed triage template pipelines is formalized common sense.

Roussev et al. [7] argue and analyze forensic triage as a real-time computation problem, which has allotted limited time and resources. One hour is considered to be an acceptable time limit for triage. The authors assume that an increase in the performance can be achieved if the acquisition and processing start and complete at almost the same time. It means that the processing should be as fast as data the cloning. The suitability of the most common open-source implementations and of most common forensic procedures to fit into the time constraints is investigated experimentally. The authors state that the triage investigation can be carried out in the field and in the laboratory. For the fieldwork, they consider 8-core workstation and for the laboratory, they consider 48-core server. The obtained results show that only a few basic methods, like file metadata extraction, crypto-hashing, and registry extraction, can fit into the time budget in the workstation triage. To increase the performance of the file acquisition, Roussev et al. [7] implement a Latency-Optimized Target Acquisition (LOTA) scheme. The main idea of this scheme is that the metadata of a filesystem is parsed to make an inverse map from blocks to files before cloning the target. This procedure allows sequential scanning of blocks and reconstructing the files. The LOTA scheme enables an improvement of a factor of two for files larger than 1 M and a factor of 100 for smaller files. It is recommended to use the scheme in the forensic environment routinely. The authors advocate employing parallel computations to obtain higher processing rates.

Lim and Lee [21] describe a unified evidence container XeBag for storing diverse digital evidence from different sources. The XeBag can be used for selective evidence collection and searching on the live system. The file structure of XeBag is based on well-known compression file formats, PKZip and WinRAR. To record forensic metadata, an Extensible Markup Language (XML) document is included additionally for each stored object. The XML format is a popular data exchange format, therefore, it enables easy access to the data. The authors provide a description of a video surveillance system to show how its digital evidence is stored and can be retrieved from the unified evidence container XeBag.

Grier and Richard III [22] introduce a new approach, called sifting collectors, for imaging of the selected regions of disk drives. The sifting collectors create a sector-by-sector, bit-for-bit exact image of disk regions that have forensic value. The forensics image is produced in an Advanced Forensics Format v3 [23], and it is fully compatible with the existing forensic tools. The selection of the regions that have forensics value is based on profiles. The authors do not expect that the examiners can prepare the profiles themselves, therefore, the profiles must be created and stored in a library. The sifting collectors firstly collect the metadata according to the defined profile. Then they interpret metadata, determine sectors of interest, and assemble them in the disk order. As a result, their methods are not suitable for unknown filesystems. If profiles are not possible to define, the alternative proposes to include a person in the scanning loop to decide what is relevant. The implemented prototype targets New Technology File System (NTFS) as a file system and uses the Master File Table as its primary source. The conducted experiment shows a speed up from 3 to 13 times in comparison to the forensic image acquisition tool Sleuthkit [24] for the test cases. The absolute values of runtimes are not provided. The accuracy of the region selection is between 54% and 95% for the considered test cases. Faster image acquisition time gives less accuracy. One important limitation of sifting collectors is their susceptibility to steganography and anti-forensics.

Penrose et al. [25] present an approach for fast contraband file detection on the device itself. The approach is based on clusters scanning, hash calculating, and comparison to the database. The cluster size is 4 KiB. A Bloom filter is used to store the cluster hashes of the contraband files. The Bloom filter reduces the size of the database of the block level Message-Digest Algorithm 5 (MD5) hashes by an order of magnitude; however, it costs a small false positive rate. The designed Bloom filter is 1 GiB in size and it uses eight hash functions. A larger Bloom filter enables faster access to the hashes of the contraband files. The performed experiment shows that the approach achieves 99.9% accuracy scanning for contraband files in minutes. Some false positives are encountered; however, the results are positive for the existence of all contraband files. The experiment was conducted in legitimate computing environment. The authors draw a conclusion that this type of case can be further investigated in a forensically sound environment.

Turnbull and Randhawa [26] describe an ontology-based approach to assist examiner-led triage. The purpose of the approach is to enable a less technically intrinsic user to run a triage tool. This is implemented by collecting low-level artifacts and inferencing hypotheses from the collected facts. The approach is oriented to automatically deriving events from the base of the forensics artefacts. A Resource Descriptive Framework (RDF) is used as the basis of the ontology. The representative feature of the approach is that the layered multiple ontologies are designed over the same dataset. The description of the ontologies used is vague. The authors find some advantages of the RDF; however, they recognize that a Web Ontology Language (OWL) could provide more possibilities. The authors suggest that the approach is applicable for the extraction of information from social networks, though, no evidence of such application can be found in the paper. The implemented system to provide a proof-of-concept consists of a knowledge base, data ingestors, reasoners, and a visualiser. The visualiser is hardcoded into the used ontology. Neither test, nor real cases are provided. To conclude, the idea of the approach is attractive, however, the description and the development are immature.

Hitchcock et al. [27] introduce a Digital Field Triage (DFT) model to offload some of the initial tasks performed in the field by forensic examiners to non-digital evidence specialists. The primary goals of the model are twofold: (i) To increase the efficiency of an investigation by providing digital evidence in a timely manner; (ii) To decrease the backlog of files at a forensic laboratory. The proposed model is based on Rogers et al. [2] and it has four phases: planning, assessment, reporting, and threshold. The DFT model has inherent risks associated with it. They are as follows: the management, training, and supporting tools. The management and ongoing training are integral parts of the success of the DFT model. The tools must support the management. For the DFT to work, there are three fundamental concepts:

1. DFT must work with a supervising examiner
2. DFT must maintain the forensic integrity of the digital evidence
3. A DFT assessment does not replace the forensic analysis

Therefore, the DFT model is not a replacement for full analysis, but is part of the overall strategy of handling digital evidence. The first version of the DFT model was implemented in Canada six years ago. The implementation achieved the goals pursued by the model; however, persistent attention needs to be turned to the risks associated with the model.

Leimich et al. [28] propose a variation of cloud forensic methodology tailored to a live analysis of Random-Access Memory (RAM) for Hadoop Distributed File System (HDFS). The aim of the methodology is to minimize the disruption to the data center after data breach. The Hadoop is a Java implemented system developed for UNIX based operating systems. It is a master/slave distributed architecture for storing and processing big data. The HDFS consists of DataNodes (slaves), which store the data, and NameNode (master) that manages the DataNodes. The methodology is oriented to the acquisition of the NameNode contents of to pinpoint the affected DataNodes. The forensic analysis of the DataNodes is out of scope of the proposed methodology. The methodology contains nine phases: preparation, live acquisition of the NameNode, initial cluster reconnaissance, checkpointing via a forensic workstation, live artefact analysis, establish 'suspect' transactions and map to data block, perform targeted dead acquisition of the DataNodes, data reconstruction, and report. To test the validity of the methodology a small HDFS cluster that has one master and three slaves, was configured with a single scenario of deleted data. The phase of data reconstruction is not carried out. The experiment confirms that the methodology enables locating the deleted data blocks. Liemich et al. [28] discuss the ability to implement the proposed methodology in forensic tool in compliance with the National Institute of Standards and Technology (NIST) Computer Forensic Tool Testing criteria.

Montasari [8] extends the Rogers et al.'s [2] model by dividing all phases into two stages and introducing new sub-tasks into the phases. The single planning activity is assigned to the first stage. The planning should be carried out before attending the site. Montasari [8] considers many models of the forensics process, not just triage models, because according to the author, the single model proposed by Rogers et al. [2] exists for the onsite triage process. The author selects activities, which would be appropriate for the triage process, from other models. Therefore, several sub-tasks are added to the model of the forensics field triage process, and the model is presented in a more detailed and categorized way. Additionally, the model is extended by a set of investigative principles joined into a group under the name of "Overriding Principles", which are an additional contribution of the paper. These principles are as follows:

1. To preserve chain of custody
2. To maintain an accurate audit trail
3. To maintain a restricted access control
4. To maintain an effective case management
5. To maintain the information flow

Peersman et al. [29] present an approach that incorporates artificial intelligence and machine learning techniques (support vector machines) to automatically label new Child Sexual Abuse (CSA) media. The approach employs two stages for labelling the unknown CSA files. The first stage uses the text categorization techniques to determine whether a file contains CSA content based on its filename. The text categorization applies the following features: predefined keywords, forms of explicit language use, expressions relating to children and family relations in English, French, German, Italian, Dutch, and Japanese. Additionally, all patterns of two, three, and four consecutive characters are extracted from the filenames. The second stage gets the files from the first level and examines the visual content of images and audio files. The second stage bases the decision on multi-modal features. The multi-modal features consist of the following representations: colour-correlograms, skin features, visual words and visual pyramids, and audio words for audio files. The conducted experiment shows a false positive rate of 20.3% after the first stage. The second stage reduces the false positive rate to 7.9% for images and 4.3% for videos. The approach is implemented into the iCOP toolkit [30] that performs live forensic analysis on a P2P network. Therefore, the proposed approach is designed for a proactive monitoring activity. To label the most pertinent candidates for the CSA media, an examiner can login to the iCOP canvas that automatically arrange the results. Additionally, the approach can be adapted to the identification of the new CSA media during a reactive investigation. The approach is implemented in the Gnutella P2P network.

Quick and Choo [31] develop the idea of data reduction introduced in [32]. The authors present the methodology to reduce the data volume using selective imaging. The methodology suggests to select only the key files and data. Windows, Apple and Linux operating systems and their filesystems are considered. A forensic examiner makes the decision to include or exclude particular file types. The decision is based on the data, contained in these file types, relevance to the case. The other possibility considered for reducing data volume is a thumbnailing of video, movie, and picture files. The thumbnailing significantly reduces large image files. Once the file types are selected and some thumbnails are loaded into the forensics software, the logical image file is created. The presented methodology can be applied using common digital forensics tools. The methodology is applied to test as well as real world data. Many results of the experiments that illustrate the viability of the methodology are provided. In general, time reductions observed are 14 min on average to collect a logical image and process in the Internet Evidence Finder, meanwhile the processing of full forensic image takes 8 h 4 min on average. The presented methodology can be applied to either write-blocked physical media or a forensic image.

3. METHODS OF POST-MORTEM TRIAGE

Marturana and Tacconi [33] summarize the research works [34,35] delivered at conferences and present a model intended for both live and post-mortem triage using machine learning techniques. The presented model consists of the following four steps: forensic acquisition, feature extraction and normalization, context and priority definition, and data classification. For such model, there are two main challenges, the definition of crime-related features and collection of a consistent set of classified samples related to the investigated crimes. The crime-related features are defined for two cases studies, copyright infringement and child pornography exchange. Guidelines for using the classifiers are provided. The attention of the experiment is mostly directed to the comparison of the classifiers used at the last stage of the model. No conclusion is made as to which classifier is best suited for the investigated cases. The presented statistical approach has proven to be valid for ranking the digital evidence related to copyright infringement and child pornography exchange. However, for this approach to be viable, it is necessary to have a deep understanding of possible relations between the crime under investigation and the potential digital evidence.

McClelland and Marturana [36] extend the research presented by Marturana and Tacconi [33]. The authors investigate the impact of the feature manipulation on the accuracy of the classification. The weights are assigned to the features. Two approaches are used for assigning weights to the features, automatic and manual. The automated feature weights are quantified using the Kullback–Leibler measure. The manual weights are determined on the basis of the surveyed digital forensic experts' contribution. The Naïve Bayes classifier is used for the experiment. The only improvement is achieved in the child pornography case.

Horsman et al. [10] extend the ideas presented in [37] and discuss a Case-Based Reasoning Forensic Triager (CBR-FT) method for retrieving the evidential data based on the location of the digital evidence in the past cases. The CBR-FT maintains a knowledge base for gathering the previous experience. Each location on the system stored in the knowledge base is assigned an evidence relevance rating (ERR), which is used as the prior probabilities in the Bayesian model to determine the priority of a particular location for searching. The model enables calculating a primary relevance figure (PRF) for each location. The search is carried out in two stages: in the first stage, only locations with a PRF above 0.5 are used, while the second stage is optional. If the examiner suspects that additional evidence can exist, s/he proceeds to the second stage. During the second stage, the examiner focuses on identifying similar patterns in cases stored in the CBR-FT knowledge base. The CBR-FT knowledge base must cover enough cases to reflect its target population correctly. That is the first restriction for application of the method. The study focuses on fraud offences and it has constructed a fraud knowledge base from 47 prior investigations. The experiment shows that the CBR-FT is more effective when compared to a commercial application EnCase Portable [38], which uses precision and recall rates. However, an additional shortcoming of this study is that it focuses only on offences of fraud.

Bashir and Khan [39] suggest a triage framework oriented to analyzing and resolving an attack. The framework contains the usual steps that belong to a general investigative process. The term "triage" refers to a certain part of the framework. The main idea of the triage framework is to create a blacklist database that contains a list of the previously known attacks with details on how to resolve. Every attack is characterized by six attributes: identifier, name, description, status, signature, and then counter measures. The key attribute is the signature that is a placeholder to store unique signatures of cyber-attacks in the form of MD5 hashes. If the signature of any of the affected files is found in the blacklist database, then it means that the attack is known. The answer to how to resolve it is in the blacklist database. However, if the attack is unknown, there is no triage process; a detailed analysis follows. The blacklist database is updated periodically on the basis of the new knowledge and new attacks.

Dalins et al. [40] introduce a crawl and search method that can be used for digital triage. The proposed method adopts the Monte Carlo Tree Search strategy that is used in games for the filesystem search, which is called Monte Carlo Filesystem Search (MCFS). The original random selection is leveraged with non-binary scoring to keep guided search. Three file scoring methods are introduced, each built on the previous one: simple scorer, type of interest scorer, and similarity-based scorer. Other customizations are made to deliver better performance: integration of domain knowledge to enhance guided search, use of proprietary Microsoft PhotoDNA algorithm to measure the similarity of images, and skin tone detection to identify exposed skin that is usual component of child pornography. The experiment is carried out on real data that was obtained from the Australian Federal Police. The data presented as forensics images are related to the possession and online trading of child pornography. The experiment shows that the proposed MCFS is an effective method for larger and complex tree structures of the file system hierarchy. The search efficiency can be improved by around a third compared to uninformed depth-first search. However, the integration of domain knowledge and skin tone detection scoring showed lower results than expected. An additional investigation is necessary to improve these customizations. In general, the improved proposed method is promising, since many performance limitations arise due to the complicated filesystem design [7].

Fahdi et al. [41] investigate the possibility of utilizing the Self-Organising Map (SOM) technique to automatically cluster notable artefacts that are relevant to the case. A SOM is a neural network that generates a mapping from the high dimensional input data into a regular two dimensional array of nodes based upon their similarity in an unsupervised manner. The approach is based on using the metadata from several sources, such as the file system, email, and Internet, as the input into the SOM clustering. Moreover, the approach is oriented at the investigation of the suspects' systems rather than the victims' systems. Several pre-processing options are employed before the application of the approach. These options include the creation of the file list, expanding compound files, data carving, entropy test for encryption, and known file search. The results of data carving are not included into the file list of the SOM. Data carving should not be deployed during triage, since data carving tends to generate a lot of data due to high false positive rates [7]. The experiment shows that the use of the approach as a triage to verify the existence of the notable files allows identifying 38.6% of notable files at a cost of 1.3% of noise files. It is possible to expand the network size to increase the percentage of the notable files, however, at the cost of picking up more noise files. Most of the analysis takes a relatively trivial amount of time for small data sets (several GB); however, it takes an hour on average to process a large data set (0.5 TB). The appeal of the approach is that the only examiner interaction required in this process is when selecting the crime category. The approach can be a building block with further research and refinement to provide a triage tool for investigating simpler and technically more trivial cases that represent a large proportion of the forensic examiners' daily activities.

4. TRIAGE OF MOBILE DEVICES

Mislan et al. [42] discuss the onsite triage process for mobile devices. The following steps are suggested for an on-scene triage investigation of mobile devices:

1. Initiate the chain of custody
2. Isolate the device from the network
3. Disable the security features
4. Extract the limited data

5. Review the extracted data
6. Preview the removable storage media.

All the steps are discussed in details. The process of the investigation should be well documented in order to validate the results. The mobile device technicians, who are less experienced as technical examiners, should perform the onsite triage. The basic requirements for the automated onsite triage tools are outlined. To present shortly, they are as follows: simplicity of use, audit trail, and access control. The legal allowances of the United States to examine mobile devices are considered as well.

Walls et al. [43] introduce an investigative tool DECODE for recovering information from mobile phones with unknown storage formats. The main idea is that the data formats from known phone models can be leveraged for recovering information from the new phone models. The evaluation focuses on feature phones, i.e., phones with less capability than that of smartphones. The DECODE takes the physical image of a mobile phone as input. It is the first limitation of the tool, because the image is not its concern. The second limitation is the assumption that the owner of the phone has left the data in plaintext format. The next shortcoming is that the extracted results are limited to address books and call log records. The contribution of the paper is a technique for an empirical mobile phone data analysis. The used technique consists of two steps—removal of known data and recovering information from the remaining data. The latter step is called an inference process. Block hash filtering accomplishes the first step. The second step adapts the techniques from natural language processing, namely the context-free grammar, and uses probabilistic finite state machines to encode typical data structures. The Viterbi algorithm treats the created finite state machines twice. Finally, the decision tree classifier is used to remove the potential false positive. The development is based on the four following models: Nokia 3200B, LG G4015, Motorola v551, and Samsung SGH-T309. The performance of DECODE's inference engine is evaluated against two metrics, recall and precision. The conducted experiment on the phones that have not been seen previously shows an average recall of 93% and precision of 52% for address books, and an average recall of 97% and precision of 80% for call logs.

Marturana et al. [34] discuss the application of machine learning algorithms for digital triage of mobile phones. The triage stage is introduced between the stages of acquisition and analysis. The extracted data are firstly preprocessed in order to clean data, remove redundant attributes, and normalize data. Several classification algorithms are used to show the ability to classify whether a mobile phone was used to commit a pedophilia crime. The attention is devoted to the performance of the classification algorithms. The research is the first step towards the post-mortem forensic triage of mobile phones.

Varma et al. [44] present a system, called LIFTR, for prioritizing the information recovered from Android phones. The initial data for the system is a forensic image extracted by a recovery engine. Three recovery engines—DECODE [43], Bulk Extractor [45], and Strings, a common UNIX utility for identifying stings of printable characters in a file —are used as the suppliers of the forensic images. Therefore, the LIFTR should operate in concert with the recovery engine, as it augments the results obtained by the engine. The basic idea is that the recovery engine returns many unrelated items to the investigated crime results, since it does not consider the semantics behind the recovered content. Varma et al. [44] explore the filesystem of the Android phones and learn the rules of storing the information. These rules learnt and the feedback from the examiner form the basis for information prioritizing. The examiner labels the relevant information units of the investigated crime at the page level. The labeling takes several times and it is performed in the cycle. All the information is ranked based on a combination of the examiner's feedback, the actual content, and the storage system locality information. To test the validity of the approach, the open-source prototype of the system LIFTR is implemented. The LIFTR's ranking algorithm is evaluated against 13 previously owned Android smart phones. Moreover, the set includes nine phones with the Yaffs filesystem [46]. To improve the results, the authors wrote a special Yaffs parser to identify the expired pages that are important to the information relevance. The experiment shows that the LIFTR ranking improves the score of standard information retrieval metric from 0.0 to an average 0.88.

Guido et al. [47] introduce a differential acquisition technique that can be used for forensic image acquisition of mobile devices for triage purposes. The advantage of the technique introduced is its runtime that is several times faster than other compared commercial tools or techniques. The main idea is to use the precomputed baseline hashes. Therefore, the hashes of the unknown blocks are only sent to the server. The prototype named Hawkeye is implemented. The Hawkeye uses MD5 algorithm for hashing. Several other improvements are implemented to obtain less runtime. They are as follows: threading (10 threads by default) and comparison function of the zero block. The Hawkeye runs on Android devices in the recovery mode. The experiment is performed with 16 GB Samsung Galaxy S3 smartphone (Samsung, Seoul, South Korea). The acquisition techniques of the tool can be applied to other platforms, such as iOS (Apple Inc., Cupertino, CA, USA) as well.

5. TRIAGE TOOLS

We only review the tools that are presented in research papers. Such tools are usually not commercial.

Roussev and Quates [12] assume that digital triage is conducted in the laboratory environment. The purpose of digital triage is to identify the most relevant artifacts. The speed and reliability are critical for digital triage; these requirements are imperative for building triage tools. The authors demonstrate the use of tools sdhash and sdhash-dd to generate similarity digests for the purposes of digital triage. The first tool, sdhash, is an open source tool [48]. The similarity digests are intended to solve the two following problems of content correlation: resemblance and containment. The overall process of the application is quite simple; the similarity digests are generated for all targets and all queries. The queries are then systematically applied to the targets. The resemblance of data is detected at the bit-stream level, and parsing or understanding of the data being processed is not necessary. For the experiment, the authors use 1.5 TB of raw data that consist of many disk images, RAM snapshots, network captures, and four Universal Serial Bus (USB) devices. The investigation of Windows machines is carried out and three cases are investigated. It takes 180 min to generate the similarity digests. The total triage time for the investigated cases is 110 min and extra 90 min for the additional file hashing to prepare the similarity digests of the targets. As we see, the preparation for the triage takes significantly more time that the triage itself. However, the digest formation can proceed in parallel with the acquisition process. The problem of the method is to have the identified targets as informative as possible, since the method is based on the content resemblance and containment.

Cantrell and Dampier [9] present the implementation of the automated phases in the partially-automated digital triage process model [13]. The implementation is carried out on the basis of series of scripts comprised of original and open source tools written in Perl. The Linux distribution CAINE [49] installed to a USB drive is chosen as the development and testing environment in order to provide some form of boot media and to incorporate full onsite capability. The Windows registry is obtained by using the open source tool RegRipper [50]. The final report is provided in the form of HyperText Markup Language (HTML) pages. The tool is implemented to search the Web browser history for Internet Explorer only. The initial testing is done on a series of 300 GB drives. The runtimes are not provided.

Lim et al. [51] introduce a Live Data Forensic System (LDFS) designed to collect and analyze live data for Microsoft Windows-based systems. The LDFS consist of two separate tools, LDFS collection and LDFS analysis. The LDFS collection system gathers volatile and non-volatile data such as: memory dump, page file, web browser artifacts, instant messaging services clients, Windows Registry, and file system metadata. The distinctive feature of the LDFS collection system is that it can decode encoded BuddyBuddy, Yahoo, and MissLee messenger clients' chat logs. The physical memory dump and dump of all active processes are performed by means of third-party applications. The focus of choosing these applications is based on the least changes to the investigated system by the tool. The XML collection report holds all the collected items with their MD5 and Secure Hash Algorithm 1 (SHA1) hash values. The LDFS collection system is tested against five different types of Windows OSs (Microsoft, Redmond, WA, USA). Several experiments are conducted to test the performance of the system. The largest collection time does not exceed 49 min. The LDFS analysis module has the capabilities for analyzing all the collected data; however, it has not been fully implemented yet. Lim et al. [51] argue that the input data and its trustworthiness are of paramount importance in the live forensics analysis. However, it is not clear whether any defense against the subversion of the collection process is implemented in the LDFS collection system.

Casey et al. [52] discuss the need for and possibilities of honing the digital forensic processes to obtain the timely results. Many tasks in the forensic processes are not resource limited, and rethinking the overall organization of the forensic processes can assure greater improvements than considering the tasks separately. Therefore, improving the complete forensic process is oriented towards two areas, namely, dismantling the barriers between the tasks of the forensic process and providing useful information to support the key decisions. The efforts discussed in this paper focus on processing data from three primary sources: (i) filesystems, (ii) malware, and (iii) network traffic. Many triage tools analyze the filesystems. The analysis reveals that the main bottleneck in this process is the disk Input/Output (I/O) speeds. Using the results of the analysis, Casey et al. [52] provide the following guidelines for the triage or forensic data extraction tools to improve efficiency:

1. A tool can simultaneously deliver data into multiple extraction operations and create the forensic duplicate

2. A tool can store extracted information in both, the XML format and SQLite database
3. A tool should provide a user-friendly interface to facilitate the viewing, sorting, and classification of files

Additionally, tool developers have to consult about each step of the development with their customers. For the malware, the main suggestion is that the tool should firstly determine whether the file has been seen before. Next, the automatic malware processing tool developed by Defense Cyber Crime Center (DC3) is presented as an illustrative example. However, no suggestions are provided for the network traffic tools. The suite of tools PCAPFAST, developed by DC3, is provided as the example of the right network traffic tool.

Garfinkel [45] extends the research work presented by Garfinkel et al. [53] and introduces a forensic tool bulk_extractor devoted to the initial part of an investigation. The base of the bulk_extractor is the analysis of bulk data. The bulk_extractor scans raw disk images or any data dump for useful patterns (emails, credit card numbers, Internet Protocol (IP) addresses, etc.). It uses multiple scanners tailored to the certain patterns and heuristics to reduce false positive results and noises. The identified patterns are stored in feature files. When processing is complete, the bulk_extractor creates a feature histogram for each feature file. To improve the speed of processing, the bulk_extractor takes advantage of available multi-core capabilities. It detects and decompresses the compressed data. A lot of attention is devoted to the decompression of data. This feature is not usual for triage tools, because it consumes a lot of processing time. However, the feature is very useful for the forensic tool. The performance of the bulk_extractor is compared to the commercial tool EnCase. The results indicate that the bulk_extractor extracts email addresses from the forensic 42 GB disk image 10 times faster than EnCase, and it takes 44 min. The processing time of the bulk_extractor is between 1 and 8 h per piece of media, depending on the size and complexity of the subject data. The processing time does not meet the triage requirements. The bulk_extractor is successfully applied to 250 GB hard disk drives in two real cases. The processing time is 2.5 h for the first case and 2 h for the second. In general, the bulk_extractor is nice-to-have; however, it is not a triage tool.

Koopmans and James [11] introduce an automated network triage (ANT) solution designed for client-server environment. The purpose of the solution is to sort the analyzed systems by their likely relevance to the investigated case. The ANT is developed on the basis of the Preboot eXecution Environment (PXE) protocol and is composed of a network server that runs various services, and the clients, which are the systems to be analyzed, in a physically isolated network. The ANT server boots a suspected computer via a network. The authors provide many technical details that explain the specific steps—what software to use and how to boot the seized computers. The interface is developed in Personal Home Page (PHP) programming language. The data for triage are as follow:

1. A list of keywords to search for
2. A list of preferred file names or extensions
3. A list of preferred directories
4. A hash database that contains the hashes of files of interest
5. A hash database index file

Three real cases of the likelihood that the suspicious computers actually pose threat are very successfully investigated; the runtimes of the three cases are within 10 min. The runtimes are very short, however, it is not clear why they are so short, and an explanation is not provided. Moreover, Horsman et al. [10] state that hashing and keyword searching approaches can limit the effectiveness of digital triage because they are too restrictive. The limitations of the ANT solution are the following: there is no possibility to boot from the external source and encrypted data could not be analysed.

Moser and Cohen [1] discuss the use of triage in quite a different context than the traditional criminal case investigation—an incident response. The authors consider the use of the GRR Rapid Response (GRR) system. It is an agent-based open source distributed enterprise forensics system. Moser and Cohen [1] overview the components of the GRR system. A more detailed description of the GRR system is available elsewhere [54]. This method lowers the total time cost of triage analysis by distributing this task to the system agents. The main attention is directed towards the reliability of agents. Constant monitoring of used resources—memory and central processing unit (CPU)— ensures the reliability of agents. The investigation consists of three phases: planning, collection, and analysis. The experiment is carried out on many corporate workstations and laptops. The GRR agents are installed on these computers. The goal of the experiment is to examine the representative cases of a typical enterprise investigation performed by an incident response team. Four cases are analyzed. The majority of agents pick up artifacts in the first few minutes after the start. Nevertheless, the GRR continues running to 24 h so, if the missing machines come back online later, the artifacts will still be detected. The case of the autorun key comparison required an extensive manual analysis, therefore, improvement is necessary for such cases.

Shaw and Browne [15] argue that a digital forensic triage has been conducted on an informal basis for several years. The authors introduce the concepts of administrative and technical triage. The administrative triage assesses the circumstances of a new case before starting an examination of the evidence. Shaw and Browne [15] discuss and summarize the weaknesses of digital triage. The enhanced previewing is suggested as an alternative to digital triage. The Linux forensic distribution CAINE [49] installed on a compact disc (CD) is chosen as a base for the implementation. The bootable CD is remastered to include the existing open source forensic tools and to add new analysis software. A high-level overview of system work is presented. The possibilities to deploy the enhanced previewing in the digital forensic laboratory are analyzed. The weaknesses of the enhanced previewing are as follows: the case management becomes more complicated and the system is not suitable to the field use at all. The authors doubt "whether the Enhanced Previewing process is a subset of technical triage or whether it is a distinct process only loosely related to technical triage". We are inclined to state that the enhanced previewing is not a subset of technical triage, because the processing time of the enhanced previewing would be quite long. We base our conclusion on the provided description of the system.

Shiaeles et al. [55] review three open source triage tools and suggest the ways to improve them. The TriageIR, TR3Secure, and Kludge tools are tested for various Microsoft Windows versions. There is currently no mature framework for practically testing and evaluating triage tools, however, the authors do not suggest a framework and evaluate the tools in their best way imagined. The first principle to assess is the access to volatile data. The next principle to assess is the adherence of tools to forensic principles ensuring the admissibility of the collected evidence to the court. An experiment shows that no single considered tool is better than others. All the tools have their strengths and weaknesses. The solution is to preferably have several tools and maintain a profile of the tool capabilities. The recommendations for improving the tools are as follows:

1. The tools should be made more adaptable, either dynamically or manually
2. Disabling Prefetch on Windows systems will result to less system alterations
3. The tools should record and undo all registry changes, which they perform to the examined system

4. The tools should collect the Internet activity artifacts that belong to all known browsers

Woods et al. [56] present an open source software for automated analysis and visualization of disk images created as part of the BitCurator project [57]. The goal of the presented software is to assist in triage tasks. The data for analysis is obtained from open source forensic tools fiwalk [58] and bulk_extractor [45]. The fiwalk tool recognizes and interprets the content of filesystems that are contained in disk images, and produces an XML report. The bulk_extractor tool reads the raw contents of the disk image and reports on various features. The BitCurator reporting tools produce Portable Document Format (PDF) reports on filesystem and for each feature separately. If data entry datasets are large, it is possible to configure the reporting tools to produce the report for a subset of the filesystem or a subset of features. The time required to manage a given disk image with forensic tools fiwalk and bulk_extractor is within the range of tens of minutes. The limiting factor in terms of time is the BitCurator reporting tools that may have to process an extremely large XML filesystem report and text feature reports. The BitCurator project freely distributes these reporting tools in a variety of ways for the practitioners and researchers to use.

Baggili et al. [59] present a five-phase, multi-threaded bootable tool Forensics2020 for forensics triage. The tool is loaded from a bootable Windows Pre-installation Environment using a USB stick. Phases proceed in sequence, however, while the tool is working, the examiner can interact with the tool to see the results up to that point and to request certain types of data. The first phase collects logical files and their metadata. The second phase analyses every image for the Exchangeable Image File Format (EXIF) data. The third phase explores and classifies each file based on its header. The fourth phase parses executable files for audit and threat purposes. The fifth phase hashes each file and takes the longest time of all the phases. The experiment is carried out to assess the efficacy and a forensic soundness of Forensics2020. In sum, 26.33 TB of data from 57 computers are analyzed. The total time required to complete the process is 10,356 s. The tool makes certain changes to the hard drive; however, the changes are greater in number than those of similar Linux-based tools. Two lessons can be learned from the development of Forensics2020. Firstly, a multi-threaded, multi-stage tool allows the examiner to interact with the evidence while the system is performing the forensics processing. Secondly, the mounting of the hard drive by a bootable tool has influence over the perception of the forensic soundness.

Haggerty et al. [60] propose an approach to automate the visualization of quantitative and qualitative email data to assist the triage of digital evidence during a forensics investigation. The quantitative information, which is retrieved from the email, refers to the network events and actor relationships. The qualitative information refers to the body of the emails themselves. The authors have developed a TagSNet software to implement the proposed approach. The software provides two views—a network of the actors and a tag of keywords that are found in the email bodies. Both views are interactive in that the forensics examiner may move the actors and text around. The experiment is carried out on the Enron email data. The average time to process and visualize email data is about 10 min. However, the visualization is not aimed at answering the investigative questions; it only aids the forensics examiner to triage email data more quickly than in the manual mode.

Vidas et al. [61] describe a free forensic tool, OpenLV, which can be deployed in the field and in the laboratory. It is noteworthy that over the past years it has been used under the name of "LiveView". The interface of the tool is oriented to the examiners with little training. The OpenLV asks for configuration and creates a virtual machine out of a forensic image or physical disk. The virtual machine enables booting up the image and gains an interactive environment without modifying the underlying image. The tool natively supports only the dd/raw image format. Other formats require third party software that can be integrated into the tool, which is Windows centric, and a limited Linux support is added. Additionally, the OpenLV aids to remove the barrier of passwords for Windows users. The authors claim that "OpenLV aims to meet the demand for an easy-to-use triage tool", however, neither an example nor a reference is provided for how OpenLV is used for triage purposes.

Conway et al. [62] discuss a development of a Virtual Crime Scene Simulator (VCSS) that can perform a live triage of digital devices. Training is important for the law enforcement officers; therefore, the tool will have a field of its application. The VCSS is an open source project, and it is implemented as game playing, where Unity3D [63] is chosen as the base platform. The virtual environment includes a three-dimensional (3D) representation of a house with four rooms, a hallway, and outside scenery. The crime scenery has a set of the following items: furniture, various hardware devices, and an avatar for interrogation. The following in-game actions are possible: live examination of the various digital devices, interrogation of the avatar, and other actions related to the crime scene. The full device interaction is implemented on Windows version only. The trainer can add new logic by modifying the existing JavaScript. The law enforcement officers from a developing country used the VCSS for training. The participants highly evaluated the educational purpose of the application.

Hegarty and Haggerty [64] present the SlackStick approach to identify files of interest for forensic examiner on the live system. The approach is based on the signatures of the files. To create the signature of the file, a block within the original file is chosen, which may be from anywhere within a file, except for the first and the last blocks. Several predetermined bytes are chosen to represent the file. The number of bytes can be chosen by balancing the tradeoff between the false positives and false negatives. The higher number of bytes decreases the likelihood of false positives. The SlackStick software written in Python under Slax operating system [65] runs from an external device. SlackStick reads the memory blocks on the target machine sequentially to generate block signatures for comparison with the signature library. If a match is found, a report that includes the matched signature and the physical location of the file in the storage media is generated. They conducted an experiment in which it took a dozen of seconds to analyze 1 GB partition that has 2 194 JPEG images. Signatures are generated by selecting 11 bytes within the second block of each target files. Neither false positives nor false negatives are found. As the number of signatures increases, no measurable impact on performance is observed.

Further, van Beek et al. [66] introduce a development of the distributed digital forensic system HANSKEN [67] that is the successor of the operating digital forensic system XIRAF [68]. The goal of HANSKEN is to speed-up the computations of big data. The three forensic drivers for the system are as follows: minimization of the case lead time, maximization of the trace coverage, and specialization of the people involved. These drivers justify the building of the distributed big data forensic platform. To mitigate the threats associated with a big data platform, the development of the system HANSKEN is based on eight design principles. They are enumerated in the order of the priority: 1. Security, 2. Privacy, 3. Transparency, 4. Multi tenancy, 5. Future proof, 6. Data retention, 7. Reliability, and 8. High availability. The first three principles are sociological; meanwhile the other five are business principles and define the system boundaries. The system uses its own forensic image format. The authors justify the need for its own format; however it could be the limitation of the system, especially for the future development. The system HANSKEN stores the data compressed and encrypted. The encryption of data ensures a restricted access to it. The process of extracting data from a forensic image starts as soon as the first bits of the image are uploaded to the system. Such approach acknowledges the right organization of the forensic processes to improve the efficiency of the forensic investigation. The authors admit that triage is a valuable approach for ordering the processing of images, not for leaving images unprocessed. Such form of triage is planned to be included into the system HANSKEN. The system is implemented on the Hadoop realization of MapReduce. The system HANSKEN was planned to be put into production at the end of the year 2015.

Koven et al. [69] further explore and develop the idea of email data visualization [60]. The authors present a visual email search tool InVEST. Firstly, the tool preprocesses the email data to create indexes for various email fields. The duplicate information and junk data are excluded from indexing. Next, the user starts the search process with defined keywords. The search results are presented in five different visual views. The visual views enable better understanding and interpreting of the search results as well as finding the relationships between the search entities. The diverse views show different relationships between search entities and present the contextual information found within these results. All the views support the possibility to refine the search results using filtering and expanding. The process of filtering and expanding is iterative until the search is successful. An experiment is carried out on the Enron email data set. Two case studies are successfully investigated. Koven et al. [69] used the term "triage" in the title of the paper. The term "triage" is used in the sense of a tool, which allows selecting a subset of the emails that are related to a particular subject from the whole email set. However, the time spent to select can be quite long. The process of selecting the subset of the email is interactive heavily involving the user. The authors present an example that "the time to make the discovery and exploration including the skimming of at least 30 of the discovered emails was approximately 1 h". Therefore, the use of the tool in triage process is quite unlikely, unless the data captured is only in form of email.

6. LESSONS LEARNED FROM THE REVIEW

To summarize the field of live triage, the noteworthy research focusses are as follows:

1. The stress of a real-time computation problem having allotted limited time and resources for triage, presented by Roussev et al. [7]. The idea is that an increase in the performance can be achieved if acquisition and processing start and complete at almost same time. The implementation of the forensic system HANSKEN [66] proves the appropriateness of the presented idea
2. The selective imaging approaches to reduce data volume, presented by Grier and Richard III [22] and Quick and Choo

[31,32]. The difference between the approaches is in selecting the regions that have a forensic value. Grier and Richard III [22] state that the profiles must be created and stored in a library. Moreover, Quick and Choo [32] suggest the idea of thumbnailing video, movie, and picture files

3. The introduction of triage template pipelines into the investigative process for the most popular types of digital crimes, presented by Overill et al. [20]. However, the authors do not enumerate these types of crimes and provide only the DDoS and P2P template diagrams without the discussion of the details

4. The artificial intelligence approaches presented by Turnbull and Radhava [26] and Peersman et al. [29]. Turnbull and Randhawa [26] describe an approach to assist a less technically intrinsic user to run a triage tool. Peersman et al. [29] present an approach to automatically label new child sexual abuse media

To summarize the field of post-mortem triage, the noteworthy research focusses are as follows:

1. Storing and using the knowledge of the past cases, presented by Horsman et al. [10,37] and Bashir and Khan [39]

2. The use of machine learning techniques, presented by Marturana and Taconi [33,34,35], McCleland and Marturana [36], and Fahdi et al. [41]. The trend is promising because such techniques are indeed valuable in many research areas; however, the presented research works are immature

To summarize the field of triage of mobile devices, the noteworthy research achievement is only single one:

The information recovery engine DEC0DE, offered by Walls et al. [1.43] and the information prioritization system LIFTR, which uses the data obtained from DEC0DE, offered by Varna et al. [44]

To summarize the field of triage tools, the noteworthy research achievements are as follows:

1. The method of similarity digests, offered by Roussev and Quates [12]

2. The online GRR Rapid Response system used for incident response, offered by Moser and Cohen [1]

3. The multi-threaded bootable tool Forensic2020, which allows interaction of the examiner, while the tool is processing data, offered by Baggili et al. [59]

4. The visualization of email data offered by Haggerty et al. [60]. Koven et al. [69] presented an approach of email data visualization, as well. However, the provided runtimes are quite long and, therefore, the tool is not suitable for triage purposes
5. The SlackStick approach to identify the files of interest, when several predetermined bytes are chosen to represent the file, offered by Hegarty and Haggerty [64]
6. The distributed digital forensic system HANSKEN that works on a big data platform, offered by van Beek et al. [66].

7. CONCLUSIONS AND FUTURE DIRECTIONS

The evolution of modern digital devices is outpacing the scalability and effectiveness of the digital forensic techniques. Digital triage is one of the solutions to this problem, as it can extract intelligence quickly at the crime scene and provides valuable information to the forensic examiner. This form of triage is known as the live triage. In a similar way, such methodology can be used in a laboratory to prioritize the analysis of digital media and to alleviate the examination backlog. This form of triage is known as the post-mortem triage. The term "forensic" should be used carefully with digital triage, because the process of digital triage does not always adhere to the rules of the forensic process. Moreover, the legitimacy of the process depends on the jurisdiction system of a specific country. Therefore, the digital triage model must be adjusted individually according to the legal system of a specific country. Live triage raises important legal concerns. Sometimes, the process of digital triage in the forensic context is an admission of failure, since important evidence can be overlooked. However, a better approach does not exist for today. To solve this problem, we have to consider digital triage as a technical process that provides information for the forensic examination and does not involve the evaluation of digital evidence.

To increase the performance of digital triage when data cloning procedure is involved, data cloning and data processing should start and complete almost at the same time. Such approach is advanced and it is applied in some forensic tools (for example, the digital forensic system XIRAF). Moreover, it means that data processing should be as fast as data cloning.

The number of mobile devices is increasing quite rapidly. Digital triage of mobile devices is harder than that of desktop computers. It also appears that due to the nature of mobile devices, many forensics procedures must inevitably involve live forensics as the device needs to be powered on. To address this problem, more effective methods for live triage and tools for mobile devices are necessary.

High-level training is required from the examiners on the field. It would be cost-effective to hire a less technically fluent specialist for such job. One possibility to accomplish this is to alter the software to make it friendlier to less skilled technical examiners or to construct specific tools that can incorporate the expert knowledge. The core concept of this approach is that the expert systems can locate and interpret the low-level computing artefacts and provide higher-level concepts. Reducing the need to locate and interpret the low-level artefacts is a still lesser-explored method for conducting digital triage.

Another direction of the research in digital triage could be incorporating intelligent technologies, i.e., techniques from artificial intelligence, computational modelling, and/or social network analysis. Almost no research works of digital triage that are directed at the vast area of the social networks are presented. Furthermore, a new developing area of the Internet of Things is left almost without attention as well.

Large volumes of data are available for forensic examination. Since every desktop computer has several processors, available resources can be applied to speed-up the computations. The methods of parallel processing are already applied; however, only in small quantities so far.

Moreover, a challenging research direction is awaiting the researchers' attention. More precisely, it is the implementation of the algorithms into hardware, because the hardware inherently performs the computations faster than the software.

ACKNOWLEDGMENTS

We express gratitude to anonymous reviewers for the valuable comments.

CONFLICTS OF INTEREST

The authors declare no conflict of interest.

REFERENCES

1. Moser, A.; Cohen, M.I. Hunting in the enterprise: Forensic triage and incident response. Digit. Investig. 2013, 10, 89–98.
2. Rogers, M.K.; Goldman, J.; Mislan, R.; Wedge, T.; Debrota, S. Computer forensics field triage process model. J. Digit. Forensic Secur. Law 2006, 1, 27–40.
3. Casey, E.; Ferraro, M.; Nguyen, L. Investigation delayed is justice denied: proposals for expediting forensic examinations of digital evidence. J. Forensic Sci. 2009, 54, 1353–1364.
4. Casey, E. Triage in digital forensics. Digit. Investig. 2013, 10, 85–86.
5. Casey, E. Differentiating the phases of digital investigations. Digit. Investig. 2016, 19, A1–A3.
6. Venčkauskas, A.; Jusas, V.; Paulikas, K.; Toldinas, J. A methodology and tool for investigation of artifacts left by the BitTorrent client. Symmetry 2016, 8, 40.
7. Roussev, V.; Quates, C.; Martell, R. Real-time digital forensics and triage. Digit. Investig. 2013, 10, 158–167.
8. Montasari, R.A. Formal Two Stage Triage Process Model (FTSTPM) for Digital Forensic Practice. Int. J. Comput. Sci. Secur. 2016, 10, 69–87.
9. Cantrell, G.; Dampier, D.A. Implementing the automated phases of the partially-automated digital triage process model. J. Digit. Forensics Secur. Law 2012, 7, 99–116.
10. Horsman, G.; Laing, C.; Vickers, P. A case-based reasoning method for locating evidence during digital forensic device triage. Decis. Support Syst. 2014, 61, 69–78.
11. Koopmans, M.B.; James, J.I. Automated network triage. Digit. Investig. 2013, 10, 129–137.
12. Roussev, V.; Quates, C. Content triage with similarity digests: The M57 case study. Digit. Investig. 2012, 9, S60–S68.
13. Cantrell, G.; Dampier, D.; Dandass, Y.S.; Niu, N.; Bogen, C. Research toward a partially-automated, and crime specific digital triage process model. Comput. Inf. Sci. 2012, 5, 29–38.
14. Hong, I.; Yu, H.; Lee, S.; Lee, K. A new triage model conforming to the needs of selective search and seizure of electronic evidence. Digit. Investig. 2013, 10, 175–192.
15. Shaw, A.; Browne, A. A practical and robust approach to coping with large volumes of data submitted for digital forensic examination. Digit. Investig. 2013, 10, 116–128.
16. Pollitt, M.M. Triage: A practical solution or admission of failure. Digit. Investig. 2013, 10, 87–88.
17. Lopez-Rojas, E.A.; Axelsson, S. Using the RetSim fraud simulation tool to set thresholds for triage of retail fraud. In Lecture Notes in Computer Science, Proceedings of the 20th Nordic Conference on Secure IT Systems, Stockholm, Sweden, 19–21 October 2015; Springer: Cham, Switzerland, 2015; Volume 9417, pp. 156–171.
18. Perumal, S.; Norwawi, N.M.; Raman, V. Internet of Things (IoT) digital forensic investigation model: Top-down forensic approach methodology. In Proceedings of the Fifth International Conference on Digital Information Processing and Communications (ICDIPC), Sierre, Switzerland, 7–9 October 2015; pp. 19–23.
19. Auberson, M.; Baechler, S.; Zasso, M.; Genessay, T.; Patiny, L.; Esseiva, P. Development of a systematic computer vision-based method to analyse and compare images of false identity documents for forensic intelligence purposes—Part I: Acquisition, calibration and validation issues. Forensic Sci. Int. 2016, 260, 74–84.
20. Overill, R.E.; Silomon, J.A.M.; Roscoe, K.A. Triage template pipelines in digital forensic investigations. Digit. Investig.2013, 10, 168–174.

21. Lim, K.-S.; Lee, C. A framework for unified digital evidence management in security convergence. Electron. Commer. Res. 2013, 13, 379–398.
22. Grier, J.; Richard, G.G., III. Rapid forensic imaging of large disks with sifting collectors. Digit. Investig. 2015, 14, S34–S44.
23. Garfinkel, S.; Malan, D.; Dubec, K.-A.; Stevens, C.; Pham, C. Advanced forensic format: an open extensible format for disk imaging. In Proceedings of 2nd International Conference on Digital Forensics, Orlando, FL, January 29-February 01 2006; Advances in Digital Forensics II. IFIP Advances in Information and Communication, vol. 222; Springer: Boston, MA, 2006; pp. 13–27.
24. The Sleuth Kit. Autopsy. Available online: http://www.sleuthkit.org/autopsy/ (accessed on 25 March 2017).
25. Penrose, P.; Buchanan, W.J.; Macfarlane, R. Fast contraband detection in large capacity disk drives. Digit. Investig.2015, 12, S22–S29.
26. Turnbull, B.; Randhawa, S. Automated event and social network extraction from digital evidence sources with ontological mapping. Digit. Investig. 2015, 13, 94–106.
27. Hitchcock, B.; Le-Khac, N.-A.; Scanlon, M. Tiered forensic methodology model for Digital Field Triage by non-digital evidence specialists. Digit. Investig. 2016, 16, S75–S85.
28. Leimich, P.; Harrison, J.; Buchanan, W.J. A RAM triage methodology for Hadoop HDFS forensics. Digit. Investig. 2016, 18, 96–109.
29. Peersman, C.; Schulze, C.; Rashid, A.; Brennan, M.; Fischer, C. iCOP: Live forensics to reveal previously unknown criminal media on P2P networks. Digit. Investig. 2016, 18, 50–64.
30. iCOP. iCOP Toolkit. Available online: http://scc-sentinel.lancs.ac.uk/icop/?q= content/icop-toolkit (accessed on 25 March 2017).
31. Quick, D.; Choo, K.-K.R. Big forensic data reduction: Digital forensic images and electronic evidence. Clust. Comput. J. Netw. Softw. Tools Appl. 2016, 19, 723–740.
32. Quick, D.; Choo, K.-K.R. Data reduction and data mining framework for digital forensic evidence: Storage, intelligence, review and archive. Trends Issues Crim. Crim. Justice 2014, 480, 1–11.
33. Marturana, F.; Tacconi, S. A machine learning-based triage methodology for automated categorization of digital media. Digit. Investig. 2013, 10, 193–204.
34. Marturana, F.; Me, G.; Berte, R.; Tacconi, S. A quantitative approach to triaging in mobile forensics. In Proceedings of the International Joint Conference of IEEE TrustCom-11, Changsha, China, 16–18 November 2011; pp. 582–588.
35. Marturana, F.; Bertè, R.; Tacconi, S.; Me, G. Triage-based automated analysis of evidence in court cases of copyright infringement. In Proceedings of the First IEEE International Workshop on Security and Forensics in Communication Systems (SFCS 2012), Ottawa, ON, Canada, 14–15 June 2012; pp. 6668–6672.
36. McClelland, D.; Marturana, F. A Digital Forensics Triage methodology based on feature manipulation techniques. In Proceedings of the International Conference on Communications, Sydney, Australia, 10–14 June 2014; pp. 676–681.
37. Horsman, G.; Laing, C.; Vickers, P. A case based reasoning framework for improving the trustworthiness of digital forensic investigations. In Proceedings of the 11th IEEE International Conference on Trust, Security and Privacy in Computing and Communications (TrustCom-2012), Liverpool, UK, 25–27 June 2012; pp. 682–689.
38. EnCase Portable. Available online: http://www.guidancesoftware.com/encase-portable (accessed on 25 March 2017).
39. Bashir, M.S.; Khan, M.N. A triage framework for digital forensics. Comput. Fraud Secur. 2015, 2015, 8–18.
40. Dalins, J.; Wilson, C.; Carman, M. Monte-Carlo filesystem search: A crawl strategy for digital forensics. Digit. Investig.2015, 13, 58–71.
41. Al Fahdi, M.; Clarke, N.L.; Li, F.; Furnell, S.M. A suspect-oriented intelligent and automated computer forensic analysis. Digit. Investig. 2016, 18, 65–76.
42. Mislan, R.P.; Casey, E.; Kessler, G.C. The growing need for on-scene triage of mobile devices. Digit. Investig. 2010, 6, 112–124.

43. Walls, R.J.; Learned-Miller, E.; Levine, B.N. Forensic triage for mobile phones with DECODE. In Proceedings of the 20th USENIX Conference on Security, San Francisco, CA, USA, 8–12 August 2011; pp. 1–14.
44. Varma, S.; Walls, R.J.; Lynn, B.; Levine, B.N. Efficient smart phone forensics based on relevance feedback. In Proceedings of the 4th ACM Workshop on Security and Privacy in Smartphones & Mobile Devices (SPSM '14), Scottsdale, AZ, USA, 3–7 November 2014; pp. 81–91.
45. Garfinkel, S.L. Digital media triage with bulk data analysis and bulk_extractor. Comput. Secur. 2013, 32, 56–72.
46. Yaffs. Available online: http://www.yaffs.net/ (accessed on 25 March 2017).
47. Guido, M.; Buttner, J.; Grover, J. Rapid differential forensic imaging of mobile devices. Digit. Investig. 2016, 18, S46–S54.
48. sdhash home. Available online: http://roussev.net/sdhash/sdhash.html (accessed on 25 March 2017).
49. CAINE. Computer Forensics Linux Live Distro. Available online: http://www.caine-live.net/ (accessed on 27 March 2017).
50. KALI Tools. RegRipper. Available online: http://tools.kali.org/forensics/regripper (accessed on 25 March 2017).
51. Lim, K.-S.; Savoldi, A.; Lee, C.; Lee, S. On-the-spot digital investigation by means of LDFS: Live Data Forensic System. Math. Comput. Model. 2012, 55, 223–240.
52. Casey, E.; Katz, G.; Lewthwaite, J. Honing digital forensic processes. Digit. Investig. 2013, 10, 138–147.
53. Garfinkel, S.; Nelson, A.; White, D.; Roussev, V. Using purpose-built functions and block hashes to enable small block and sub-file forensics. Digit. Investig. 2010, 7, S13–S23.
54. Cohen, M.; Bilby, D.; Caronni, G. Distributed forensics and incident response in the enterprise. Digit. Investig. 2011, 8, S101–S110.
55. Shiaeles, S.; Chryssanthou, A.; Katos, V. On-scene triage open source forensic tool chests: Are they effective? Digit. Investig. 2013, 10, 99–115.
56. Woods, K.; Lee, C.A.; Misra, S. Automated analysis and visualization of disk images and file systems for preservation. In Proceedings of the 10th IS & T Archiving Conference, Washington, DC, USA, 2–5 April 2013; pp. 239–244.
57. BitCurator. Available online: https://www.bitcurator.net/ (accessed on 25 March 2017).
58. Fiwalk. Available online: http://www.forensicswiki.org/wiki/Fiwalk (accessed on 25 March 2017).
59. Baggili, I.; Marrington, A.; Jafar, Y. Performance of a logical, five-phase, multithreaded, bootable triage tool. Advances in Digital Forensics X. In IFIP Advances in Information and Communication Technology; Springer: Berlin/Heidelberg, Germany, 2014; pp. 279–295.
60. Haggerty, J.; Haggerty, S.; Taylor, M. Forensic triage of email network narratives through visualisation. Inf. Manag. Comput. Secur. 2014, 22, 358–370.
61. Vidas, T.; Kaplan, B.; Geiger, M. OpenLV: Empowering investigators and first-responders in the digital forensics process. Digit. Investig. 2014, 11, S45–S53.
62. Conway, A.; James, J.I.; Gladyshev, P. Development and Initial User Evaluation of a Virtual Crime Scene Simulator Including Digital Evidence. In Lecture Notes of the Institute for Computer Sciences Social Informatics and Telecommunications Engineering, Proceedings of the 7-th International ICST Conference on Digital Forensics and Cyber Crime (ICDF2C), Seoul, South Korea, October 6–8, 2015; Springer: Cham, Switzerland, 2015; Volume 157, pp. 16–26.
63. Unity. Available online: https://unity3d.com/ (accessed on 27 March 2017).
64. Hegarty, R.; Haggerty, J. SlackStick: Signature-based file identification for live digital forensics examinations. In Proceedings of 2015 European Intelligence and Security Informatics Conference, Manchester, UK, 7–9 September 2015; pp. 24–29.
65. Slax Linux. Available online: https://www.slax.org/ (accessed on 27 March 2017).

66. Van Beek, H.M.A.; van Eijk, E.J.; van Baar, R.B.; Ugen, M.; Bodde, J.N.C.; Siemelink, A.J. Digital forensics as a service: Game on. Digit. Investig. 2015, 15, 20–38.
67. Hansken. Available online: https://www.forensicinstitute.nl/ products_and_services /forensic_products/hansken.aspx (accsessed on 27 March 2017).
68. Bhoedjang, R.A.F.; van Ballegooij, A.R.; van Beek, H.M.A.; van Schie, J.C.; Dillema, F.W.; van Baar, R.B.; Ouwendijk, F.A.; Streppel, M. Engineering an online computer forensic service. Digit. Investig. 2012, 9, 96–108.
69. Koven, J.; Bertini, E.; Dubois, L.; Memon, N. InVEST: Intelligent visual email search and triage. Digit. Investig. 2016, 18, S138–S148.

CHAPTER 17

The Psychopath as an Irrational Agent: A Canadian Criminal Law Perspective by Kevin Moustapha

Kevin Moustapha*

LLD Candidate and lecturer at the University in Montreal (Faculty of law), Canada

ABSTRACT

The purpose of this study is to look at the capacity of psychopaths to act in a rational manner within the meaning of section 16 of the Canadian *Criminal Code*. We therefore focus on certain neuroscientific studies in order to see whether they can inform us about the disorders that psychopaths suffer from, and whether those disorders could have an impact on their ability to act rationally. If they do, then psychopathy could be an ailment that triggers the section 16 defence of not criminally responsible owing to a mental disorder and also the implications of the Supreme Court of Canada decision in *Bouchard-Lebrun*.

INTRODUCTION

Thanks to the advanced analyses and diagnoses that science has provided us with in recent years, more and more questions are being raised regarding the precise nature of psychopathy. Initially considered to be a personality disorder in international classifications, psychopathy is now a major subject of scientific discussion on brain dysfunctions. In the field of neuroscience, for example, the studies by Kent Khiel, R.J.R Blair and Adrien Raine have made it possible to corroborate a number of symptoms linked with psychopathy. According to them, the dysfunction of certain areas of the brain in psychopathic individuals could be the direct origin of their antisocial behaviour.

Among those studying the Canadian insanity defence, this theory has sparked interest in that it could provide a partial explanation of why psychopaths sometimes behave irrationally. Nonetheless, although the plenitude of these various studies allows us to glean increasingly relevant information about the nature of psychopathy, taken as a whole, these medical and neuroscientific theories are insufficient to establish with certainty that dysfunction in certain regions of the human brain inexorably leads the individual to commit criminal offences. However, developments in medical science, coupled with the valuable teachings of neuroscience, open the way to a debate on the possible application of the Canadian insanity defence to psychopathy. Section 16 of the *Criminal Code* excuses individuals who are unable to understand their environment and interact with others in a rational manner. In Canadian criminal law, the inability to make a rational choice can flow from a whole range of mental disorders. As the justices in *Bouchard-Lebrun* stated, criminal law presumes that every person is independent and rational, and that his or her acts and omissions incur liability [1]. In this regard, we think that psychopathy is a disorder that could overturn that presumption, referred to in *Bouchard-Lebrun*. We view emotional and volitional problems as responsible for mental disturbances that deprive a psychopath of the ability to assess, rationally, what he or she is doing. Psychopaths' clear failures on the moral level (circuit of emotions) and the difficulties they may encounter when making decisions (volition) lead them perceive a bad action as good or justifiable. In order to explore this, we will begin by analysing the link between the notions of psychopathy and irrationality (I). Next, we will see that, thanks in particular to the contributions of neuroscience, it is now possible to state that psychopaths are unable to act rationally within the meaning of Canadian criminal law. Nonetheless, we have to ask whether considering them not criminally responsible might not be such a good idea (II).

the link between the notions of psychopathy and irrationality

A) Rationality, a fuzzy, even elusive concept

In the strict sense of the term, rationality refers to the nature of what is rational [2]. The concept of rationality comes into play "every time we view human behaviour from the angle of action, more specifically, action that obeys reasons" [3]. Above all, the principal characteristic of rationality is that it confers a form of intelligibility and predictability to all of human behaviour[1]. Finally, rationality seems to have an explanatory function[2]. The term "rational" provides us with further information. From the Latin rationalis, the term "rational" refers to all that is specific to reason [4]. This therefore implies that, to be rational, an action or form of behaviour must appear logical, reasonable and in accordance with good sense [4]. While making a rational choice often refers to economic, or even political or sociological, reasons [5], we have to admit that rational choice theory has gradually been exported to more distant lands, such as criminology [6] and criminal law [7]. Beyond academic studies of the notion of rationality, it is useful to note that it is difficult to unravel this concept with precision, no matter what the discipline in the human and social sciences [8]. Law, and more specifically criminal law, is no exception to this because neither case law nor doctrine seems to provide a definition of rationality in the legal sense of the term. Stephen J. Morse also refers to these difficulties when he tried to decipher the concepts of rationality and irrationality: "Rationality and irrationality are continuum concepts. The degree of irrationality required [to conclude that an agent is not responsible] might vary from context to context. For example, the amount of irrationality that would permit a finding of incompetence to contract need not be the same as the amount necessary to avoid criminal liability [9]. These highly useful comments by Morse confirm that the concepts of rationality and irrationality remain fuzzy and almost elusive. However, it is above all the very relative dimension of these notions that stands out. Depending on the area of law in question, these terms have very different meanings. From this observation thus flows the fact that the notions of rationality and irrationality are normative, certainly, but above all contingent. This idea can also be found in Morse's writings: If human beings were not creatures who could understand and follow the rules of their society, who could not be guided by reasons, the law and all other systems, such as morality, that regulate conduct by reasons and rules would be powerless to affect human action. Rule followers must be creatures who are capable of properly using the rules as premises in practical reasoning. It follows that a legally responsible agent is a person who is so capable according to some contingent, normative notion of both rationality itself and how much capability is required. For example, legal responsibility might require the capacity to understand the reason for an applicable rule, as well as the rule's narrow behaviour command. These are matters of moral, political, and, ultimately, legal judgment, about which reasonable people can and do differ. There is no uncontroversial definition of rationality or of what kind and how much is required for responsibility. Finally, these difficulties resulting from the definition of the concept of rationality in criminal law, and more specifically with respect to the insanity defence, give judges some latitude since the concept is flexible. The flexibility seems to be a real advantage for psychopathic individuals, who have rarely been associated with any form of irrationality in criminal law, despite the seriousness but especially the impact of their disorders with regard to their antisocial behaviour. This observation has to be taken into account in the structure of our reflection since we hold that neuroscience can explain in part why psychopaths suffer from a form of irrationality, and that this can be used as a basis for showing that psychopaths meet the second criterion of section 16 of the *Criminal Code*. The dysfunction of areas linked with emotions has a significant effect on psychopaths' ability to distinguish well from bad, and thereby make rational choices in accordance with that distinction. As philosopher Samuel Lepine says with exemplary clarity, "psychopaths certainly may wield the concepts of good and bad, but those concepts are devoid of their evaluative and prescriptive value in their eyes. In other words, psychopaths seem to be indifferent with respect to moral rules very simply because they do not consider that things are really morally good or bad, but only that they are considered good or bad by other people". It seems that the conflict between reason and emotions when determining volition has now become anachronistic. The anachronism flows obviously from recent contributions by neuroscience, which, instead of placing reason and emotions in opposition, associates them in moral judgment, as well as in decision making. Rather than seeing reason and emotions as enemies in analysis of psychopathy, some neuroscientists make them allies in order to show that they are essential and above all complementary in an individual's decision-making process.

Today, we can also assess the impact of neuroscientific disciplines against the measure of studies on the complementarity between reason and emotions, and their impact on the moral judgment indispensable in an individual's decision-making process. Indeed, it is very interesting to note that most neuroscientists studying the issue base their work on psychopathic dysfunctions to illustrate the very essence of their respective reflections [10]. Can we consider that it is irrational for an individual to be amoral? Finally, is the principle of rationality a function of our ability to use our reason or of our ability to use our emotions? In order to best develop the concept of irrationality in psychopathy, we need to identify the principal questions that have to be raised to understand how reason and emotion are hinged together, and especially whether an individual's inability to analyse his or her environment on the moral level flows exclusively from an emotional deficit, or is instead related to reasoning disorders.

B) The neuroscientific approach to psychopaths' rationality: between reason and emotions

One of the first neuroscientists to describe the brain as a sophisticated device indispensable for an individual to have free choice is Michael

Gazzaniga. He sees it as an instrument that makes it possible for an individual to interact with his or her environment so as to be able to learn how to determine his or her own way of reacting, and as an evolving system for making decisions [11]. However, it is Antonio Damasio who places the faculty of reason at the centre of biological regulation mechanisms involving, notably, emotions. According to Damasio, our moral judgment, and general, social and moral decision-making occurs thanks to the collaboration of different levels of our brain [12]. He replaces emotion at the centre of the individual's decision process, and makes it the perfect ally of reason.

Thus, Damasio highlights the fact that emotion plays a role in reason and assists in the reasoning process, instead of necessarily disturbing it, as was supposed by a certain Cartesian line of thought, of which one of the best examples is none other than René Descartes himself. Damasio's line of thought reminds us once again to what point emotions are essential in an individual's adaptive process. His research is all the more remarkable in that it makes it possible to establish the foundations for an analysis of psychopathy from the point of view of an allegedly unbreakable link between reason and emotions. In his work, *Descartes' Error*, he bases his arguments on the concrete example of a nineteenth-century psychopath named Phineas Gage. In order to fully understand the structure of Damasio's thought and see how it is linked with psychopathic disorders, we need to take a short look at this case. Phineas Gage, foreman of a company specializing in building railroads and respected by his peers for his abilities, skillfulness and good manners, was the victim of a terrible work accident. Following an error, a 6-kilo, 3 centimetre thick, 110-centimetre long iron bar was blown right through his head. Against all expectations, Gage survived his injuries. However, the most important aspect of the event is to be seen in regard to the extraordinary change in personality that the man subsequently experienced. His behaviour changed radically. Whereas he used to be jovial and peaceful, he became irritable and violent to such a point that his employer had to let him go [13]. At the time, science did not have the tools necessary to diagnose an antisocial personality disorder, or even a psychopathic disorder. It was thus only a century and a half later, under the influence of researchers such as Damasio, that neuroscientists, aided by the formidable technological progress in tools for observing the human brain, were able to locate the damaged brain areas that were at the origin of Gage's change in personality and his lack of morality. Damasio uses the theory of somatic markers to explain Gage's new behaviour [12]. The somatic marker hypothesis entails that emotions participate in reason and that they can assist the reasoning process, instead of hindering it. His explanation of this idea is very clear: Because the feeling is about the body, I gave the phenomenon the technical term somatic state ("soma" is Greek for body); and because it "marks" an image, I called it a marker. Note again that

I use somatic in the most general sense (that which pertains to the body) and I include both visceral and nonvisceral sensation when I refer to somatic markers. What does the somatic marker achieve? It forces attention on the negative outcome to which a given action may lead, and functions as an automated alarm signal which says: Beware of danger ahead if you choose the option which leads to this outcome. The signal may lead you to reject, immediately the negative course of action and thus make you choose among other alternatives. The automated signal protects you against future losses, without further ado, and then allows you to choose from among fewer alternatives. There is still room for using a cost/benefit analysis and proper deductive compe¬tence, but only the automated step drastically reduces the number of options. Somatic markers may not be sufficient for nor¬mal human decision-making since a subsequent process of reason¬ing and nal selection will still take place in many though not all instances. Somatic markers probably increase the accuracy and e ciency of the decision process. Their absence reduces them [14]. Thus, when the circuit of emotions (of the brain) is affected, as in the case of Phineas Gage, the entire decision process is affected. Given that emotion and reason work together in establishing moral judgment, an individual who suffers emotional dysfunctions may begin to make inconsistent decisions that will gradually take him or her down the path of irrationality. A series of antisocial forms of behaviour adopted by Gage after his accident provides a perfect illustration of this idea, thereby confirming the idea that emotional problems resulting from brain problems can foster antisocial behaviour. Moreover, it is important to note that in the most extreme cases, such as that of Gage, the individual's behaviour is not only antisocial but also completely irrational. This is what Damasio very rightly reminds us when discussing psychopathic disorder; developmental sociopaths or psychopaths are well known to all of us from the daily news. They steal, they rape, they kill, and they lie. They are often smart. The threshold at which their emotions kick in, when they do is so high that they appear unflappable, and are, from their self-reports, unfeeling and uncaring. They are the very picture of the cool head we were told to keep in order to do the right thing. In cold blood and to everybody's obvious disadvantage including their own, sociopaths often repeat their crimes. They are in fact yet another example of a pathological state in which a decline in rationality is accompanied by diminution or absence of feeling. It is certainly possible that developmental sociopathy arises from dysfunction within the same overall system which was impaired in Gage, at cortical or subcortical level. But rather than resulting from blunt macroscopic damage occurring in adulthood, however, the impair¬ment of developmental sociopaths would come from abnormal circuitry and abnormal chemical signaling and begin early in devel¬opment. Psychopathy, which is translated precisely by the dysfunction of an individual's circuit of emotions, therefore seems to be perfectly consistent with the canons of the theory of somatic markers developed by Damasio. Indeed, with respect to the theory of somatic markers, how can we say for certain that a psychopath has the ability to interact with his or her environment in a rational manner when his or her emotional faculties are defective? Psychopathy is the perfect disorder to support Damasio's research because it is one of the only disorders that are described mainly by a major failure at the level of emotions. If our system of reasoning is efficient only because it is an extension of our emotional system, as some neuroscientists say, then it seems inconceivable from a scientific point of view to consider a psychopath as a rational being. Since brain imaging shows that a psychopath absolutely does not have the ability to feel the emotions that he or she expresses (although there seems to be some compensation of affective failures through mimetism) [15], it seems difficult to say with certainty that psychopaths' decision-making process is not deficient. Not only is emotion the perfect complement of reason when establishing a moral judgment, but it is also one of an individual's essential mechanisms for making rational choices. Psychopathic disorders are, in the end, only a striking example of the fact that emotions play a role in making all individuals into rational beings, able to interact with their environment. Deep analysis of a psychopath's rationality leads to questioning of the positions traditionally adopted in philosophy. Confined to analysis flowing from the opposition between rationalism and sentimentalism, the study of rational behaviour now takes the royal road drawn by neuroscience. Contrary to rationalist affirmations, to be rational does not necessarily mean cutting oneself off from one's emotions. In contradiction with what sentimentalists might say, emotions are not the only things that come into play in the establishment of moral judgments: they must be associated with reason. Moral judgment and all the mechanisms at the origin of an individual's decision process thus follow from this association between reason and the emotions. Neuroscientific studies of rational behaviour have made it possible for us to conceptualize the principle of a psychopath's irrationality and to see what the implications in law could be with regard to section 16 of the *Criminal Code* and the jurisprudence established by *Bouchard-Lebrun.*

The psychopath's inability to behave rationally

A) Possible application of the Canadian insanity defence to psychopathic disorders

In line with our analysis above, emotional disorders have significant influence on the immorality of psychopaths. However, to an even stronger degree, such emotional disorders prevent psychopaths from behaving in a rational manner. Regarding the problems related in particular to brain dysfunctions, we think that psychopaths are unable to behave rationally. This hypothesis, though far from being unanimously accepted, has nonetheless been suggested a number of times by some authors, who draw a parallel between psychopathy and irrationality. They include Paul Litton and Heidi Maibom, who have studied in detail the difficulties that psychopaths encounter when trying to behave rationally. These authors base all of their argumentation on the famous inability of the psychopath to behave in a rational manner. According to them, this incapacity is what prevents psychopaths from assessing their actions in relation to public standards set by the state. In order to analyse all of the components of a psychopath's irrationality, we will focus on the various arguments developed by these two authors. To begin with, according to Professor Litton, psychopathic disorders cannot be summarized as simply a deficiency on the moral level. According to him, psychopaths' actions lack consistency, which would explain in part their inability to behave rationally. That inability allegedly flows from the fact that psychopaths cannot evaluate, decipher or analyse signs in their environment accurately. It is not merely that the psychopath has not internalized any moral standards; he has a weakened capacity for possessing any kind of evaluative standards, and that weakened capacity is related to signs of irrationality [...] Individuals with psychopathy may not exhibit the kind of disordered thinking prevalent in psychoses, but agents can be irrational in other ways. Young children are not responsible due to their lack of rational powers, and in many ways, persons with psychopathy are similar [16].

intoxicating substances. Yet, like Paul Litton, we think that psychopathic individuals demonstrate a form of irrationality. According to him, it is above all the absence of morality that determines the irrational actions that psychopaths perform. In the end, what we draw from his arguments is that it is because psychopaths are totally immoral that they are irrational. The fact that they are the complete opposite of what the community perceives as moral or immoral makes them completely irrational. As Litton puts it very clearly: "Indeed, their actions in pursuit of excitement and pleasure, including their immoralities towards those ends, are not unintelligible to us in the same way that, say, a mere desire to place green books all over one's roof is unintelligible". Litton is not, however, the only one to ask questions about psychopaths' ability to behave rationally. For many years, Heidi Maibom has also been studying the impact of immorality on the lack of rationality of psychopaths' actions. According to her, in contrast, psychopaths know how to tell the difference between good and bad. Their immense problem lies in their complete inability to apply that knowledge: As opposed to the criminally insane, psychopaths exhibit knowledge of right and wrong. Nevertheless, their understanding is deviant and incomplete [17]. Maibom corroborates Litton's arguments by pointing out, once again, the difficulties psychopaths have adopting rational behaviour: Psychopathic individuals suffer from principled difficulties that non-psychopathic individuals do not. Compared to the normal population, they have impaired practical rationality. They have problems willing the necessary and sufficient means to their ends, making sure that specific intentions are internally coherent and consistent with the underlying intention, and determining whether foreseeable outcomes of their actions are compatible with their ultimate aim [18]. Thus, both authors come to the same conclusion: psychopaths do not have the ability to behave rationally. While the conclusion is the same, the two authors put forth different reasons explaining the irrationality. Litton places the source of psychopaths' irrationality in their immorality alone, insisting on their inability to follow moral rules set by the community. In contrast, Maibom thinks that psychopaths can tell right from wrong. The only difficulty lies in their inability to apply that knowledge rationally. Thus, in general, the two professors' positions are in the same direction: psychopaths do not have the ability to behave rationally. Psychopaths are unable to internalize the criminal norm and, more broadly, the values protected by society. Since moral judgments are based on emotional reactions that cannot take place without some rational reflection, it seems difficult to say that complete amorality allows an individual to behave rationally. This seems, in fact, to echo the position of Robert Hare, who strengthens our arguments by adding that the case of psychopaths is absolutely incomparable since they are unable to really comprehend what motivates other people, and they very often adopt inconsistent behaviour: One of our subjects, who scored high on the Psychopathy Checklist, said that while walking to party he decided to buy a case of beer, but realized that he had left his wallet at home six or seven blocks away. Not wanting to walk back, he picked up a heavy piece of wood and robbed the nearest gas station, seriously injuring the attendant [19]. That inconsistency is above all a manifestation of the deficiency of psychopaths' emotional systems. Psychopathic disorders indisputably show that the participation of emotion in the reasoning process is absolutely indispensable. While there used to be discord regarding the roles of reason and emotions in the mechanisms at the origin of an individual's decision process, we have to observe that thanks to neuroscientific studies of psychopathy, this is no longer necessarily the case today because it has been proven that these two entities work together.

Thus, given all of these assertions, how should we analyse the case of psychopaths with regard to the Canadian insanity defence?

Under section 16 of the Canadian *Criminal Code*, the criminal responsibility of an individual cannot be attracted if that individual demonstrates that he or she was unable to assess the nature and quality of the action in question, or know whether the action was wrong [20]. Above all, since the clarifications brought by *Bouchard-Lebrun*, we know that Canadian criminal law excuses individuals who are unable to interact with their environment in a rational manner[3].

It is from this point of view that criminal responsibility is assigned with regard to schizophrenic individuals, and people suffering from dementia or paranoid disorders. The defence has also been rejected in a number of cases with respect to psychopathic disorders [21]. Unlike the other disorders, psychopathy is not the direct product of a deficiency at the level of reason alone, but rather a deficiency that breaks the link between reason and emotions.

Yet, as we have seen through the detailed study of neuroscience, this link is fundamental because reason uses emotion to fine tune the decision process and thereby make it possible for an individual to behave rationally. Without emotion, reason cannot express itself fully, and, alone, remains insufficient to guarantee that the actions performed by the individual will be rational. Since an individual's actions are indisputably linked with his or her emotional faculties, it is evident that emotion is one of the most essential manifestations of human free will. All of these statements provide us with strong support in our reflection, and further weaken the perception that the justices of the Supreme Court have of psychopathy. Owing to the specific nature of the disorder, a psychopath could be considered irrational. Psychopathy is unique because it is the only disorder that reveals so clearly the role that emotion plays with respect to reason in the decision process and the capacity of any individual to behave rationally. While psychopaths know perfectly well what the terms good and bad mean, they are absolutely unable to apply that knowledge rationally. They do not have the ability to apply the knowledge of right and wrong rationally because they are unable to behave in compliance with all of the moral principles and values familiar to us. It is in this sense that Hervey Cleckley speaks of a "mask of sanity" when referring to the behaviour of psychopaths in society. That expression highlights the fact that psychopaths claim to feel the emotions that we feel every day, whereas they are unable to really understand them, and especially how to apply them [22]. Psychopaths thus have special status, for, although they can distinguish between right and wrong, they are absolutely unable to use that knowledge rationally. Psychopaths do not have the ability to behave rationally because psychopathic disorders undermine the very essence of the links between reason and emotions. Their thoughts and irrational actions do not flow from misunderstanding of what is good and bad, but indeed from an inability to apply that distinction to concrete situations. If an individual is unable to apply his or her knowledge of right and wrong rationally, to distinguish right from wrong in practical terms, and to adapt his or her actions in light of the moral principles promoted by our society, then we cannot decently consider that individual as criminally responsible. That individual is not criminally responsible because he or she does not have the level of rationality required to attract criminal responsibility. We therefore consider that, under section 16 of the *Criminal Code*, psychopathy, in particular at its most extreme stage, renders an individual unable to understand that the actions that he or she performs are deeply wrong in the sense of *Bouchard-Lebrun*.

B) Psychopaths' lack of criminal responsibility justified by their irrationality: not such a good idea?

The recent teachings of neuroscience have been very useful in thinking about psychopathy. Moreover, they could have a strong influence on possible application of the insanity defence to psychopathy. Nonetheless, the idea of incorporating psychopathy into the insanity defence needs to be tempered. In order to protect the public from the dangerousness of individuals who are often repeat offenders, the legislator and the judge seem inclined to analyse the social and societal consequences of possibly making psychopathy a disorder creating eligibility for the insanity defence. This pragmatism can be explained by the need to protect the public from the danger that psychopaths can entail owing to the disorder they suffer, and also because there is no viable treatment for psychopathy. In all democratic legal systems, fundamental principles govern life in society. The principles take different forms related to notions of justice and morality, or target social objectives pursued by the authorities. Those objectives include public safety [23] which is one of the most important issues with respect to smooth functioning of the state, as well as justice. The concept of public safety echoes the principles of criminal responsibility and the stakes involved in possibly making sufferers of psychopathic disorders eligible for the insanity defence [24]. In the context of our reflection on the very special case of the psychopath, the concept of public safety is indispensable, especially when we know that psychopaths have among the highest repeat offender rates [25]. Deeply disturbed by emotional and volitional disorders, a psychopath is an individual who tends to tirelessly repeat the same criminal actions. Public safety is impacted by this kind of individual, and both the legislator and the judge have to be able to provide appropriate legal and judicial responses [26]. Let us look at an example to illustrate this: in section 753 of the Canadian *Criminal Code*, the legislator has provided that an individual must be declared a dangerous offender if "the offence for which the offender has been convicted is a serious personal injury offence . . . [there is] a pattern of repetitive behaviour by the offender, of which the offence for which he or she has been convicted forms a part, showing a failure to restrain his or her behaviour . . . [or there is] a substantial degree of indifference on the part of the offender respecting the reasonably foreseeable consequences to other persons of his or her behaviour . . ." [27]. When we turn our attention to the principal criteria identified in section 753, we see that there is great similarity between the notion of a dangerous offender and the antisocial behaviour adopted by psychopaths. Thus, we can wonder whether this aspect of criminal law was not designed specifically for this type of individual. Indeed, a dangerous offender is unable to refrain from repeating his or her actions, and demonstrates marked indifference to the reasonably predictable consequences that his or her actions can have on others, which is also the case with individuals suffering from psychopathy. Is this a coincidence? We do not think so because we think that this legislation on dangerous offenders shows a desire to control psychopaths over the long term so as to meet public safety objectives. In substance, is it not specifically their inability to comply with social norms, and above all the determinism that afflicts them, which makes it impossible to align their situation with that of the insane, with respect to both criminal responsibility and sentencing?

The second element of the reflection underlying this principle of public safety is the fact that there is no viable treatment for psychopathy [28]. This information is fundamental from our point of view because the fact that there is no viable treatment and psychopaths' inability to adopt behaviour in compliance with social norms inevitably lead judges to classify them as dangerous offenders and to make them ineligible for the insanity defence. We consider that the determinism that they suffer makes it impossible for them to employ the insanity defence because the judge would be forced to place them in specific structures [29]. In Canadian criminal law, all decisions regarding an accused who is not criminally responsible must be taken in accordance with section 672.54 of the *Criminal Code* [30]. The procedure provided for in that section is, moreover, summarized in clear terms in *Winko*: "The task of the court or Review Board is to determine whether the evidence discloses a "significant" or real risk to the community should the NCR accused be released. If that risk exists, the NCR accused remains under supervision, either subject to conditions while living in the community or detained in a hospital. It does mean, though, that however difficult the task, the court or Review Board cannot avoid the responsibility of making that determination. Just as a jury is obliged to come to a conclusion on an accused's guilt or innocence, so is the court or Review Board required determining whether or not releasing the NCR accused would pose a "significant risk" to public safety. If, at the end the day, the court or Review Board cannot so conclude, the legal justification for confinement is absent and the NCR accused must be released . . ." [31].

In *Swain*, Lamer J. referred to the importance of therapeutic follow up of individuals declared not criminally responsible on account of mental disorder [32]. These two decisions thus highlight the importance of treatment in steps taken to rehabilitate individuals suffering from mental disorders in Canada. However, it is very complicated to provide psychopaths with care. Indeed, statistics show that psychopaths regularly fail when placed in a care facility because they resist treatment [33], and there is abundant research suggesting that they often manipulate those around them [34]. The present literature in Canada suggests also that psychopathy is a stable disorder for most individuals and many researchers remain sceptical about supposed changes in a personality disorder such as psychopathy with time [35]. For some scholars, current methods of treatment for psychopaths show no efficacy and many of them show detrimental effects [35].

Although psychopathic offenders are a challenging population to treat, some researchers consider that psychopathy is not untreatable. For them, these theories are largely based on outdated and poorly designed studies [36]. Moreover, this therapeutic nihilism on treating psychopathy [36] is inconsistent with important recent advances, which have been made in assessing psychopathy and recidivism risks. For Stephen Wong and Mark E. Olver for example, treatment programs in line with the model have been in operation, and the results of early outcome evaluations are encouraging [36]. Unfortunately, recents research from correctional service Canada seems to confirm that psychopaths tend to exploit unstructured programs, masking their resistance with verbal skills [37]. These results are major obstacles for correctional workers trying to provide appropriate intervention [38] and will certainly justify the severity of the judges.

REFERENCES

1. Rc Bouchard-Lebrun (2011) CSC 58: 575.
2. Rationalité
3. Mongin P (2002) Le principe de rationnalité et l'unité des sciences sociales. Revue économique 53: 301-323.
4. Rationnel, Rationnelle.
5. Rule J B (2002) Les leçons du choix rationnel Sociologie et societies 34; Gagne L (1997) Jon Elster et les limites de la rationalité: Choix rationnel et normes dans l'explication des comportements individuels M.A. thesis submitted to the Department of Political Science, Faculty of Graduate Studies at Laval University.
6. Cusson M (2015) Choix rationnel et pensée stratégique Dictionnaire de criminologie en ligne: Poupart J (2002) Choix rationnel et criminologie : limites et enjeux Sociologie et societies 34:133-145.
7. Morse SJ (2003) Diminished rationality, diminished responsibility Ohio State Journal of Criminal Law; Morse SJ (2000) Rationality and Responsibility Southern California Law Review 74; Ulen T.S (1999) Rational choice theory in law and economics Encyclopedia of law and economics 1:790-793.
8. Morse SJ (2003) Diminished rationality, diminished responsibility Ohio State Journal of Criminal Law.
9. Morse SJ (2002) Uncontrollable urges and irrational people. Faculty Scholarship. p. 648.
10. Blair JR (2003) Neurobiological basis of psychopathy. The British Journal of Psychiatry 182: 5-7 ; Raine A, Glenn A (2014) Psychopathy: An introduction to biological findings and their implications. NYU Press.
11. Gazzaniga M (2005) Ethical brain. Dana Press p.101.
12. Damasio AR (1994) Emotion, reason and the human brain. Avon Books.

13. Damasio AR (1994) Emotion, reason and the human brain. Avon Books. p.7-10.
14. Damasio AR (1994) Emotion, reason and the human brain. Avon Books p.173.
15. Cleckley H (2003) The mask of sanity: An attempt to clarify some issues about the so called psychopathic personality. American Journal of Psychiatry 113: 92.
16. Litton P (2008) Responsibility status of the psychopath: On moral reasoning and rational self-governance. Rutgers Law Journal 39: 375.
17. Maibom H Moral (2005) Unreason: The case of psychopathy. Mind & Language 20: 241.
18. Maibom H Moral (2005) Moral unreason: The case of psychopathy. Mind & Language 20: 244-245.
19. Hare R (1993) Without conscience: The disturbing world of the psychopaths among Us. New-York, Guilford Press 125: 40-41.
20. 20Code criminel, LRC (1985) C-46, art 16.
21. Kjeldsen v. La Reine, [1981] 2 RCS 617 ; R. v. Simpson, (1977) 35 C.C.C. (2d) 337.
22. Cleckley H (2003) The mask of sanity: An attempt to clarify some issues about the so called psychopathic personality.
23. Divito v.Canada (Sécurité publique et Protection civile), 2013 CSC 47, [2013] 3 R.C.S. 157.
24. Petrosino AJ, Petrosino C (1999) The public safety of Megan's law in Massachusetts: An assessment from a sample of criminal sexual psychopaths. Crime and Delinquency 45.
25. Barbaree HE (2005) Psychopathy, treatment behaviour and recidivism: an extended follow-up of Seto and barbaree. Journal of interpersonal violence 20: 1115-31.
26. Ortega RJ, Fitch L (2000) Law and the confinement of psychopaths behavioral. Sciences and the law 18.
27. Code criminel, LRC (1985)C-46, art.753.
28. Hart SD (1998) The role of psychopathy in assessing risk for violence: conceptual and methodological issues. Legal and criminological psychology 3: 121-37.
29. Levy K (2011) Dangerous psychopaths: Criminally responsible but not morally responsible, subject to criminal punishmentaAnd to preventive detention. San Diego law review 48 .
30. Code criminel, LRC 1985 , C-46, art 672.54.
31. Winko v. British Columbia (Forensic Psychiatric Institute), [1999] 2 SCR 625.
32. 32 Rv. Swain [1991] 1 R.C.S. 933.
33. Pham TH, Cotte G (2000) Psychopathie : théorie et recherche. Presse universitaire du septentrion.
34. Pham TH, Saloppe X, Leistedt S (2012) La manipulation et le mensonge pathologique de la psychopathe comme défi majeur pour la recherche de la vérité judiciaire. Annales Médico-Psychologiques, 170: 141-142.
35. Shaw J and Porter S (2012) Forever a psychopath? Psychopathy and the Criminal Career Trajectory in Helinä Häkkänen-Nyholm and Jan-Olof Nyholm, Psychopathy and Law: A Practitioner's Guide p. 204.
36. Wong SCP, Olver ME (2015) Risk reduction treatment of psychopathy and applications to mentally disordered offenders. CNS Spectr. 20: 303-310.
37. FORUM on Corrections Research.

CHAPTER 18

Pharmacokinetics of Heroin and Its Metabolites in Vitreous Humor and Blood in a Living Pig Model

Andre´ Gotta˚s [1,2] • Marianne Arnestad[1] • Per Steinar Halvorsen[3] • Liliana C. Bachs[1] • Gudrun Høiseth[1,4]

[1]*.Division of Forensic MedicineNorwegian Institute of Public HealthOsloNorway*

[2]*.Department of PharmacologyOslo University HospitalOsloNorway*

3.The Intervention CentreOslo University HospitalOsloNorway

[4]*.Center for PsychopharmacologyDiakonhjemmet HospitalOsloNorway*

ABSTRACT

Vitreous humor (VH) is an alternative matrix for drug analysis in forensic toxicology. However, little is known about the distribution of xenobiotics, such as opioids, into VH in living organisms. The aim of this study was to simultaneously measure heroin and metabolite concentrations in blood and VH after injection of heroin in a living pig model. Six pigs were under non-opioid anesthesia during the surgical operation and experiment. Ocular microdialysis was used to acquire dialysate from VH, and a venous catheter was used for blood sampling. Twenty milligrams of heroin was injected intravenously with subsequent sampling of blood and dialysate for 6 h. The samples were analyzed by ultra-performance liquid chromatography–tandem mass spectrometry. Heroin was not detected in VH; 6-monoacetylmorphine (6-MAM) and morphine were first detected in VH after 60 min. The morphine concentration in VH thereafter increased throughout the experimental period. For 6-MAM, C_{max} was reached after 230 min in VH. In blood, 6-MAM reached C_{max} after 0.5 min, with a subsequent biphasic elimination phase. The blood and VH 6-MAM concentrations reached equilibrium after 2 h. In blood, morphine reached C_{max} after 4.3 min, with a subsequent slower elimination than 6-MAM. The blood and VH morphine concentrations were in equilibrium about 6 h after injection of heroin. In conclusion, both 6-MAM and morphine showed slow transport into VH; detection of 6-MAM in VH did not necessarily reflect a recent intake of heroin. Because postmortem changes are expected to be small in VH, these experimental results could assist the interpretation of heroin deaths.

Keywords: Heroin Pharmacokinetic Vitreous humor 6-MAM Ocular microdialysis Pigs

INTRODUCTION

Heroin is a highly lipid soluble opioid prodrug with negligible affinity and efficacy to μ-opioid receptor [1, 2, 3]. After intake, heroin is rapidly and completely converted to 6-monoacetylmorphine (6-MAM) spontaneously or by serum and tissue cholinesterases, and by tissue carboxyesterases, among others. 6-MAM is further metabolized to morphine that is in turn metabolized in the liver to the inactive morphine-3-glucuronide (M3G) and the active morphine-6-glucuronide (M6G), as well as several other minor compounds (for reviews see [4, 5, 6]). In man, the blood elimination half-life for heroin is less than 5 min, for 6-MAM around 15–20 min, for morphine 2–3 h, and for its glucuronides, 4–6 h [7]. The most intense phase of heroin effects, the "rush", takes place shortly after intake. At this stage 6-MAM is the predominant metabolite in both blood and brain [7, 8, 9, 10]. Later, 1–2 h after intake, morphine and M6G are the main metabolites able to mediate heroin's effects.

In forensic toxicology, results from postmortem body fluids or tissues are used to interpret the role of drugs in each death. Postmortem redistribution of drugs is a major challenge in this interpretation, making the estimation of the true concentrations at the time of death difficult, especially in cases with long postmortem intervals and potentially substantial drug redistribution. Vitreous humor (VH) is used as an alternative specimen for toxicological interpretation, and it is assumed that this medium is less prone to postmortem redistribution, making the detected concentration more representative for the one present at the time of death. When trying to estimate a blood concentration from the one detected in VH, toxicologists fall short, because no studies have investigated the kinetics of heroin and metabolites in both blood and VH in vivo. Therefore, the penetration and detection times for drugs and metabolites in VH as compared to those in blood are unknown.

In fatal cases where heroin metabolites are detected, interpretation of the toxicological findings from different specimens is especially important, because the presence of metabolites and their concentrations can provide important information not only about the cause of death, but also about the time between intake of heroin and death. How heroin and its metabolites distribute into different body compartments after administration has previously been well described for blood and urine in man [6]; the distribution of heroin and its metabolites into the brain has been also reported in animal models [8, 9, 11]. However, the distribution into VH has been studied less thoroughly, and such knowledge is important when interpreting the death involving heroin.

Most previous studies investigating the distribution of heroin and metabolites in both blood and VH have used postmortem samples [12, 13]; antemortem in vivo kinetics of drugs in VH have been studied for only a very few drugs related to forensic toxicology in living animal models using rabbits [14, 15, 16]. A study by Crandall et al. [17] compared concentrations of morphine in VH and blood for one time point in a porcine model. One of the main problems in the study of the kinetics of drugs in VH is the limited amount of VH available for sampling. This limits the number of time points and forces the use of different animals for the different time points increasing the spreading of the data. Some of the challenges associated with sampling from VH can be overcome by the use of ocular microdialysis, and this technique has previously been used to investigate drug distribution into the aqueous and vitreous body of the rabbit eyes [18, 19, 20, 21] (for review see Boddu et al. [22]).

With more antemortem knowledge about both blood and VH kinetics of heroin and metabolites, the interpretation of postmortem cases is easier and more exact. A better understanding of the pharmacokinetics of these substances in VH might further open new perspectives about the utility of VH as a complementary sample in forensic toxicology. The aim of this study was, therefore, to investigate the concentration–time profiles of heroin, 6-MAM, morphine, and M3G in both VH and blood after intravenous (i.v.) administration in the living organism using ocular microdialysis. This could further elucidate the usefulness of VH as a forensic material. Pigs were chosen as the preferred specimen for this study, because their physiology resembles that of humans [23]. It is previously indicated that formation of higher amounts of M3G could be expected in pig as compared to those in humans, and formation of M6G is not expected [24, 25, 26].

MATERIALS AND METHODS

Chemicals and reagents

Heroin hydrochloride, heroin-d_3, 6-MAM, 6-MAM-d_3, 6-MAM-d_6, morphine-d_3, M6G, M6G-d_3, M3G, and M3G-d_3 were obtained from Lipomed (Lipomed GmbH, Arlsheim, Switzerland); morphine from NMD (NMD Grossisthandel AS, Oslo, Norway); and morphine-d_6 from Toronto Research Chemicals Inc. (Toronto, Ontario, Canada). Standard compounds were stored according to supplier recommendations. HPLC-grade methanol and acetonitrile were purchased from Labscan Ltd. (Poch SA, Gliwice, Poland); analytical grade ammonium formate and formic acid from Merck (Whitehouse Station, NJ, USA). All water used was provided by a MilliQ A10 purification system (Merck, Darmstadt, Germany). Stock and working solutions were prepared as described previously [11].

Animals and conditions

Six Norwegian Landrace pigs (*Sus scrofa domesticus*) of either sex (weight 45 ± 5 kg) supplied from the Centre for Comparative Medicine, Oslo University Hospital, Rikshospitalet were used. They were housed separately at standard housing conditions (08:00–20:00 lights on), with free access to food and water. The experimental protocol was approved by the Norwegian National Animal Research Authority (Authorization Number 6425) and carried out in accordance to Norwegian regulations and international standards.

In vivo experiment

Anesthesia was induced with intramuscular (i.m.) injection of ketamine (20 mg/kg), azaperone (3 mg/kg), and atropine (0.02 mg/kg). The anesthesia was maintained with i.v. ketamine (1–2 mg/kg) and pentobarbital (1–3 mg/kg) until a tracheotomy through a neck midline incision was performed, and the pig was mechanically ventilated by an anesthesia machine (Leon *plus*, Heinen + Löwenstein; Bad Ems, Germany). Thereafter, general anesthesia was continued by an infusion of ketamine (2–2.5 mg/kg/h) together with inspired isoflurane at a concentration of 1.0–1.5 % using a Tec 7 gas analyzer (Heinen + Löwenstein). The inspired oxygen fraction was set to 0.35 and ventilation adjusted to keep expired end-tidal carbon dioxide concentration between 4.5 and 5.5 kPa. The right internal jugular vein was cannulated with a triple lumen central venous catheter (CVC) and advanced to vena cava superior for pressure registration and for heroin administration and blood sampling. The administration of heroin and collection of blood samples were obtained from different lumens of the CVC to avoid cross contamination of samples. The right carotid artery was cannulated for continuous blood pressure registration. A bladder catheter was placed by a mini-laparotomy to extract urine samples and measure diuresis and core temperature. Additional monitoring of the pig included electrocardiography, heart rate, and SpO_2. Fluid administration during the experiment was made with i.v. Ringer's acetate at a rate of 20 mL/kg/h.

A microdialysis guide cannula (AT6.14.iC, AgnTho´s, Lidingö, Sweden) was carefully inserted into the eye through a 19-gauge syringe needle. The needle was carefully inserted about 3 mm below the corneal scleral limbus through the pars plana; the needle was then retracted and the guide cannula inserted in its place. The guide dummy was then retracted and the microdialysis probe (AT6.14.4, AgnTho´s) was placed through the guide into the vitreous body. The probe position was adjusted, by visual inspection through the lens, such that the 4 mm long microdialysis membrane (outer diameter, 0.6 mm) resided well suspended in the mid-VH, and was then secured with surgical tape (Figure 1).

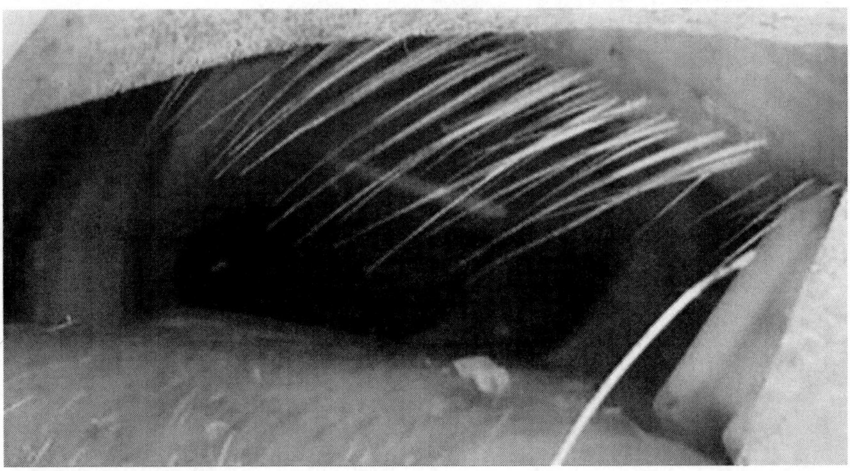

Figure 1. Microdialysis guide penetrating the pars plana with the probe extending through the guide shaft into the vitreous chamber of the eye

Before implantation of the microdialysis probe, the probe was connected by fluorinated ethylene propylene-tubing to a syringe containing a physiological Ringer´s solution with deuterated recovery calibrators (6-MAM-d_3, morphine-d_6, M3G-d_3) for each compound under study. The syringe containing the perfusion fluid was placed in a syringe microinfusion pump (CMA 400; CMA Microdialysis, Solna, Sweden). The pump perfusion flow was set to 0.7 μL/min. After implantation of the microdialysis probe the pig received an i.v. bolus injection (2 mL) of 20 mg heroin through the CVC, followed by flushing with 10 mL physiological saline solution. Dialysate was collected at 10 min intervals, into vials containing 5 μL internal standards (ISs), during the entire experimental session. That is, 7 μL dialysate was collected per vial, giving a total sample volume of 12 μL including the ISs. The vials were placed in a fraction collector (CMA/470; CMA Microdialysis) cooled to 6 °C during collection. Concurrently, 0.1 mL blood samples were taken through the CVC at 1, 2, 5, 7, 10, 15, 30, and then at every 30 min up to a maximum of 360 min after the injection of heroin. In one pig, samples from the living pig were only taken until 180 min after the injection, and postmortem samples were collected for another 150 min (postmortem data is not included in the succeeding data processing). In the remaining pigs, samples from the living animals were collected for 300–360 min after injection of heroin. The exact time point for each sample was recorded.

Two samples from the VH, in the contralateral eye of the one holding the microdialysis probe, were collected by syringe aspiration at different time points during the experiment. After euthanasia, and removal of the microdialysis probe, an additional aspiration of VH was performed from the eye holding the probe. The aspirated samples were taken to compare to the microdialysis samples (dialysates).

The blood samples were prepared in accordance with a previously established method with some modifications [9, 27]. In brief, blood samples were drawn using a syringe and transferred to microcentrifuge tubes containing 100 μL ammonium formate buffer (5 mM, pH 3.1) with 4 mg/mL NaF and 17.8 IE/mL heparin sodium, handled on ice, and immediately frozen on a freezing block at −80 °C (placed in a −80 °C freezer overnight and placed on ice in a Styrofoam box just before sampling). The block was switched with a new block from the −80 °C freezer after the first 60 min).

Each pig was finally euthanized by an i.v. injection of 1 g pentobarbital and 100 mmol potassium chloride through the CVC.

Chemical analysis

Dialysate required no further sample preparation; the VH sample analysis was done according to our previously established method with minor modifications [11]. Briefly, the samples were analyzed on a Waters Xevo TQ-S ultra-performance liquid chromatography (UPLC)–tandem mass spectrometry (MS/MS) system equipped with an Aquity HSS T3-column (Waters, Milford, MA, USA). This system was more sensitive than the system used in Gottas et al. [11]. We, therefore, determined new lowest levels of detection (LODs) and lowest levels of quantification (LLOQs) for 6-MAM, morphine, and M3G. Each LOD was determined by evaluation of the signal/noise (S/N) ratio for diluted series of the lowest standard, with an acceptance criteria of $S/N > 3$.

Sample preparation of the aspirated VH samples was performed according to a previously published method [27]; however, the chromatographic conditions were the same as in the previously published method for urine [28]. Briefly, a 100-μL aliquot of VH was added 50 μL IS solution (0.5 μM in water), and liquid-liquid extraction with 500 μL acetonitrile/methanol (85:15) was performed. Blood samples were prepared for analysis similarly to a previously published method [27]. Briefly, 50 μL IS solution (0.5 μM in water) was added to the blood samples before protein precipitation plus liquid–liquid extraction with 500 μL acetonitrile/methanol (85:15). High-performance liquid chromatography (HPLC)-MS/MS analysis was performed with a Waters Quattro Premier coupled to a XE MS/MS system equipped with a Xterra® MS C18 column (150 × 2.1 mm internal diameter, 3.5 μm particle size) (Waters). Chromatographic conditions were in accordance with a previously established method [9].

Statistical analyses

Kinetica version 5.1 (Thermo Fisher Scientific Inc., Waltham, MA, USA) was used for processing of the data. Results are presented from fitting curves from this program, in addition to individual data in some figures. The compartment model was selected based on the best goodness of fit as determined by the lower Akaike Information Criterion for each drug.

The calculated fitted results were used in a noncompartmental analysis to calculate the area under the concentration–time curve from time zero to last sample time (AUC_{Last}), the estimated maximum concentration (C_{max}) and the time, T_{max}, to reach C_{max} for all the different substances. The half-lives ($t_{1/2}$) were calculated for heroin, 6-MAM and morphine in blood. Additionally, $t_{1/2}$ was calculated for 6-MAM in VH. The AUC calculations were based on a mixed log linear model implemented in the Kinetica software, with an i.v. bolus model using the extrapolated zero concentration functions for heroin in blood after i.v. administration, and an extravascular model for the metabolites in blood and all compounds in VH. All data below LLOQ were discarded from the dataset before processing in Kinetica.

RESULTS

Validation of the analytical method

Although most analytical conditions were almost the same as described previously [11], the UPLC-MS/MS system is somewhat different and more sensitive. In the previous report, the dialysate samples of rat brain extracellular fluid were analyzed; in the present study, the dialysate samples of pig VH were dealt with.

To check the sensitivity of the new instrumental system, the lowest reference standard solutions were diluted down to 6.0 pM; at this concentration the S/N ratios were as high as 13.5, 58.3, and 6.0 for 6-MAM, morphine and M3G, respectively. The performance of the lowest reference standard (100 pM) determined the LLOQ. The variation (% relative standard deviation) and accuracy (% bias) were 0.4 and 0.7 % for 6-MAM, 1.6 and 2.9 % for morphine, and 9.6 and 6.0 % for M3G, respectively, at this concentration, well within the widely accepted criteria of ±20 % [29].

Recoveries for each microdialysis probes in VH were calculated by retrodialysis using isotope analogs as described by Gottas et al. [11]. This procedure was used for calculating the concentration of unbound analyte in the VH. The average recovery values for the deuterated recovery calibrators were 38 ± 9 % [mean ± standard deviation (SD)] for 6-MAM-d_3, 31 ± 6 % for morphine-d_6 , and 28 ± 10 % for M3G-d_3 (n = 5 each).

The validation data for blood samples has been described in our previous report [9].

Blood pharmacokinetics

The concentration-versus-time profiles of heroin and its metabolites in blood and VH of pigs are shown in Figure 2, and the calculated pharmacokinetic parameters are listed in Table 1. The heroin level in blood declined below LLOQ already after about 15 min, with a biphasic concentration-versus-time curve (Figure 2a). The initial alpha phase (ultra-rapid distribution, between 0 and 2 min) had an apparent $t_{1/2}$ of 0.4 min and the terminal elimination phase (between 5 and 15 min) a $t_{1/2}$ of 1.9 min. The extrapolated C_0 for heroin was 116 µM. Based on the population-calculated data, 6-MAM in blood reached a C_{max} at 2.29 µM and T_{max} after 0.5 min. The decline in 6-MAM blood concentration curve was also biphasic, with an initial alpha phase (between 0.5 and 5 min) with an apparent $t_{1/2}$ of 2.0 min, and a terminal elimination phase

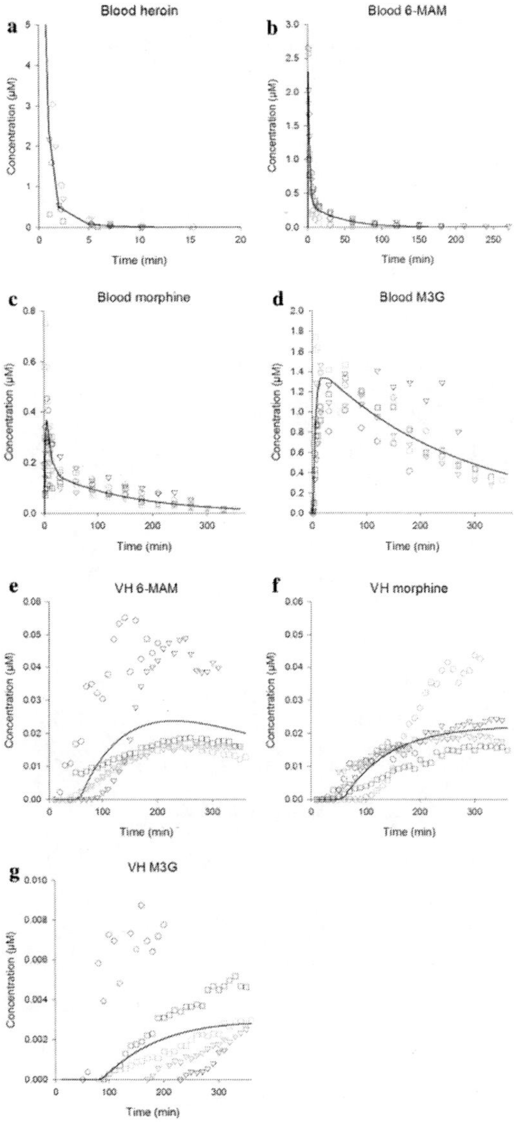

Figure 2. Blood and vitreous humor (VH) concentrations of heroin, 6-monoacetylmorphine (6-MAM), morphine, and morphine-3-glucuronide (M3G) after intravenous administration of 20 mg heroin. *Points* represent the observed concentrations (the *different symbols* representing the different animals), *lines* show the fitted values calculated in the Kinetica software. **a–d** Observed and fitted values for each separate compound in blood ($n = 6$) (for heroin, $n = 5$). **e–g** Observed and fitted values for each separate compound in VH ($n = 5$)

Table 1. Pharmacokinetic parameters for blood and vitreous humor (VH) dialysate after intravenous injection of 20 mg heroin to each pig

Drug	Parameter	Blood	VH dialysate
Heroin	AUC_{Last} [μM × min]	26.4	–
	C_0 (μM)	116[a]	–
	T_{max} (min)	0.0	–
	$T_{1/2}$ alpha-phase (min)	0.4	–
	$T_{1/2}$ elimination-phase (min)	1.9	–
6-MAM	AUC_{Last} [μM × min]	18.7	5.9
	C_{max} (μM)	2.29	0.0238
	T_{max} (min)	0.5	230
	$T_{1/2}$ alpha-phase (min)	2.0	–
	$T_{1/2}$ elimination-phase (min)	27.2	336
Morphine	AUC_{Last} [μM × min]	26.1	4.9
	C_{max} (μM)	0.37	0.0217[b]
	T_{max} (min)	4.3	360[b]
	$T_{1/2}$ elimination-phase (min)	109	–
M3G	AUC_{Last} [μM × min]	285	0.5
	C_{max} (μM)	1.36	0.0028[b]
	T_{max} (min)	20.2	360[b]

All parameters are based on population simulation data in Kinetica

[a] C_0 calculated in Kinetica must be considered as uncertain

[b] C_{max} and T_{max} at the end of experiment presented

(between 30 and 270 min) with a $t_{1/2}$ of 27.2 min (Figure 2b). In the alpha phase, heroin was still present and, therefore, influenced the concentration curve for 6-MAM, whereas virtually all heroin was cleared from the blood during the terminal elimination phase. The morphine concentrations increased gradually, with a C_{max} of 0.37 μM after 4.3 min (T_{max}) (Figure 2c). Morphine further had a $t_{1/2}$ of 109 min in the terminal elimination phase between 270 and 360 min, i.e. when no 6-MAM was detected in blood. The blood levels of M3G increased even slower than morphine, with a C_{max} of 1.36 μM after 20.2 min (T_{max}) (Figure 2d).

Ocular/vitreous humor pharmacokinetics

The concentration-versus-time profiles in VH are also shown in Figure 2, and the calculated pharmacokinetic parameters are listed in Table 1. Heroin was not detected in VH dialysate. 6-MAM appeared in VH dialysate 56.8 min (range 20–80 min) after injection of heroin. Thereafter the 6-MAM concentrations increased gradually and reached a C_{max} of 0.0238 μM after 230 min (T_{max}) (Figure 2e). The apparent $t_{1/2}$ for 6-MAM was 336 min. Similarly to 6-MAM, morphine was first detected in VH 57.2 min (range 30–100 min) after the heroin injection. Thereafter, the concentration increased gradually throughout the experimental procedure to reach a concentration of 0.0217 μM after 360 min (Figure 2f). M3G first appeared in VH after about 85.9 min (range 60–240 min), thereafter the levels increased slightly during the total experimental procedure to reach a concentration of 0.0028 μM after 360 min (Figure 2g).

To show the microdialysis technique superiority over traditional aspiration, aspirated VH samples were also collected. A total of 18 VH samples were collected by aspiration. Only eight of these did not show too much contamination by blood, making the results reliable. For these samples, the mean concentration ratio of aspirated samples to dialysates at the same time point were 2.4 ± 1.8 (mean ± SD) for morphine and 2.1 ± 3.0 for 6-MAM. The results from aspirated samples were not used any further, because of the possible blood contamination.

Relationship between drug concentrations in blood and vitreous humor

The blood/VH concentration ratio for morphine according to the fitted curve was 160 after 1 h, 7 after 2 h and decreased to approximately 1 after 6 h (Figure 2c, f). The blood/VH concentration ratio for 6-MAM according to the fitted curve was 64 after 1 h, 1 after 2 h and ≤1 for the remaining experimental period (Figure 2b, e). The blood/VH concentration ratio for M3G was >100 for the total experimental period (Figure 2d, g).

DISCUSSION

This study described the transfer of heroin metabolites into VH. To our knowledge, this is the first pharmacokinetic study on heroin metabolites in both blood and VH in the living animal. We observed a slow distribution from blood to VH for all of the heroin metabolites studied; heroin, however, was not detected in the VH. Importantly, the interval when 6-MAM can be detected was much longer in VH as compared to that in blood.

The strength of the present study is the tracing of the concentrations of heroin metabolites in VH in detail in a living animal. This was made possible by the use of the microdialysis technique, without regular aspiration demanding a large number of animals and incurring a high probability of blood contamination of the samples. Macha and Mitra [30] stated that microdialysis was superior to direct sampling techniques and could be applied to study ocular drug pharmacokinetics without rendering considerable physiological changes to the ocular barriers and the intraocular environment. We chose the pig model because it resembles the human being in both size and physiology [23, 31], and we found that the blood pharmacokinetics were comparable to data previously published on humans [6]. From the literature, it has also been indicated that glucuronidation and esterase-mediated hydrolysis in pigs is expected not to differ substantially from those of humans [24, 25, 32]. One weakness of this model is the lack of formation of M6G, which is also used in the interpretation of postmortem cases. There was also some inconsistency in the experimental period with sampling from the living animal, because it was prolonged as compared with the handling of postmortem samples, after we discovered the slow transfer to VH in the first pig. In the first pig, where postmortem samples were also collected for 150 min after death, the blood/VH concentration ratios of 6-MAM, morphine, and M3G continued to fall during the postmortem period (data not shown). This needs to be studied in more detail. The results in the present study were quite consistent among the animals, with only one pig showing higher concentrations of 6-MAM and M3G in VH (shown by dark blue circles in Figure 2e, g), but not in blood (Figure 2b, d). These findings might reflect the fact that individual differences in penetration could be expected, but their prolonged detection times in VH were still to be expected.

Our results are in accordance with the limited previous studies on kinetics of drugs of abuse in VH. In rabbit models, equilibrium between blood and VH for ethanol after its i.v. administration was attained after 10–20 min [14], for diazepam 60–120 min after its i.m. administration [15], and for MDMA 30 and 120 min after its i.m. administration [16]. Crandall et al.

[17] reported that morphine appeared in the aspirated VH 20 min after i.v. administration of a fatal dose (about 120 mg) of morphine using a pig model. However, in our experience, it was difficult to avoid blood contamination of the aspirated VH samples, and we believe that it cannot be excluded that the results presented by Crandall et al. [17] are influenced by blood contamination. We also observed early and high levels of both morphine and M3G in our aspirated blood-contaminated VH samples (data not shown).

Although the pharmacokinetics of opioids in VH in vivo has not previously been studied in detail, a number of previous studies have investigated on the postmortem relationship between blood and VH concentrations for different drugs. Our results could explain the findings in two previous studies regarding heroin metabolite ratios of blood to VH. Bogusz et al. [12] compared concentrations in femoral blood versus VH in 12 human cases where deaths were attributed to intoxication with heroin. The concentrations of morphine were generally lower in VH than in blood, and they found a mean ratio of blood/VH of 4.2 for morphine; 6-MAM was most often not detected, or only trace amounts were found in either blood or VH (LOD 0.5 ng/mL). They concluded that lower concentrations of morphine and its glucuronides in VH as compared to those in femoral blood might reflect a slower distribution of drugs to VH, which was verified in our study (Figure 2c, d, f, g) [12]. Wyman and Bultman [13] investigated 25 autopsy cases identified as heroin exposures. They also found concentrations of morphine higher in femoral blood than in VH, with a mean blood/VH concentration ratio of 3.5. For 6-MAM, however, they found that its concentrations in VH, in general, were higher than in femoral blood, with a blood/VH concentration ratio of 0.2. They concluded that the possibility of detecting 6-MAM in postmortem specimens could be greatly improved when VH was included as a specimen for analysis in such cases [13]. This conclusion was strengthened by the experimental data in the present study (Figure 2b, e), because the concentration of 6-MAM in VH exceeded the one in blood at a relatively early time point for 6-MAM, as compared with morphine as shown in Figure 2b, e.

The present results raise questions about the physiology of transfer of drugs from blood into VH. The entry of systemic administered drugs into the VH is prevented by different barriers, e.g., the blood-aqueous barrier and the blood-retina barrier (BRB) [22, 33]. The penetration of drugs through membranes, including the BRB, depends on their size, shape, charge, and lipophilicity [34, 35]. The flow is further governed by hydrostatic and osmotic pressure, as well as the concentration differences across the membrane [34]. The differences in penetration into VH observed between drugs have been attributed also to protein-binding of the drugs studied [36]. VH is low in proteins, with only 1–3 % of the total serum protein concentration [16], and we assume that there are no relevant differences between humans and pigs in this regard. The protein binding of morphine in pig and human plasma does not show significant differences [37].

As seen in Figure 2, a delay in diffusion was observed for all of the opioids under study, and especially for 6-MAM, its concentration in VH still increased after the concentration in blood became very low (Figure 2b, e). Even though the mechanisms for this finding are not fully understood, some theories can be launched. It has been suggested that the cornea may act as a reservoir for drugs [21]; furthermore, some drugs might have an affinity for melanin pigment present outside the retina [38, 39, 40], possibly both delaying and prolonging the distribution of drugs into different compartments of the eye (for reviews see Gaudana et al. [41] and Bévalot et al. [42]). In a study by Pitkanen et al. [40], the most lipophilic beta blockers showed the highest permeability coefficients through the outer BRB; moreover, the longest VH diffusion time (permeation lag time) was seen for the most lipophilic drugs. They also suggested that this might be due to an increasing drug-melanin binding with increasing lipophilicity, because these drugs showed the greatest binding to melanin [39]. Whether the theories mentioned above also apply to the opioids in this study is only speculative; however, the binding of opioids to melanin in hair has been documented [43]. Regarding the prolonged detection times for 6-MAM, it was previously suggested that esterases in VH showed lower activities and were more saturable as compared with those in blood [44, 45]. The mechanisms behind the decline in concentrations of morphine in VH have not previously been described, but it should be noted that prolonged detection times were observed for all of the metabolites.

How can our results assist in the practical interpretation of postmortem cases, considering the fact that drug concentrations in VH are expected to be subject to fewer postmortem changes [36, 46]? If the approximate time frame between intake and death is known, the true antemortem blood concentration could be calculated based on the concentration in the VH. Often, a very high blood concentration of morphine is seen together with a low concentration in VH, which might represent the true antemortem situation if death occurred within the first hour after intake of heroin; however, if some hours had elapsed, which is often the case with heroin deaths, the true antemortem blood morphine concentration would be expected to be closer to or lower than the morphine concentration observed in the VH.

Based on the final measured concentrations and the half-lives of 6-MAM, in the present study, the expected detection time for 6-MAM in VH could be more than 20 h in a living subject. Assuming limited postmortem changes in VH [36, 46], the detection of 6-MAM in VH postmortem could indicate that death occurred as many as 20 h or more after an intake of heroin. The dose and the LLOQ may, of course, influence this estimate. In postmortem cases, a negative 6-MAM in blood together with a positive 6-MAM in VH is often observed. In light of our findings, this could be explained by the considerably longer detection times for 6-MAM in VH as compared to those in blood. Because the present study also indicated longer detection times for morphine in VH, as compared to those in blood, it can be assumed that the detection times for other drugs are also longer in VH, as compared to those in blood.

It has previously been reported that a high blood/VH concentration ratio of drugs suggest rapid death [42]. The results from the present study could argue that the detection of 6-MAM or morphine in VH indicates that death did not occur within the first hour after intake of heroin. This finding is, however, complicated by the fact that heroin is often used on a daily basis, and therefore, detections in VH could be caused by intake, for instance, on the day before.

CONCLUSIONS

To our knowledge, this is the first report dealing with a pharmacokinetic study on heroin metabolites in VH using a living pig model. The pig resembles the human in both size and physiology except from minor metabolic species differences. To get pharmacokinetic data from VH samples, the present microdialysis sampling using the pig model is most recommended. Such data of heroin metabolites obtained from the blood and VH samples collected from living pigs gave an important clue for interpretation of antemortem heroin poisoning in human victims by the postmortem analysis of heroin metabolites in both blood and VH samples.

ACKNOWLEDGMENTS

The authors are grateful to Elisabeth Øiestad, Jørg Mørland and Gunhild Heide for helpful comments to the manuscript.

COMPLIANCE WITH ETHICAL STANDARDS

Conflict of interest

There are no financial or other relations that could lead to a conflict of interest.

Ethical approval

All applicable international, national, and/or institutional guidelines were followed for the care and use of the animals. This study did not involve samples collected from human participants.

REFERENCES

1. Gianutsos G, Cohen SD, Carlson G, Heyman R, Salva P, Morrow G, Hite GJ (1986) Alteration of in vivo and in vitro effects of heroin by esterase inhibition. Toxicol Appl Pharmacol 82:14–18
2. Inturrisi CE, Schultz M, Shin S, Umans JG, Angel L, Simon EJ (1983) Evidence from opiate binding studies that heroin acts through its metabolites. Life Sci 33(Suppl 1):773–776
3. Selley DE, Cao CC, Sexton T, Schwegel JA, Martin TJ, Childers SR (2001) μ Opioid receptor-mediated G-protein activation by heroin metabolites: evidence for greater efficacy of 6-monoacetylmorphine compared with morphine. Biochem Pharmacol 62:447–455
4. Salmon AY, Goren Z, Avissar Y, Soreq H (1999) Human erythrocyte but not brain acetylcholinesterase hydrolyses heroin to morphine. Clin Exp Pharmacol Physiol 26:596–600
5. Maurer HH, Sauer C, Theobald DS (2006) Toxicokinetics of drugs of abuse: current knowledge of the isoenzymes involved in the human metabolism of tetrahydrocannabinol, cocaine, heroin, morphine, and codeine. Ther Drug Monit 28:447–453
6. Rook EJ, Huitema AD, van den Brink W, van Ree JM, Beijnen JH (2006) Pharmacokinetics and pharmacokinetic variability of heroin and its metabolites: review of the literature. Curr Clin Pharmacol 1:109–118
7. Rook EJ, van Ree JM, van den Brink W, Hillebrand MJ, Huitema AD, Hendriks VM, Beijnen JH (2006) Pharmacokinetics and pharmacodynamics of high doses of pharmaceutically prepared heroin, by intravenous or by inhalation route in opioid-dependent patients. Basic Clin Pharmacol Toxicol 98:86–96
8. Andersen JM, Ripel A, Boix F, Normann PT, Morland J (2009) Increased locomotor activity induced by heroin in mice: pharmacokinetic demonstration of heroin acting as a pro-drug for the mediator, 6-monoacetylmorphine, in vivo. J Pharmacol Exp Ther 331:153–161
9. Gottås A, Øiestad EL, Boix F, Vindenes V, Ripel Å, Thaulow CH, Mørland J (2013) Levels of heroin and its metabolites in blood and brain extracellular fluid after i.v. heroin administration to freely moving rats. Br J Pharmacol 170:546–556

10. Gottås A, Boix F, Øiestad EL, Vindenes V, Mørland J (2014) Role of 6-monoacetylmorphine in the acute release of striatal dopamine induced by intravenous heroin. Int J Neuropsychopharmacol 17:1357–1365
11. Gottas A, Oiestad EL, Boix F, Ripel A, Thaulow CH, Pettersen BS, Vindenes V, Morland J (2012) Simultaneous measurement of heroin and its metabolites in brain extracellular fluid by microdialysis and ultra performance liquid chromatography tandem mass spectrometry. J Pharmacol Toxicol Methods 66:14–21
12. Bogusz MJ, Maier RD, Driessen S (1997) Morphine, morphine-3-glucuronide, morphine-6-glucuronide, and 6-monoacetylmorphine determined by means of atmospheric pressure chemical ionization-mass spectrometry-liquid chromatography in body fluids of heroin victims. J Anal Toxicol 21:346–355
13. Wyman J, Bultman S (2004) Postmortem distribution of heroin metabolites in femoral blood, liver, cerebrospinal fluid, and vitreous humor. J Anal Toxicol 28:260–263
14. Fernandez P, Lopez-Rivadulla M, Linares JM, Tato F, Bermejo AM (1989) A comparative pharmacokinetic study of ethanol in the blood, vitreous humour and aqueous humour of rabbits. Forensic Sci Int 41:61–65
15. Teixeira HM, Reis F, Proença P, Ramos P, Quintela O, López-Rivadulla M, Marques E, Vieira DN (2004) Vitreous humour as a complementary sample to blood for the detection/confirmation of diazepam: ante-mortem and post-mortem studies in an animal model. Hum Exp Toxicol 23:571–577
16. De Letter EA, De Paepe P, Clauwaert KM, Belpaire FM, Lambert WE, Van Bocxlaer JF, Piette MH (2000) Is vitreous humour useful for the interpretation of 3,4-methylenedioxymethamphetamine (MDMA) blood levels? Experimental approach with rabbits. Int J Legal Med 114:29–35
17. Crandall CS, Kerrigan S, Aguero RL, Lavalley J, McKinney PE (2006) The influence of collection site and methods on postmortem morphine concentrations in a porcine model. J Anal Toxicol 30:651–658
18. Waga J, Ohta A, Ehinger B (1991) Intraocular microdialysis with permanently implanted probes in rabbit. Acta Ophthalmol (Copenh) 69:618–624
19. Waga J, Nilsson-Ehle I, Ljungberg B, Skarin A, Stahle L, Ehinger B (1999) Microdialysis for pharmacokinetic studies of ceftazidime in rabbit vitreous. J Ocul Pharmacol Ther 15:455–463
20. Macha S, Mitra AK (2001) Ocular pharmacokinetics of cephalosporins using microdialysis. J Ocul Pharmacol Ther 17:485–498
21. Katragadda S, Gunda S, Hariharan S, Mitra AK (2008) Ocular pharmacokinetics of acyclovir amino acid ester prodrugs in the anterior chamber: evaluation of their utility in treating ocular HSV infections. Int J Pharm 359:15–24
22. Boddu SH, Gunda S, Earla R, Mitra AK (2010) Ocular microdialysis: a continuous sampling technique to study pharmacokinetics and pharmacodynamics in the eye. Bioanalysis 2:487–507
23. Forster R, Bode G, Ellegaard L, van der Laan JW (2010) The RETHINK project on minipigs in the toxicity testing of new medicines and chemicals: conclusions and recommendations. J Pharmacol Toxicol Methods 62:236–242
24. Higashi E, Ando A, Iwano S, Murayama N, Yamazaki H, Miyamoto Y (2014) Hepatic microsomal UDP-glucuronosyltransferase (UGT) activities in the microminipig. Biopharm Drug Dispos 35:313–320
25. Matal J, Jancova P, Siller M, Masek V, Anzenbacherova E, Anzenbacher P (2008) Interspecies comparison of the glucuronidation processes in the man, monkey, pig, dog and rat. Neuro Endocrinol Lett 29:738–743
26. Lykkegaard K, Lykkesfeldt J, Lauritzen B, Svendsen O (2008) Morphine reduces spinal c-fos expression dose-dependently during experimental laparotomy in pigs: a combined pharmacokinetic and surgical study. Res Vet Sci 84:457–464
27. Karinen R, Andersen JM, Ripel A, Hasvold I, Hopen AB, Morland J, Christophersen AS (2009) Determination of heroin and its main metabolites in small sample volumes of

whole blood and brain tissue by reversed-phase liquid chromatography-tandem mass spectrometry. J Anal Toxicol 33:345–350

28. Berg T, Lundanes E, Christophersen AS, Strand DH (2009) Determination of opiates and cocaine in urine by high pH mobile phase reversed phase UPLC-MS/MS. J Chromatogr B 877:421–432
29. Peters FT, Drummer OH, Musshoff F (2007) Validation of new methods. Forensic Sci Int 165:216–224
30. Macha S, Mitra AK (2001) Ocular pharmacokinetics in rabbits using a novel dual probe microdialysis technique. Exp Eye Res 72:289–299
31. Douglas W (1972) Of pigs and men and research: a review of applications and analogies of the pig, sus scrofa, in human medical research. Sp Life Sci 3:226–234
32. Hosokawa M, Maki T, Satoh T (1990) Characterization of molecular species of liver microsomal carboxylesterases of several animal species and humans. Arch Biochem Biophys 277:219–227
33. Hughes PM, Olejnik O, Chang-Lin JE, Wilson CG (2005) Topical and systemic drug delivery to the posterior segments. Adv Drug Deliv Rev 57:2010–2032
34. Schmidt A, Ganßman B, Skopp G, Klinder K, Pötsch L, Aderjan R, Lutz R, Mattern R (1997) An in vitro experiment for postmortem vascular permeation. The passage of morphine and morphine glucuronides across a vascular wall. J Forensic Sci 42:486–491
35. Cunha-Vaz JG (2004) The blood-retinal barriers system. Basic concepts and clinical evaluation. Exp Eye Res 78:715–721
36. Holmgren P, Druid H, Holmgren A, Ahlner J (2004) Stability of drugs in stored postmortem femoral blood and vitreous humor. J Forensic Sci 49:820–825
37. Baggot JD, Davis LE (1973) Species differences in plasma protein binding of morphine and codeine. Am J Vet Res 34:571–574
38. Cheruvu NP, Amrite AC, Kompella UB (2008) Effect of eye pigmentation on transscleral drug delivery. Invest Ophthalmol Vis Sci 49:333–341
39. Leblanc B, Jezequel S, Davies T, Hanton G, Taradach C (1998) Binding of drugs to eye melanin is not predictive of ocular toxicity. Regul Toxicol Pharmacol 28:124–132
40. Pitkanen L, Ranta VP, Moilanen H, Urtti A (2005) Permeability of retinal pigment epithelium: effects of permeant molecular weight and lipophilicity. Invest Ophthalmol Vis Sci 46:641–646
41. Gaudana R, Ananthula HK, Parenky A, Mitra AK (2010) Ocular drug delivery. AAPS J 12:348–360
42. Bévalot F, Cartiser N, Bottinelli C, Fanton L, Guitton J (2016) Vitreous humor analysis for the detection of xenobiotics in forensic toxicology: a review. Forensic Toxicol 34:12–40
43. Kronstrand R, Forstberg-Peterson S, Kagedal B, Ahlner J, Larson G (1999) Codeine concentration in hair after oral administration is dependent on melanin content. Clin Chem 45:1485–1494
44. Behar-Cohen FF, Gauthier S, El Aouni A, Chapon P, Parel JM, Renard G, Chauvaud D (2001) Methylprednisolone concentrations in the vitreous and the serum after pulse therapy. Retina 21:48–53
45. Mains J, Tan LE, Zhang T, Young L, Shi R, Wilson C (2012) Species variation in small molecule components of animal vitreous. Invest Ophthalmol Vis Sci 53:4778–4786
46. Drummer OH, Gerostamoulos J (2002) Postmortem drug analysis: analytical and toxicological aspects. Ther Drug Monit 24:199–209

CHAPTER 19

Stature Prediction from Anthropometry of Extremities among Jordanians

Amal A Mashali*[1], Omar El taweel[2] and Elerian Ekladious[3]

[1] *Professor of Forensic Medicine & Clinical Toxicology, Faculty of Medicine, Alexandria University, Egypt*

[2] *Student, Mutah University, Resident at Princess Basma Teaching Hospital, Irbid, Jordan*

[3] *Professor of Anatomy, Faculty of Medicine, Ein Shams University, Egypt*

ABSTRACT

Stature of an individual has an important role in identification, which is often required in medico-legal practice. The estimation of stature is an important step in the identification of dismembered remains or when only a part of a skeleton is only available as in major disasters or with mutilation. There is no published data on anthropological data among Jordanian population. The present study was designed in order to find out relationship of stature to some anthropometric measures among a sample of Jordanian population and to determine the most accurate and reliable one in predicting the stature of an individual.

A cross sectional study was conducted on 314 adult healthy volunteers, free of bone diseases, nutritional diseases and abnormalities in the extremities after taking their consent. Students of Faculty of Medicine, Mutah University helped in collecting the data. The anthropometric measurements (anatomically defined) were stature, humerus length, hand length and breadth, foot length and breadth, foot index and knee height on both right and left sides of the body. The measurements were typical on both sides of the bodies of the studied sample. All the anthropometric data showed significant relation with age except the knee height. There was a significant difference between male and female measurements. There was a significant positive correlation between the different measures and the stature of the individuals. The most sensitive measure for prediction of a stature from the measurements in the entire studied sample and in males was found to be the humerus length, while in females it was the knee height. Three sets of equations, each of three equations, were developed for estimation of stature from the different measurements, after excluding all the nonsignificant parameters with a high rate of accuracy. The equations were verified by using the same partcipants' data with a high rate of accuracy. These formulae are applicable to the Jordanian population only, due to inherent variations in these dimensions which may be attributed to biological and environmental factors.

INTRODUCTION

Forensic anthropology is the specialized sub-discipline of physical anthropology that applies the techniques of osteology and skeletal identification to problems of legal concern [1]. Ascertaining major biological characteristics such as age, sex, race/ ethnicity and stature are often the first pieces of data that help focus the investigation on specific group characteristics. When remains are complete, determining sex, race/ethnicity and stature can be done without difficulty [2].

Stature provides insight into various features of a population including nutrition, health and genetics. An experienced physical anthropologist can estimate the height of an individual if complete long bones are present. This is done by accurately measuring the long bones and comparing these measurements to charts based on regression formulae that have been developed after many measurements of skeletal remains [1]. The best skeletal elements to use are the long bones of the arms and legs. One of the pioneers in the study of stature estimation was Dr. Mildred Trotter in World War II [3]. Different formulas have been developed using long bones for reconstruction of stature from skeletal / dismembered remains as Steele's formula and Bayesian approaches [2,4].

With the increasing frequency of mass disasters, homicides, air plane crashes, train and road accidents etc., there is always need for such studies which help in identifying the deceased from fragmentary and dismembered human remains. Many authors have successfully tried to estimate stature from percutaneous body measurements [5-10]. Other studies correlated foot dimensions with stature and with sex differences [11,12]. Stature was estimated from various measurements of the foot like foot length and breadth based on statistical equations and formulae [13-17]. Many studies were successfully performed on this topic despite a wide range of ethnics and races through the populations. Most of the studies about stature estimation are concentrated on upper and lower extremities [18,19].

In Jordan, there were no published studies regarding anthropometric measurements. Therefore, the present study was designed in order to find out relationship of stature to some anthropometric measures among a sample of Jordanian population and to determine the most accurate and reliable one in predicting the stature of an individual.

MATERIALS AND METHODS

Approval of the ethics committee of Faculty of Medicine, Mutah University was obtained after reviewing the proposal.

Informed consent was obtained from every subject with explanation of the purpose and nature of the study.

The study is a cross-sectional one on a sample of 314 normal healthy volunteering adults, Jordanians, above age of 18 up to 45.

Exclusion criteria were bone disease, nutritional or developmental disorders and orthopedic deformities.

Demographic data were obtained including the name of the family, occupation and the governance of residence.

Anthropometric measures were recorded in centimeters, using Stadiometer and Sliding Calipers, according to the techniques described by Vallois [20]. The anthropometric measurements were hand length, hand breadth, foot length, foot breadth, humerus length and knee height were taken independently on the left and right side of each individual. The dimensions of hand and foot (Figure 1) were measured following the International Biological Program Protocol [21,22]. The measurements were done by the group of students assisting in the research and redone by researchers. The mean values were taken in the study.

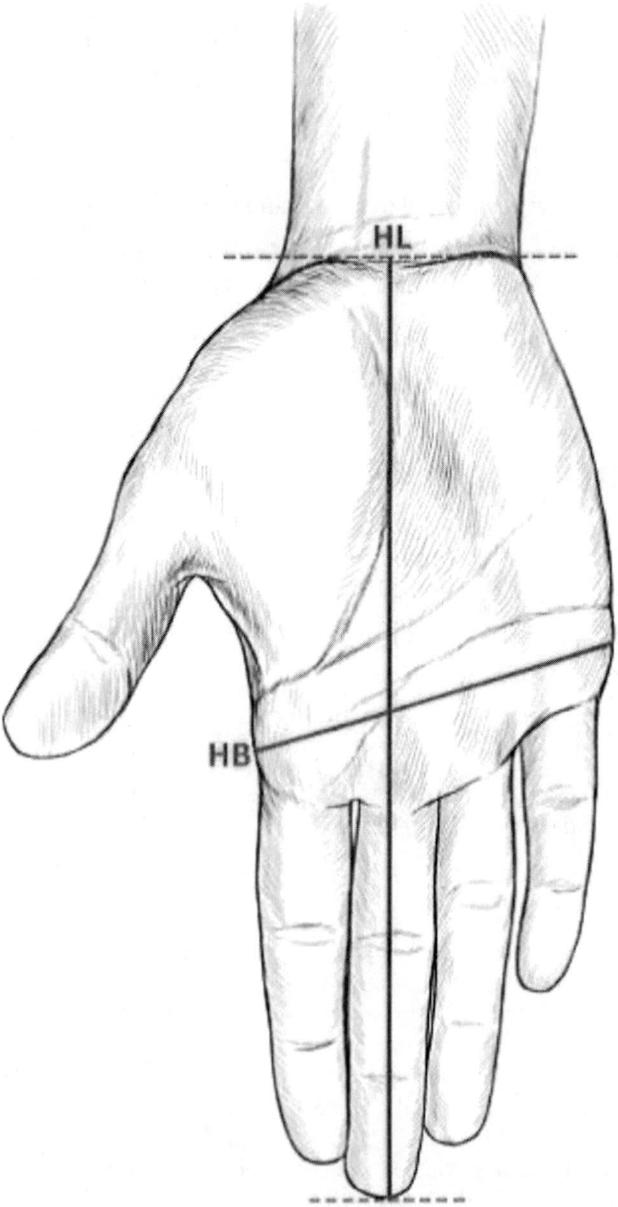
HL
HB

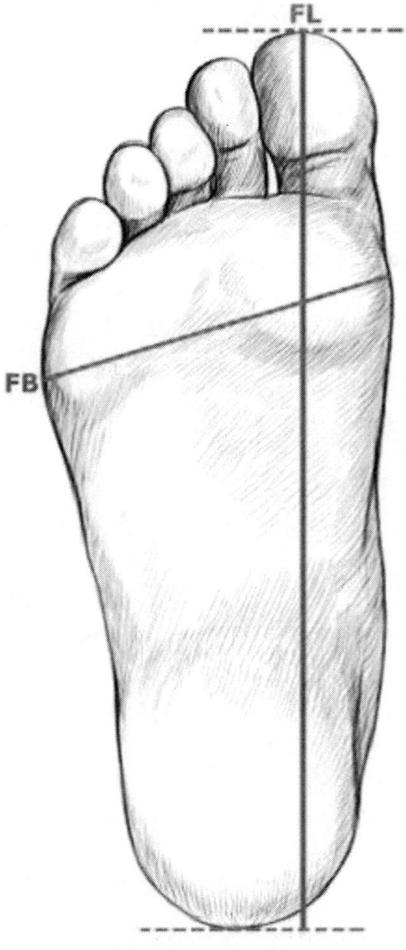

Figure 1: Diagram showing the measurements of the length and breadth of hand and the length and breadth of foot

a) Stature: Height-vertex (stature): the vertical distance between the point vertex and the floor. Vertex is the highest point on the head is held in (Frankfurt Horizontal) plane. The subject was made to stand in a standing posture and measurement was taken without any wear on head and foot

b) b) Hand length: The distance between the points inter-stylion and the tip of the third finger. Inter-stylion: is the middle point of the line connecting the point stylion radial (the most distal point on the styloid process of radius) and stylion ulnare (the most distal point on the styloid process of ulna) (right and left).

c) Hand breadth: The distance between the most prominent point, outside of the lower epiphyses of the 2nd metacarpal (metacarpal radiale) to the most prominent inside point of the lower epiphyses (metacarpal ulnare) of the 5th metacarpal (right and left).

d) Foot length: The straight distance between the most backward point on the heel of the foot (pternion) and the most forward placed point on the longest toe of the foot (acropodian) when the foot is fully stretched (right and left).
e) Foot breadth: The distance between the points of the anterior epiphyses (distal) of the 1st metatarsal, the most prominent of the inner side of the foot (metatarsal-tibiale), and the joint of the anterior epiphyses of the 5th metatarsal, the most prominent of the outer side (metatarsal-fibulare) (right and left).

f) Foot index: Was calculated individually for both feet in males and females by using the formula: Foot index = (foot breadth/foot length) X 100.

g) Humerus length : From just below the most prominent point of the shoulder to the midpoint between the lateral and medial epicondyles (right and left).

h) Knee height: From the midpoint between the 2 condyles of femur to the floor (right and left).

Statistical analysis

Data were subjected to statistical analysis using Statistical Package for Social Sciences (SPSS) version 16. Results were expressed as arithmetic mean and standard deviations. Student t-test and F-test were used to establish any significant differences. Linear regression analysis was undertaken to derive predictive equations for estimation of stature using the different parameters, except the non-significant parameters. The regression formulae were calculated for various combinations to reach the best estimate possible. The level of significance was set at $p \leq 0.05$.

RESULTS

The study included 314 healthy Jordanian adults, 144 females and 170 males. Their ages ranged between 18 and 45 years.

Table 1 show that the stature ranged between 144 and 190 cm, with a mean of 167.74 ± 9.82 cm.

Table 1: Distribution of the studied sample regarding their stature (n = 314)

Stature	Number	Percent
144- 150	9	2.9
151-160	76	24.2
161-170	99	31.5
171-180	101	32.2
181+	29	9.2
Total	314	100.0
Range	144-190	
Mean	167.74	
S.D.	9.82	
Median	171.0	

Table 2 compared the means of the bilateral measurements of the studied sample. There was no significant statistical difference for all parameters on the left and right sides. However, the right side presented higher mean values in most measurements.

Table 3 shows descriptive statistics for all measurements with age. There was a statistically significant relation for all parameters except for knee height, where $F = 2.491$, $P = 0.060$.

Table 2: Comparison between right and left measurements of different parts in the studied sample

	Mean	Std. Deviation	Std. Error Mean	P
Knee height R	33.16	3.182	.180	0.136
Knee height L	33.12	3.173	.179	
Foot breadth R	8.97	.970	.055	0.116
Foot breadth L	8.91	.942	.053	
Foot length R	18.856	1.6729	.0944	0.419
Foot length L	18.867	1.6657	.0940	
Hand length R	25.36	2.209	.125	0.376
Hand length L	25.30	2.403	.136	
Hand breadth R	10.79	1.351	.076	0.335
Hand breadth L	10.74	1.550	.087	
Length of humerus R	46.067	5.6143	.3168	0.404
Length of humerus L	46.151	5.1903	.2929	
Foot index R	47.706	5.0808	.2867	0.102
Foot index L	47.382	4.9758	.2808	

Table 4 shows that stature of males ranged between 154 and 190 cm with a mean of 174.68 ± 6.624 and that of females ranged between 148 and 182 cm with a mean of 159.55 ± 5.893 . This depict a statistically significant difference between statures of males and females, where t = 449.792 and P = 0.0001.

All measurements in both sexes showed that males were having greater mean values of all measurements with a significant statistical difference, except for foot index where t = 1.380 and P = 0.241.

Table 3: Relation between age and different measurements (mean value of right and left side)

		N	Min.	Max.	Mean	S.D.	F	p
Stature	< 20	25	148	178	159.7	8.1	11.194	0.0001**
	20 - 30	233	148	190	169.4	9.7		
	30 - 40	36	148	184	165.6	9.2		
	40+	20	152	176	162.5	7.4		
Knee height	< 20	25	27	40	31.9	2.9	2.491	0.060
	20 - 30	233	27	42	33.1	3.2		
	30 - 40	36	29	44	34.0	3.1		
	40+	20	27	40	33.8	3.3		
Foot breadth	< 20	25	7	10	8.2	0.7	7.404	0.0001**
	20 - 30	233	7	12	9.0	0.9		
	30 - 40	36	7	11	9.2	1.0		
	40+	20	7	12	9.2	1.2		
Foot length	< 20	25	14.0	21.0	17.6	1.6	6.057	0.001*
	20 - 30	233	15.3	23.0	19.0	1.5		
	30 - 40	36	16.5	22.2	19.2	1.5		
	40+	20	7.5	22.0	18.5	3.1		
Hand length	< 20	25	20	28	23.8	1.8	5.008	0.002*
	20 - 30	233	16	32	25.5	2.2		
	30 - 40	36	20	31	25.6	2.2		
	40+	20	22	30	24.8	2.2		
Hand breadth	< 20	25	7	12	9.7	1.4	6.329	0.0001**
	20 - 30	233	6	15	10.9	1.3		
	30 - 40	36	8	14	10.8	1.5		
	40+	20	8	16	11.1	2.0		
Length of humerus	< 20	25	32.0	55.0	42.9	5.5	3.626	0.013*
	20 - 30	233	24.5	56.0	46.5	5.3		
	30 - 40	36	33.0	53.0	45.8	4.5		
	40+	20	38.0	60.0	46.4	6.2		
Foot index	< 20	25	39.5	51.6	46.4	3.1	5.314	0.001**
	20 - 30	233	38.1	58.6	47.3	3.5		
	30 - 40	36	34.6	55.6	48.0	4.2		
	40+	20	41.2	106.7	51.5	13.9		

* Significant statistically

Table 4: Relation between sex and different measurements (mean value of right and left side)

		N	Min.	Max.	Mean	S.D.	*t*-test	Sig.
Stature	Males	170	154	190	174.68	6.624	449.792	.0001**
	Females	144	148	182	159.55	5.893		
Knee height	Males	170	28	42	33.67	3.125	10.825	.001**
	Females	144	27	44	32.51	3.115		
Foot breadth	Males	170	8	12	9.38	0.775	109.974	.0001**
	Females	144	7	12	8.41	0.862		
Foot length	Males	170	16.0	23.0	19.634	1.2240	106.916	.0001**
	Females	144	7.5	21.9	17.949	1.6581		
Hand length	Males	170	16	32	26.49	1.821	148.268	.0001**
	Females	144	20	31	23.97	1.840		
Hand breadth	Males	170	6	14	11.05	1.207	16.929	.0001**
	Females	144	6	16	10.42	1.515		
Length of humerus	Males	170	38.0	58.0	49.284	3.6056	225.825	.0001**
	Females	144	24.5	60.0	42.361	4.5535		
Foot index	Males	170	38.1	56.2	47.849	3.4767	1.380	.241
	Females	144	34.6	106.7	47.189	6.2839		

* Significant statistically

Table 5 illustrates the correlation between stature and the mean of different measurements (left and right, as there was no difference between both sides) in males and females. There was a significant positive correlation between all measures and stature in females except for foot index, where r = 0.037 and -0.06 respectively. In males there was a significant positive correlation between all variables measured except the foot breadth (r = 0.086, P = 0.263).

This correlation was important to predict the most significant variable which has effect on the equation in stature calculation.

The best measurement for predicting stature for all the sample studied and for males was the length of humerus, where r = 0.738 and 0.504 respectively. In females the best measurement for predicting stature was the knee height (r = 0.586).

Three sets of equations, each of three equations, were developed for estimation of stature from the different measurements, after excluding all the nonsignificant parameters, as shown in table 6.

Table 5: Correlation between different measurements (as mean of right and left)

Stature #	Male		Female		Total	
	r	p	r	p	r	p
Knee height	0.385	0.001**	0.586	0.001**	0.436	0.001**
Foot breadth	0.086	0.263	0.476	0.001**	0.537	0.001**
Foot length	0.425	0.001**	0.539	0.001**	0.649	0.001**
Hand length	0.482	0.001**	0.575	0.001**	0.711	0.001**
Hand breadth	0.202	0.008*	0.375	0.001**	0.350	0.001**
Length of humerus	0.504	0.001**	0.497	0.001**	0.738	0.001**
Foot index	-0.264	0.001*	-0.06	0.477	0.037	0.513

*Significant positive correlation

Table 6: The three sets of linear regression equations developed to estimate the stature from the measurements of the studied sample

Subjects	Parameters	Equation	Accuracy
Whole subjects (n =314)	Using all measurements (except nonsignificant ones)	Stature = 69.42 + (0.142 x knee height) + (0.521 x Foot breadth) + (1.452 x foot length) + (1.239 x hand length) – (0.474 x hand breadth) + (0.764 x length of humerus)	87.1%
	Using upper limb measurements (except nonsignificant ones)	Stature = 83.89 + (1.686 x hand length) + (0.072x hand breadth) + (0.875 x length of humerus)	845%
	Using lower limb measurements (except nonsignificant ones)	Stature = 88.433 + (0.251 x knee height) + (1.806 x foot breadth) + (2.908 x foot length)	85.0%
For males (n = 170)	Using all measurements (except nonsignificant ones)	Stature = 107.031 + (0.403 x knee height) + (1.027 x foot length) + (0.855 x hand length) – (0.417 x hand breadth) + (0.609 x length of humerus) - (0.296 x foot index)	83.8%
	Using upper limb measurements (except nonsignificant ones)	Stature = 134.865 + (0.949 x hand length) + (0.339x hand breadth) + (0.646 x length of humerus) – (0.437 x foot index)	80.0%
	Using lower limb measurements (except nonsignificant ones)	Stature= 126.437 + (0.464 x knee height) + (1.661 x foot length)	79.5%
For females (n = 144)	Using all measurements (except nonsignificant ones)	Stature= 102.935+ (0.493 x knee height) – (0.159 x Foot breadth) + (0.822 x foot length) + (0.643 x hand length) + (0.168 x hand breadth) + (0.227 x length of humerus)	83.0%
	Using upper limb measurements (except nonsignificant ones)	Stature = 111.518 + (1.171 x hand length) + (0.430x hand breadth) + (0.366 x length of humerus)	78.1%
	Using lower limb measurements (except nonsignificant ones)	Stature = 112.640 + (0.733 x knee height) + (0.423 x foot breadth) + (1.088 x foot length)	79.0%

The equations were verified by using the same partcipants' data with a high rate of accuracy.

Accuracy of the linear equations for estimating stature from measurements of all cases studied and males of studied sample was highest using all data followed by using measurements of upper limbs then using those of upper limbs. Accuracy of the linear equation for estimating stature from measurements of all females of studied sample was higher using measurements of all data followed by using those of lower limbs then those of upper limb.

DISCUSSION

In medico-legal autopsies, establishing personal identity of the victims is often required. Estimation of stature from extremities and their parts plays an important role in identifying the dead in forensic examinations [14].

The study included 314 healthy Jordanian adults, 144 females and 170 males. The stature ranged between 144 and 190 cm, with a mean of 167,74 ± 9.82 cm.

The present study did not show any significant statistical difference for all parameters on the left and right sides. However the right side presented higher mean values in most measurements. Similar results were obtained by different researches [14].

A statistically significant relation was found between age and stature of the studied sample. The stature obtained by different researchers such as Duyar., *et al.* Nath., *et al.* and Narde and Dongre varies slightly. This might be due to geographical variations, which play an important role in human development and growth [23-25].

In the current study, males were having greater mean values of all measurements with a significant statistical difference, except for foot index. These results are in accordance with many other works in different researches. These statistically significant differences may be attributed to the early maturity of girls than boys. These were in accordance with results of Ozaslan., *et al.* and Krishan and Sharma [6,14].Hormonal, genetic and environmental factors are effective in occurring of physical characteristics between sexes; Age of puberty being 2years later in males as compared with females giving them extra time for growth [26,27].

In the present study there was a significant positive correlation between all measures and stature in females except for foot index. In males there was a significant positive correlation between all variables measured except the foot breadth. This correlation was important to predict the most significant variable which has effect on the equation in stature calculation. These results are in accordance with Krishan and Sharma and Ozaslan., *et al.* [6,14].

The study revealed that best measurement for predicting stature for the sample studied and for males was the length of humerus, while in females it was the knee height. Ozaslan., *et al.* found that the best correlation coefficient was determined in males for stature estimation as follows; in hand dimensions for hand length and in foot dimensions for foot length. Also in females; hand length presented the best correlation between hand dimensions and stature [6].

Krishan and Sharmain their study found that the dimensions of hands and feet can successfully be used for estimation of stature and recommended to use them by law enforcement agencies and forensic scientists. The only precaution which must be taken into consideration is that these formulae were applicable to the population from which the data have been collected due to inherent population variations in these dimensions which may be attributed to genetic and environmental factors like climate and nutrition [14].

Three sets of equations, each of three equations, were developed for estimation of stature from the different measurements, after excluding all the nonsignificant parameters. The equations were verified by using the same partcipants' data with a high rate of accuracy.

Similar to our results, Sanli., *et al.* also stated that the multiple linear regression models is best fitted than simple linear regression model for estimating height from foot length and hand length [15].

Qamra., *et al.* explored from his study that foot length appeared to be a suitable measurement than foot breadth for stature estimation [28].

Giles., *et al.* presented linear regressions for determining height from foot length for males and females [29].

CONCLUSION

It is concluded that males have greater mean value of stature as compared to that of females.

The results of the present study show that the dimensions of hands and feet can successfully be used for estimation of stature by law enforcement agencies and forensic scientists. These formulae are applicable to the Jordanian population only, due to inherent variations in these dimensions which may be attributed to biological and environmental factors. These equations may be helpful to obtain approximate stature of an individual when there is difficulty in obtaining direct measurement such as in fragmented remains of body.

ACKNOWLEDGEMENT

The authors are grateful to the students of fourth and fifth year Medicine, Faculty of Medicine, Mutah University, Jordan, in 2011 for their assistance in the study in many data collection and tabulation.

REFERENCES

1. 1 Ellis Kerley (1978) In: Warren MW, Walsh-Haney HA, Freas L E (2008) The Forensic Anthropology laboratory, CRC Press: Taylor & Francis: London, New York 5-18.
2. Pickering R, Bachman D (2009) The use of Forensic Anthropology (2nd Ed), CRC Press, Taylor & Francis, London, New York 7-17.
3. Ellis Kerley (1978) In: Warren M W , Walsh-Haney H A , Freas L E. The Forensic Anthropology laboratory (2008) CRC Press, Taylor & Francis, London, New York 5-18.
4. Gelman A, Carlin JB, Stern HS, Rubin DB (2004) Bayesian Data analysis. Chapman & Hall, CRC Press, Boca Raton, Fl 319-329.
5. Abdel-Malek AK, Ahmed AM, Sharkawi SAA, Hamid NMA (1990) Prediction of stature from hand measurements. Forensic Sci Int 46: 181-7.
6. Ozaslan A, Iscan MY, Ozaslan I, Tugcu H, Koc S (2003) Estimation of stature from body parts. Forensic Sci Int 132: 40-5.
7. Kumar A, Srivastava AK, Verma AK (2010) Estimation of Stature by Percutaneous Measurements of Distal Half of Upper Limb (Forearm & Hand). J Indian Acad Forensic Med 32: 325-8.
8. Jain P, Kaur S, Nath S (1996) Reconstruction of stature from hand and foot dimensionsamong male Brahmins of Kumaon (India). J Ind Acad Forensic Sci 35: 22-9.
9. Sen J, Ghosh S (2008) Estimation of stature from foot length and foot breadth among the Rajbansh: an indigenous population of North Bengal. Forensic Sci Int 181: 1-6.
10. Zeybek G, Ergur I, Demiroglu Z (2008) Stature and gender estimation using foot measurements. Forensic Sci Int 181: 1-5.
11. Ozden H, Balci Y, Demirustu C, Turgut A, Ertugrul M (2005) Stature and sex estimate using foot and shoe dimensions. Forensic Sci Int 147: 181-4.
12. Sigh JP, Meena MC , Rani Y, Sharma GK (2013) Stature Estimation from the Dimensions of Foot in females. Antrocom Online Journal of Anthropology 9: 237-41.
13. Agnihotri AK, Purwar B, Googoolybe K, Agnihotri S, Jeebun N (2007) Estimation of stature by foot length. J Forensic Leg Med 14: 279-83.
14. Krishan K, Sharma A (2007) Estimation of stature from dimensions of hands and feet in North Indian population. J Forensic Leg Med 14: 327-32.
15. Sanli SG, Kizilkanat ED, Boyan N, Ozsahin ET, Bozkir MG, et al. (2005) Stature estimation based on hand length and foot length. Clin Anat 18: 589-96.
16. Saxena SK (1984) A study of correlations and estimation of stature from hand length, hand breadth and sole length. Anthropol Anz 42: 271-6.

17. Quamara SR, Jit I, Deodhar SD (1980) A model for construction of height from foot measurements in an adult population of North west India. Ind J Med Res 71: 77-83.
18. Ozaslan A, Karadayi B, Ahsen Kaya A, Afsin H (2012) Predictive role of hand and foot dimensions in stature estimation. Rom J Leg Med 20: 41-46.
19. Celbis O, Agritmis H (2006) Estimation of stature and determination of sex from radial and ulnar bone lengths in a Turkish corpse sample. Forensic Sci. Int 158: 135-9.
20. Vallois HV (1965) Anthropometric techniques. Curr Anthropol 6: 127-44.
21. Weiner JS, Lourie JA (1969) Human Biology: A guide to field methods, Blackwell Scientific Publications, Oxford.
22. Olivier G (1969) Practical Anthropology, C.C. Thomas, Springfield, IL.
23. Duyar I, Pelin C, Zagyapan R (2006) A new method of stature estimation for forensic anthropological application. Anthrop Sci 114: 23-27.
24. Nath S, Jain P (2005) Use of lower limb measurements in estimation of the stature. Forensic-Anthropology Science and Medicine. (1st Ed) Delhi: Kamla-Raj Enterprises: 37-43.
25. Narde AL, Dongre AP (2013) Body Height Estimation Based on Foot Length and Foot Breadth. J Indian Acad Forensic Med 35.
26. Hasegawa I, Uenishi K , Fukumaga T, Kimura R, Osawa M (2009) Stature estimation formulae from radiologically determined limb bone length in a modern Japanese population. Leg Med 11: 260-66.
27. Malina RM, Little BB, Stern MP, Gaskill SP, Hazuda HP (1983) Ethnic and social class differences in selected anthropometric characteristics of Mexican American and Anglu adults: The San Antonio heart study. Hum Biol 55: 867-83.
28. Qomara SR, Jit I, Deodhar SD (1980) A model of reconstruction of height from foot measurement in an adult population of Northwest India. Indian J Med Res 71: 77-83.
29. Giles E, Vallendigham PH (1992) Height estimation from foot and shoeprint length. J Forensic Sci May 36: 1134-1151.

CHAPTER 20

Variation in the Incidence of Self-Immolation According To Culture and Income Level: A Literature Review

Sarah L[1,2] and Prat SS*[3,4,5]

[1] *Unité de consultation et de soins ambulatoires de la Maison d'Arrêt de Tours, Centre Hospitalo-Universitaire, Tours, France*

[2] *Urgences psychiatriques, Centre Hospitalo-Universitaire, Tours, France*

[3] *Department of Psychiatry and Behavioural Neurosciences, McMaster University, Hamilton, Canada*

[4] *Forensic Psychiatry Program, St Joseph's Healthcare, Hamilton, Canada*

[5] *E.A.2114 Psychologie des âges de la vie, Université François Rabelais, Tours, France*

ABSTRACT

Self-immolation is most commonly observed in the Middle East and Western Asia. It is far less common in the Western world. Documented cases in the West indicate that victims are more likely to be immigrants, indicating that there may be cultural influences that motivate this behavior. We have performed a literature review on self-immolation in order to better understand the cultural differences behind self-immolation and the psychopathology related to this specific behavior.

INTRODUCTION

Self-immolation is an especially violent form of suicide because of the degree suffering during the act and the level of physical damage it causes. The impact on public opinion has allowed it to become a powerful tool in mobilizing mass movements such as in the "Jasmine Revolution" [1]. This form of suicide has historically been used as a way of expressing political dissent: in 1963 in Saïgon by Bonz Thich Quang-Duc, in 1969 in Prague by Jan Palach, and in 1998 in Rome by Alfredo Ormando [2,3]. Immolation has been carried out as an isolated act by one person or even a group of people, sometimes through replication or what has been called the "copycat phenomenon" [4]. The epidemic nature of this phenomenon is often facilitated by the media, and has recently been documented in India, Sri Lanka and Iran [5-7]. However, it has also gained traction in countries like the UK where the rate of this method of suicide has increased over a one-year period [6].

Self-immolation is most commonly observed in the Middle East and Western Asia [5]. It is far less common in the Western world, where estimates place this method at 0.06% to 1.5% of all suicides and suicide attempts [8,9]. By contrast, estimates place the rate at as high as 40% in certain developing countries [5,7,10-12]. Despite the low overall rate in the Western world, self-immolation may represent 1 to 10% of the cases in Burn Units [9,13]. Documented cases in the West indicate that victims are more likely to be immigrants, indicating that there may be cultural influences that motivate this behavior [6,14-19].

We have performed a literature review on self-immolation in order to better understand the cultural differences behind self-immolation and the psychopathology related to this specific behaviour.

MATERIALS AND METHODS

For our review, we carried out a systematic search of the Medline, Pubmed and PsychInfo databases. Search terms included "auto-immolation", "self-burning", "self-immolation", and "suicide by fire". We selected studies and case reports and included papers dealing with both living and dead victims. We excluded papers dealing with minor burning-related self-harm behaviours that could not be linked to a suicide behavior, such as self-harm with cigarettes. We did not limit our search by date of publication.

From these initial results, we extracted sociodemographic data (sex, mean age, marital status, socio-professional environment), psychiatric data (diagnosis, inpatient admission, treatment), and medico-legal data surrounding the suicide act (location, precipitating factor, substance abuse, chemical substance use). The main objective was to highlight ethnic and geographical differences among victims of self-immolation and to offer a potential explanation for such differences.

Cultural and ethnic factors may be also be related to income level in different countries. Peck et al. suggested sorting data into low and medium-income countries for a better understanding of factors behind self-immolation [20,21]. Thus, we grouped our data based on income level (low-income countries vs high-income countries) according to the World Bank Classification (http://data.worldbank.org/about/country-and-lending-groups accessed on March 9th, 2015).

RESULTS

We found 57 papers on self-immolation that contained information related to the victims' ethnic and geographic backgrounds. The papers were published between 1973 and 2015 [5,22]. We sorted our data based on the level of income of countries in order to make comparisons based on cultural background. The separation between low and high-income countries approximates the separation between Asian and African nations and countries from the Western world.

The majority of the studies contained epidemiological data. Sample sizes varied from 10 to 242 [23,24]. In order to be able to compare the studies, despite small sample sizes, we extrapolated the results in the form of percentages.

This literature review allowed the description of the sociodemographic data (ethnic and demographic; social instability and isolation), psychopathological data (psychiatric history; diagnosis; premeditation, impulsivity and intentionality) and data surrounding this type of suicidal behaviour (precipitating factor; substance abuse; location; chemical use). We have summarized the data in a table (Table 1). Results are detailed and commented on in the Discussion section. There did not appear to be any clear differences between low and high-income countries with regards to premeditation, lethal intentionality, and social vulnerability. However, we decided to include this data due to its relevance.

Table 1: Summary of the data from the literature according to self-inflicted burns (y-o: years old; SA: suicide attempt; med: medication) [1]Female victims are usually older than male victims

		High-income countries/ Western countries	Low-income countries/ Africa – Eastern countries
General profile		Male 50 y-o Psychiatric disorder	Female Young adult Illiterate, poor
Epidemiological aspects	Sex	Mostly male	Mostly female
	Age	35 – 50 y-o	25 y-o
	Marital status	Unmarried (75 – 81%)	Young adult married female
	Unemployement	25 – 80 %	
	Homeless	6 – 15 %	
	Unspecified psychiatric disorder	58 – 83 %	
	SA	20 – 69 %	
	Psychotropic med	> 50 %	
	Admission in psychiatry	50 %	
	Diagnosis	Depression Psychosis (3 – 44 %) Personnality disorder (60%) Addiction (25 – 59 %)	Adjustment disorder Depression PTSD
Behavioral aspects	Lethal intentionnality	20 – 41 %	
	Impulsivity	17 – 46 %	
	Premeditation	22 – 26 %	
Precipitating factor		Spouse conflict Financial issues (unemployment 65%)	Spousal conflict Political and social demands
Drugs	Alcohol Other drugs	13 – 50% 43%	
Location		Isolated place	In public
Chemical use		Gas	Gas Alcohol Paraffin

DISCUSSION

This literature review allowed us to construct two broad profiles of victims of self-immolation:

a) Males around 50 years of age, with a clear psychiatric disorder in the high-income countries
b) Young females with low levels of education for whom self-immolation is a response to circumstances in low-income countries.

These two profiles highlight specific sociodemographic, psychopathological, and circumstances surrounding self-immolation.

Sociodemographic characteristics

Ethnic and epidemiological aspects

Public self-immolation, which may be politically and socially motivated, is more frequent in Asia and the Middle East [5,7,10,22,25,26]. As stated earlier, self-immolation appears to be more common in those regions of the world. Higher rates by these same racial and ethnic groups may also been seen in immigrant populations [6,14-17]. There appears to be a higher rate of self-immolation committed by immigrants in England and South Africa, compared to the non-immigrant population. However these racial and ethnic differences have not been observed by studies conducted in the United States [18,19,27]. This difference might be interpreted as a result of variations in integration and whether ethnic community traditions persist more strongly in certain countries than in others.

In high-income countries, victims of self-immolation are mainly male, aged 37.4 to 50.2 years old [8,28–34]. The mean age is lower for self-immolation as compared to other methods of suicide. However, female victims are usually older than males.

In low-income countries, victims are typically young women or female adolescents [7,24,35-38].

Social instability and isolation

Social instability and isolation seem to be common criteria of all the victims, regardless of their country of origin. Unemployment is described in 25 to 80% of cases, and is sometimes associated with a lower level of education [9,18,23,32,39-41]. This appears to be more common with male victims. Isolation has also been described in 6 to 15% of cases, with higher rates found in homeless or incarcerated populations [18,32].

In high-income countries, 75 to 81% of victims lived alone [8,18,27,40].

In low-income countries, social vulnerability applies mainly to women, because of their low education level and illiteracy that contribute to their dependence on their spouses or in-laws [10,11,42,43]. The victims are usually young married women, and the suicide act is committed around 5 to 6 years after marriage [10]. The birth-order of the victims might also be linked to the act of committing suicide. According to Ahmadi et al. both the eldest or the youngest siblings are at a higher risk of suicide [11]. The hypothesis is tied to familial pressures of having a successful marriage. Domestic violence and dependence on a partner due to lack of education are commonly described. On the other hand, being married with several children seems to reduce the risk of suicide for such women.

PSYCHOPATHOLOGY

Psychiatric history

The majority of the data related to the psychiatric history of victims has been collected from surveys of high-income countries.

In high-income countries, psychiatric history is present in 60 to 83% of victims and familial psychiatric history in 60% [8,27,28,33]. Victims have received psychiatric medication and have been admitted to a psychiatric facility in more than half of cases [31,33,44]. Previous suicide attempts prior to the immolation event are found in 20 to 69% of cases, but the specific act of immolation is often the first time attempt for victims [8,18,32,33]. This raises the question of the escalation of methods in committing suicide. No study describes a case of a second-time attempt of immolation after a non-lethal first attempt by immolation.

In low-income countries, as we described with the typology of the female victim, we hypothesize that the circumstances surrounding marriage might trigger mental instability. Therefore, we could assume that there is no significant psychiatric history prior to the event, in contrast to the typical victims in the high-income countries. However, it may be valuable to explore the influence of marriage in more depth in order to better understand which victims are more or less vulnerable in such situations. Zor., *et al.* described a small sample of ten victims of self-burning, involved in the military, who all had psychiatric history and a diagnosis of personality disorder [23].

Psychiatric diagnosis

The psychiatric diagnoses of the victims differ according to their backgrounds, which may shed light on the reasons behind their suicide attempts.

In high-income countries, major depressive episode is the most common diagnosis, with or without psychotic features and substance abuse [8,10,18,27,29,31,32,45-47]. Other diagnoses include psychotic episodes in 3 to 44%, personality disorder in 60% and substance abuse in 25 to 59% [8,9,18,19,23,27,29,41,45].

In low-income countries, a psychiatric diagnosis is less often described. The lack of accessibility to care due to the isolation may be a reason for the absence of previous diagnosis and history of treatment. When victims are diagnosed, it is usually with adjustment disorder in depressive episode or PTSD [11,20,21]. Male victims in low-income countries seem to have higher rates of substance abuse disorder, especially alcoholism [11].

Lethal intentionality, premeditation and impulsivity

Our review has not been able to identify any clear information about intentionality, premeditation and impulsivity toward the suicidal behaviour in low-income countries. However, such factors are worth further investigation due to their potential impact on the mental status assessment and ability to predict the risk of this type of self-harm.

Self-harming behaviour with violent methods is not necessarily linked to a true lethal intent. Often this intent can also fluctuate according to the emotion dysregulation of personality disorder. A true intent to die is described in only 20 to 41% of cases [19,27,32,48-50]. Impulsivity as the reason behind the decision to commit suicide is present in 17 to 46% of cases, whereas premeditation is found in 22 to 26% of cases [27,40,51,52].

CIRCUMSTANCES SURROUNDING SELF-HARMING BEHAVIOR

Ethnic and epidemiological aspects

Emotional dysregulation and economic concerns appear to be the main motives behind self-immolation. Emotional dysregulation is related to issues, such as break-ups, infidelity, or spousal violence [11,29,32,33,35,37,38]. Motives such as social isolation are also described [18,27]. Intra-familial conflict and financial issues are described in both low and high-income countries [7,11,20,21,29,32,33,35,52].

In high-income countries, economic concerns (approximately 65% of victims found to be unemployed) are the main motives [18,29,33].

In low-income countries, we identified arranged marriages, spousal violence and political reasons as the motives [20,21,37,38,53].

Substance abuse

The majority of studies describing drug abuse and self-immolation have been conducted in high-income countries [13,19,29,32,33]. Alcohol use is found in 13 to 50% of cases and toxic levels of drugs are found in 25 to 43% [19,29,32,33,40,52]. According to some studies, more than half of the victims were found to be on psychiatric medication at the time of the event, explaining the victims' easy access to such substances [31,33,44].

Location

Home is the main location for immolation, described in 50 to 65% of the cases [8,17,28,29,31-33,38]. Some psychopathological characteristics may influence this choice of location. Many victims meet the criteria for a cluster C personality disorder, and this may explain their preference to commit immolation in a public area [23]. Furthermore, compared to in high-income countries, in low-income countries the event may take place in public areas as a way of expressing political dissent [7,10,20,21]. We must also bear in mind that the influence of isolation and social vulnerability, which people may experience when they are hospitalized or incarcerated. These factors has been observed to motivate victims to use fire to commit suicide, in both low and high-income countries [31,40,45,54]. In controlled environments, where many items are banned for safety, fire may be the only means available for suicide, and therefore exists as the default tool for suicide attempts because of lack of access to less lethal methods such as medications for overdose.

Chemical use to facilitate the ignition

Using an inflammable product has an impact on the gravity of injuries and lethality. The use of such products is frequently described in the literature, in 30 to 77% of the victims [8,13,19,26,28-30,32,33,35,39,55,56]. According to Wiechman *et al.*, inflammable product use is linked to suicide and lethal intentionality, compared to self-harming behavior where cigarette and matches are used more commonly instead [49].

In high-income countries, gas is the most common product used [8,13,19,28-30,32,33].

In low-income countries, gas, alcohol and paraffin use is described [17,19,35,37].

CONCLUSION

This literature review allowed us to compare the differences between the victims of immolation in high and low-income countries. As described above, motives and sociodemographic characteristics are the main differences. There is little information on differences in psychopathology and this element needs to be studied in more depth [52]. Major burns may be a result of crime, suicide or accident. It is important to identify individuals with this type of suicidal ideation and behavior to adapt treatment and prevent future harm [8,26,57]. The immigration status and ethnic background of victims may represent an important clinical consideration in identifying the triggers behind such suicide attempts. Medical teams need to be able to tailor their care to the backgrounds and specific needs of patients, in order to successfully treat and rehabilitate victims of immolation.

REFERNCES

1. Lim M (2013) Framing Bouazizi: "White lies", hybrid network, and collective /connective action in the 2010-11 Tunisian uprising. Journalism 14: 921-941.
2. Biggs M (2005) Dying without killing: self-immolations, 1963-2002. In: Gambetta D (editor). Making sense of suicide missions. Oxford, Oxford University Press 173-208.
3. Boylan A (2013) Unresolved commemorations: memorials to victims of homophobic violence in Italy. The Italianist 33: 138-157.
4. Greenberg B, Strous RD (2012) Werther's syndrome: copycat self-immolation in Israel with a call for responsible media response. Isr Med Assoc J 14: 467-469.
5. Rezaeian M (2015) Self-immolation as a proxy measure for unmet needs among the vulnerable. Burns 41:17-418.
6. Ashton JR, Donnan S (1981) Suicide by burning as an epidemic phenomenon: an analysis of 82 deaths and inquests in England and Wales in 1978-9. Psychol Med 11: 735-739.
7. Laloë V (2004) Patterns of deliberate self-burning in various parts of the world. Burns 30: 207-215.
8. Gauthier S, Reisch T, Bartsch C (2014) Self-burning –a rare suicide method in Switzerland and other industrialised nations – a review. Burns 40: 1720-1726.
9. Palmu R, Isometsä E, Suominen K, Vuola J, Leppävuori A, et al. (2004) Self-inflicted burns: an eight year retrospective study in Finland. Burns 30: 443-447.
10. Poeschla B, Combs H, Livingstone S, Romm S, Klein MB (2011) Self-immolation: Socioeconomic, cultural and psychiatric patterns. Burns 37: 1049-57.
11. Ahmadi A (2007) Suicide by self-immolation: comprehensive overview, experiences and suggestions. J Burn Care Res 28: 30-41.
12. Badiye A, Kapoor N, Ahmed S (2014) An empirical analysis of suicidal death trends in India: a 5 year retrospective study. J Forensic Leg Med 27: 29-34.
13. Franchitto N, Faurie C, Franchitto L, Minville V, Telmon N (2011) Self-inflicted burns: the value of collaboration between medicine and law. J Forensic Sci 56: 638-642.
14. Davidson TI, Brown LC (1985) Self-inflicted burns: a 5-year retrospective study. Burns 11: 157-160.

15. Prosser D (1996) Suicides by burning in England and Wales. Br J Psychiatry 168: 175-182.
16. Soni Raleigh V, Balarajan R (1992) Suicide and self-burning among Indians and West Indians in England and Wales. Br J Psychiatry 161: 365-368.
17. 17Sukhai A, Harris C, Moorad RGR, Dada MA (2002) Suicide by self-immolation in Durban, South Africa: a five-year retrospective review. Am J Forensic Med Pathol 23: 295-298.
18. Pham TN, King JR, Palmieri TL, Greenhalgh DG (2003) Predisposing factors for self-inflicted burns. J Burn Care Rehabil 24: 223-227.
19. Squyres V, Law EJ, Still JM Jr (1993) Self-inflicted burns. J Burn Care Rehabil 14: 476-479.
20. Peck MD (2012) Epidemiology of burns throughout the World. Part II: intentional burns in adults. Burns 38: 630-637.
21. Peck MD (2011) Epidemiology of burns throughout the world. Part I: distribution and risk factors. Burns 37: 1087-1100.
22. Topp DO (1973) Fire as a symbol and as a weapon of death. Med Sci Law 13: 79-86.
23. Zor F, Deveci M, Bozkurt M, Dikkatli S, Duman H (2005) Psychological evaluation of self-inflicted burn patients: suicide or parasuicide? Burns 31: 178-181.
24. Ambade VN, Godbole HV, Kukde HG (2007) Suicidal and homicidal deaths: a comparative and circumstantial approach. J Forensic Leg Med 14: 253-260.
25. Singh SP, Santosh PJ, Avasthi A, Kulhara P (1998) A psychosocial study of "self-immolation" in India. Acta Psychiatr Scand 97: 71-75.
26. Forster NA, Nuñez DG, Zingg M, Haile SR, Künzi W, et al (2012) Attempted suicide by self-immolation is a powerful predictive variable for survival of burn injuries. J Burn Care Res 33: 642-8.
27. Daniels SM, Fenley JD, Powers PS, Cruse CW (1991) Self-inflicted burns: a ten-year retrospective study. J Burn Care Rehabil 12: 144-147.
28. Cave Bondi G, Cipolloni L, Parroni E, Cecchi R (2001) A review of suicides by burning in Rome between 1947-1997 examined by the Pathology Department of the Institute of Forensic Medicine, University of Rome "La Sapienza". Burns 27: 227-231.
29. Makhlouf F, Alvarez J-C, de la Grandmaison GL (2011) Suicidal and criminal immolations: an 18-year study and review of the literature. Leg Med 13: 98-102.
30. Malic CC, Karoo ROS, Austin O, Phipps A (2007) Burns inflicted by self or by others--an 11 year snapshot. Burns 33: 92-97.
31. Castellani G, Beghini D, Barisoni D, Marigo M (1995) Suicide attempted by burning: a 10-year study of self-immolation deaths. Burns 21: 607-609.
32. Squires TJ, Busuttil A (1996) Suicide by fire in Scotland: 1980-1990. J Clin Forensic Med 3: 81-85.
33. Rothschild MA, Raatschen HJ, Schneider V (2001) Suicide by self-immolation in Berlin from 1990 to 2000. Forensic Sci Int 124: 163-166.
34. Nakae H, Zheng YJ, Wada H, Tajimi K, Endo S (2003) Characteristics of self-immolation attempts in Akita Prefecture, Japan. Burns 29: 691-696.
35. Mzezewa S, Jonsson K, Aberg M, Salemark L (2000) A prospective study of suicidal burns admitted to the Harare burns unit. Burns 26: 460-464.
36. Ganesamoni S, Kate V, Sadasivan J (2010) Epidemiology of hospitalized burn patients in a tertiary care hospital in South India. Burns 36: 422-429.
37. Natarajan M (2014) Differences between intentional and non-intentional burns in India: Implications for prevention. Burns 40: 1033-1039.
38. Rasool IA, Payton JL (2014) Tongues of fire: women's suicide and self-injury by burns in the Kurdistan Region of Iraq. Sociol Rev 62: 237-254.
39. Ali SN, Soueid A, Rao K, Moiemen N (2006) Self-inflicted burns, outcome and cost. Burns 32:63-466.
40. Seoighe DM, Conroy F, Hennessy G, Meagher P, Eadie P (2011) Self-inflicted burns in the Irish National Burns Unit. Burns 37: 1229-1232.

41. Kamolz L-P (2013) Attempted suicide by self-immolation is a powerful predictive variable for survival of burn injuries. J Burn Care Res 34: e271.
42. Raj A, Gomez C, Silverman JG (2008) Driven to a fiery death--the tragedy of self-immolation in Afghanistan. N Engl J Med 358: 2201-2203.
43. Das Gupta SM, Tripathi CB (1984) Burnt wife syndrome. Ann Acad Med Singapore 13: 37-42.
44. Mulholland R, Green L, Longstaff C, Horner B, Ross E, et al (2008) Deliberate self-harm by burning: a retrospective case controlled study. J Burn Care Res 29: 644-649.
45. O'Donoghue JM, Panchal JL, O'Sullivan ST, O'Shaughnessy M, O'Connor TP, et al (1998) A study of suicide and attempted suicide by self-immolation in an Irish psychiatric population: an increasing problem. Burns 24: 144-146.
46. Horner BM, Ahmadi H, Mulholland R, Myers SR, Catalan J (2005) Case-controlled study of patients with self-inflicted burns. Burns 31: 471-475.
47. Thombs BD, Bresnick MG, Magyar-Russell G (2007) Who attempts suicide by burning? An analysis of age patterns of mortality by self-inflicted burning in the United States. Gen Hosp Psychiatry 29: 244-250.
48. Wallace KL, Pegg SP (1999) Self-inflicted burn injuries: an 11-year retrospective study. J Burn Care Rehabil 20: 191-194.
49. Wiechman SA, Ehde DM, Wilson BL, Patterson DR (2000) The management of self-inflicted burn injuries and disruptive behavior for patients with borderline personality disorder. J Burn Care Rehabil 21: 310-317.
50. Henderson A, Wijewardena A, Streimer J, Vandervord J (2013) Self-inflicted burns: a case series. Burns 39: 335-340.
51. Swenson JR, Dimsdale JE (1990) Substance abuse and attempts at suicide by burning. Am J Psychiatry 147: 811.
52. Hahn AP, Jochai D, Caufield-Noll CP, Hunt CA, Allen LE, (2014) Self-inflicted burns: a systematic review of the literature. J Burn Care Res 35: 102-119.
53. Boostani D, Abdinia S, Rahimipour Anaraki N (2013) Women victims of self-immolation: a "Grounded Theory" study in Iran. Qual Quant 47: 3153-65.
54. Andreasen NC, Noyes R (1975) Suicide attempted by self-immolation. Am J Psychiatr 132: 554-6.
55. Rashid A, Gowar JP (2004) A review of the trends of self-inflicted burns. Burns 30: 573-6.
56. Krummen DM, James K, Klein RL (1998) Suicide by burning: a retrospective review of the Akron Regional Burn Center. Burns 24: 147-9.
57. Laborderie S, Prat SS (2014) Aspect suicidaires, criminels et accidentels des décès par immolations : une revue de la littérature. Rev Med Leg 5: 148-59.

CHAPTER 21

Application of High-Resolution Mass Spectrometry to Determination of Baclofen in a Case of Fatal Intoxication

Paweł Szpot[1,2] • Agnieszka Chłopas´[2] • Grzegorz Buszewicz[2] • Grzegorz Teresin´ski

[1].Department of Forensic MedicineWroclaw Medical UniversityWrocławPoland

[2].Department of Forensic MedicineMedical University of LublinLublinPoland

ABSTRACT

The study focused on the application of high-resolution mass spectrometry (HRMS) to postmortem toxicological analysis. Fast and simple sample preparation involved precipitation with acetonitrile, removal of phospholipids using special columns and filtration. Qualitative and quantitative analyses were performed using ultra-performance liquid chromatography coupled with quadrupole time-of–flight mass spectrometry. The method was validated by determining the limit of quantification, precision, recovery and matrix effect. The use of a high-resolution spectrometer allowed us to determine the precise masses of the fragments of interest and to suggest the fragmentation pathway of baclofen. The usefulness, effectiveness and assets of the procedure were confirmed by an authentic case of a 25-year-old woman fatally intoxicated with baclofen who was found dead in her apartment. Toxicological analysis of postmortem blood samples demonstrated that the baclofen concentration was 30.7 μg/mL. In only one published case describing fatal baclofen intoxication were no other xenobiotics (that could interact with baclofen) found. To our knowledge, this is the first report dealing with analysis of baclofen by HRMS.

Keywords: Baclofen Fatal intoxication QTOF LC–MS HRMS

INTRODUCTION

Baclofen [(±)-4-amino-3-(4-chlorophenyl) butanoic acid, β-(aminomethyl)-4-chlorobenzenpropane acid, Figure 1], with commercial names of Atrofen®, Baclofen®, Lioresal®and Lyflex®. Baclofen is a drug affecting the central nervous system (CNS). The mechanism of action of baclofen, which is a derivative of γ-aminobutyric acid (GABA), involves stimulation of $GABA_B$-ergic receptors located pre- and postsynaptically [1]. The substance decreases the skeletal muscle tone by inhibiting the mono- and polysynaptic reflexes at the spinal cord level, but does not affect the neuromuscular conduction [2]. Baclofen is used for the treatment of spinal cord diseases, cerebral stroke and cerebrospinal meningitis. It is also applied in patients with severe chronic spasticity in multiple sclerosis [3, 4]. Moreover, the drug is known to minimize the symptoms of alcohol craving [5, 6, 7].

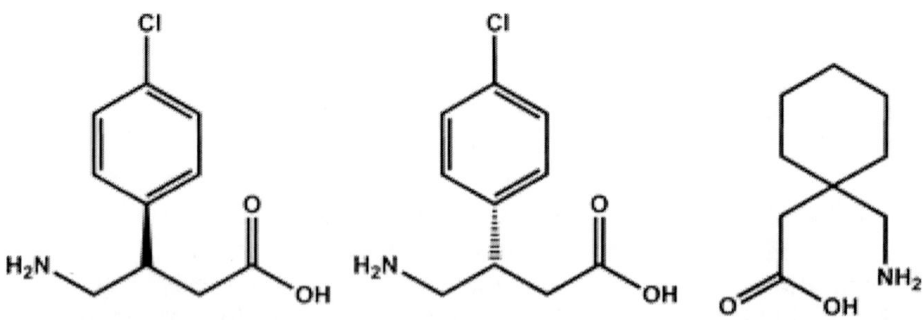

R-Baclofen S-Baclofen Gabapentin

Figure 1. Structures of enantiomers of baclofen and the internal standard (IS) gabapentin

Baclofen occurs as a mixture of two biologically active enantiomers (R and S), that differ in the arrangement of substituents around a chiral carbon atom. It has been found that the R(−) isomer of baclofen exhibits about 100-fold more potent pharmacological action than its S(+) form [8]. Wuis et al. [9] compared the pharmacokinetics of racemic baclofen and its two enantiomers, and they have found that the biological half-life ($t½$) of R(−) baclofen is 4.5 h, which is longer than the half-life of its racemic mixture (3.8 h) as well as S–baclofen (2.9 h). These findings suggest the existence of stereoselective renal elimination pathways [9]. After its oral administration, baclofen is rapidly and completely absorbed. It reaches the maximum blood concentration after 2–3 h. The biological half-life is 2.5–4 h [10]. It is bound to blood proteins at a rate of about 30 %. Due to deamination and oxidation, 15 % of the absorbed dose of baclofen is metabolized in the liver to β-(*p*-chlorophenyl)-γ-hydroxybutyric acid whereas 85 % is excreted intact into urine [11]. The oral therapeutic dose for adults should be tailored individually and ranges between 15–80 mg/day [3]; significant complications, and life-threatening cases have been reported with doses as low as 300 mg [12].

The literature reports of acute and fatal baclofen intoxication are rare, as compared to other CNS-affecting drugs [13, 14]. In some cases, the symptoms of overdose occur already during the use of the therapeutic dose and include decreases in muscle tone, dizziness, sedation, seizures, loss of consciousness, hypothermia, inhibition of the respiratory function, apnoea and coma [15]. Long-term administration followed by withdrawal can lead to psychotic, manic and paranoid anxiety disorders [16].

To date, baclofen has been determined in various biological materials using the techniques based on gas [17, 18] or liquid [19, 20, 21] chromatography, liquid chromatography–tandem mass spectrometry [22, 23, 24, 25] as well as capillary electrophoresis [26]. However, the majority of these methods require time-consuming sample preparation. As far as medico-legal toxicology is concerned, it is essential to design a simple and fast analytical procedure, providing reliable results.

Substantial advances in the development of special devices for identification and determination of chemical compounds in biological material have recently been observed. Thanks to equipment, increasingly more substances at markedly lower concentrations can be detected. However, considering the interpretation of results and established norms regarding, e.g. validation of methods, much more attention should be paid to toxicological analysis; therefore, numerous recent descriptions contain detailed data regarding analytical problems associated with examinations of the postmortem material [27, 28, 29]. In this manuscript, high-resolutions mass spectrometry (HRMS) was used for the determination of baclofen in postmortem blood. The technique based on liquid chromatography–hybrid quadrupole time-of-flight-mass spectrometry (LC–QTOF-MS) enables comprehensive targeted forensic screening. The use of this detector allows one to determine the exact weight and elemental composition of an unknown compound. To our knowledge, this is the first paper describing the determination of baclofen in postmortem blood by LC–QTOF-MS. The aim of the study was to develop and validate a highly specific, simple and quick method of determination of baclofen in biological material, to analyze fragmentation of this substance and to review the literature concerning baclofen intoxication cases.

MATERIALS AND METHODS

Chemicals and reagents

Water (Chromasolv® LC–MS), acetonitrile (Chromasolv® LC–MS), methanol (Chromasolv®LC–MS), gabapentin and formic acid were purchased from Sigma-Aldrich (Steinheim, Germany); (±)–baclofen was purchased from Polpharma SA (Starogard Gdański, Poland). Standard solutions of baclofen were prepared in water. Gabapentin [internal standard(IS)] was prepared in methanol. The standard solutions were stored in a refrigerator at −20 °C.

Blank human blood material

Four blank samples of postmortem human blood were derived from the Chair and Department of Forensic Medicine. Blank samples did not contain any anticoagulant and were screened prior to spiking to ensure that they were free from baclofen and gabapentin.

Chromatographic conditions

Chromatographic analysis was performed using an ultra-high performance liquid chromatograph (UHPLC Infinity 1290, Agilent Technologies, Waldbronn, Germany). The separation was done employing a Poroshell 120 EC-C18 column (3.0 × 100 mm); 2.7 μm (Agilent Technologies, Santa Clara, NM, USA).

A mixture of 0.1 % formic acid in water (A) and 0.1 % formic acid in acetonitrile (B) was used as a mobile phase. The gradient elution was carried out at constant flow rate of 0.4 mL/min in the following gradient mode: 0 min (95 % A/5 % B); 0.5 min (70 % A/30 % B), 5 min (30 % A/70 % B), 7 min (5 % A/95 % B) and 8 min (95 % A/5 % B). Return to the initial conditions (95 %A) was achieved 2.5 min after the 8-min long analysis. The injection volume was 5 μL.

Mass spectrometry

Detection of the study compounds was achieved using a quadrupole time-of-flight (QTOF) mass spectrometer (QTOF 6540, Agilent Technologies, Santa Clara, NM, USA). The spectrometer was equipped with an electrospray ionization (ESI) source. The source conditions were as follows: gas temperature 350 °C; gas flow 10 L/min.; nebulizer (N_2) 35 psi; sheath gas temperature 400 °C; sheath gas flow 12 L/min.; capillary voltage 3000 V; fragmentor voltage 140 V; skimmer voltage 65 V; oct 1 RF Vpp 750 V; mass range *m/z* 50–1000. Quantitative and qualitative analyses were carried out in the MS mode. To ensure accuracy of mass measurements, reference mass correction was used. Masses at *m/z* 121.0509 and 922.0098 were used as reference ions. The mass tolerance used for the extracted ion chromatograms was ±20 ppm. The mass error for the reference masses used prior to analysis was lower than 2.7 ppm.

The analysis of product ion formation of baclofen at a concentration of 5 μg/mL was carried out in the tandem QTOF mode with spectral parameters: mass range *m/z* 50–1000; acquisition rate 1.5 spectra/s; isolation width *m/z* ~4; and collision energies (CEs) 5, 20 and 35 V. Retention times for baclofen and gabapentin were 2.02 and 1.96 min, respectively. Positive ionization was performed. Quantitative and qualitative ions for determinations of baclofen and gabapentin were *m/z* 214.0629, 151.0309 and *m/z* 172.1332, 154.1226, respectively.

Sample preparation

A 200-μL volume of postmortem blood was poured into a 2-mL Ependorff tube; then 20 μL of gabapentin (IS) at a concentration of 10 μg/mL was added and mixed. The blood prepared in such a way was precipitated by adding drops of 400 μL of frozen 0.1 % formic acid in acetonitrile (constantly mixed) and the mixing was continued for 30 s. The samples were centrifuged for 15 min at 20,627*g* at 5 °C (2–16 K, Sigma, Osterode am Harz, Germany). Subsequently, 400 μL of the supernatant was loaded onto a column of Phree Phospholipid Removal 1 mL (Phenomenex, Torrance, CA, USA) coupled with a polytetrafluoroethylene syringe filter (pore size 0.22 μm, FilterBio, Nantong, China) and filtered using a Visiprep vacuum manifold (Supelco, Bellefonte, PA, USA) under pressure of 10 mmHg. A 100-μL volume of each of the extracts was transferred to the inserts and analyzed using LC–QTOF-MS.

Method validation

Method validation was carried out following the EURACHEM guidelines [30]. The method was validated by determining the following parameters: specificity, precision, recovery, limit of quantification (LOQ) and matrix effect. The postmortem blood was chosen as a target matrix for method validation. The specificity of the method was evaluated by analyzing a blank sample and a blank sample spiked with the LOQ levels of baclofen and IS. Calibration linearity was verified by analyzing each of four different blank blood samples with the additions of baclofen at the concentrations of 1, 2, 5, 10, 20 and 50 μg/mL. Precision was expressed using % relative standard deviation (%RSD) for each calibration curve point analyzed in four repeats. Recovery and the matrix effect were determined for four points of the calibration curve: 1, 5, 10 and 20 μg/mL. Recovery was determined by adding 20 μL of baclofen and IS, and 400 μL of frozen 0.1 % formic acid in acetonitrile to 180 μL of water. The solution of standards prepared in this way was analyzed using LC–QTOF-MS, and the data were compared with each of the four calibration curves in postmortem blood. The LOQ in this method was the blood concentration of baclofen for which both quantitative and qualitative ions could be observed and the standard deviation (SD) of the concentration did not exceed 20 %. To calculate the matrix effect, the formula by Chambers et al. [31] was applied:

\% Matrix effect =(Response extracted sampleResponse standard−1)×100.\% Matrix effect =(Response extracted sampleResponse standard−1)×100.

The negative result evidences suppression while the positive result evidences enhancement of analyte signals. The validated method was used for analysis of blood samples sent to the Chair and Department of Forensic Medicine to determine the cause of death.

CASE HISTORY

A 25-year-old woman was found dead in her apartment. On the day of death, the woman consumed an undetermined amount of drug(s) of unknown origin, which most likely caused her death. Near the deceased, the following containers and baclofen tablets were found: one white tablet, 5 mm in diameter, one empty Baclofen 25 mg container for 50 tablets, one full container of the same drug with 51 tablets and one container with 12 and 1/2 tablets. The above suggested that the woman took 87 and 1/2 tablets (maximal potential dose 2187.5 mg). The case investigation revealed that the woman abused alcohol, often had arguments and caused various problems. Before her death, she wrote to her boyfriend informing him that she took 100 tablets of baclofen. During the prosecutor's investigation it was established that multiple times after arguments with her boyfriend, the woman had taken various drugs, had drunk alcohol and subsequently vomited.

External examination and autopsy did not reveal any pathological changes. There was no ethyl alcohol present in her blood. Autopsy results excluded death due to mechanical trauma, acute alcohol intoxication and diseases.

In this case, postmortem venous blood samples from the femoral vein were collected via a chemically pure plastic tube, imblood samples indicated that there were impure compoundsmediately frozen at −22 °C, and stored until analysis. No anticoagulants were added to the sample.

RESULTS AND DISCUSSION

The use of HRMS confers high specificity and reduces the risk of potential interferences related to the complexity of the matrix. The absence of chromatographic peaks at the same retention times as those of target substances in blank blood samples (Figure 2a) indicated that there were no impure compounds that might give false positive signals in these blank samples. The LOQ for the method was 1 µg/mL, whereas the coefficient of determination (R^2) for the calibration curve was 0.996. Table 1 presents concentrations, SDs and precisions for individual calibration curve points. %RSD for the method was 5.1–15.3 %.

Table 1. Calculated concentrations in the calibration curve for baclofen in postmortem blood

Nominal concentration (µg/mL)	Level	Calculated concentration (µg/mL)[a]	SD	Precision (%RSD)
1	1	0.91	0.14	15.3
2	2	2.12	0.20	9.82
5	3	4.99	0.54	10.8
10	4	9.97	0.51	5.08
20	5	20.0	1.21	6.06
50	6	50.4	4.67	9.27

SD standard deviation, *RSD* relative standard deviation

[a] Mean of four repeats

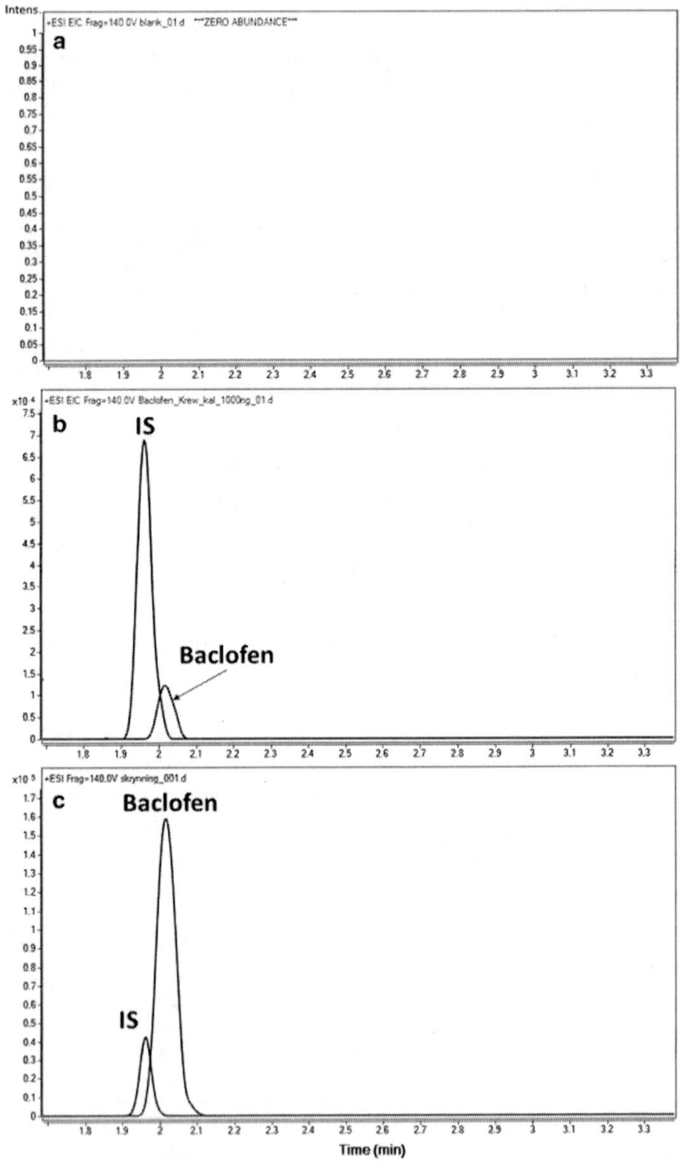

Figure 2. High-resolution extracted ion chromatograms obtained by liquid chromatography–quadrupole time-of-flight-mass spectrometry for a blank blood sample (**a**), blank blood sample spiked with 1.0-µg/mL baclofen and 1.0-µg/mL IS (**b**), and the authentic postmortem blood sample in the present case spiked with 1.0-µg/mL IS (**c**) using the precursor ions of baclofen and IS at a collision energy (CE) of 5 V

Table 2 shows the mean percent recovery rates and mean matrix effect at four different concentrations of baclofen in postmortem blood.

Table 2. Recovery rates and matrix effect of baclofen in postmortem blood

Nominal concentration (μg/mL)	Recovery rate (%)[a]	Matrix effect rate (%)[a]
1	104	−40.9
5	90.3	−50.5
10	97.3	−50.3
20	109	−38.2

[a] Mean of four repeats

The method described enabled detection of baclofen at a level lower than the LOQ, i.e. 1 μg/mL. The quantitative and qualitative ions were also observed at the concentration of 0.5 μg/mL; however, at such a wide range of concentrations (0.5–50 μg/mL), the concentration calculation error for this point was over 20 %; moreover, reduced linearity was observed. Therefore, for concentrations up to 0.5 μg/mL, a separate calibration curve should be constructed. It can be assumed that the concentration at 0.5 μg/mL of baclofen in blood was a limit of detection (LOD). The recovery rates in samples were 90.3–109 %, which confirms that the use of gabapentin as an IS was the right decision (used earlier for determination of baclofen in rat materials by Kim et al. [32] with recoveries of 93.2–100.4 %). The negative matrix effects evidenced suppression of analyte signals, which could not be completely removed despite the use of precipitation, sorbent-removing proteins and phospholipids as well as filtration during sample preparation.

The presented method has a limitation especially for quantification of baclofen in postmortem blood. Gabapentin is clinically used for treating epilepsy; as such, there is potential influence from originally occurring gabapentin. For this reason, it is important to first analyze the sample without the IS–gabapentin when using this method.

Qualitative analysis was based on the automatic comparison of all ions from the total ion chromatogram to our database. Crucial criteria included retention time (±0.5 min.) and mass error (±2 ppm). The qualitative analysis showed a positive result for baclofen, which was confirmed by comparison of examined blood samples with the reference standard of baclofen by product ion mass spectra, with parameters of the precursor ion at *m/z* 214.0629, fragmentor voltage at 140 V and CE at 20 V. The concentration of baclofen determined in the postmortem blood was 30.7 µg/mL ($n = 4$, SD of 1.8 µg/mL; %RSD of 5.94; Figure 2c). Tested blood was screened prior to analysis in order to ensure that it was free from gabapentin. Screening testing did not indicate other compounds in the biological material.

The application of a QTOF detector enabled obtaining high-resolution mass spectra (MS/MS). The MS/MS analysis allowed us to investigate the mechanism of product ion formation from baclofen, which, to our knowledge, has not been described before. The obtained product ions of baclofen may be useful in the determination of this compound by less accurate detectors such as a single quadrupole or a triple quadrupole. Depending on the CE, baclofen underwent product ion formation to three major ions at *m/z* 197.03638 (CE 5 V), 151.0309 (CE 20 V) and 116.06205 (CE 35 V), which can be used as confirmation ions (Figure 3).

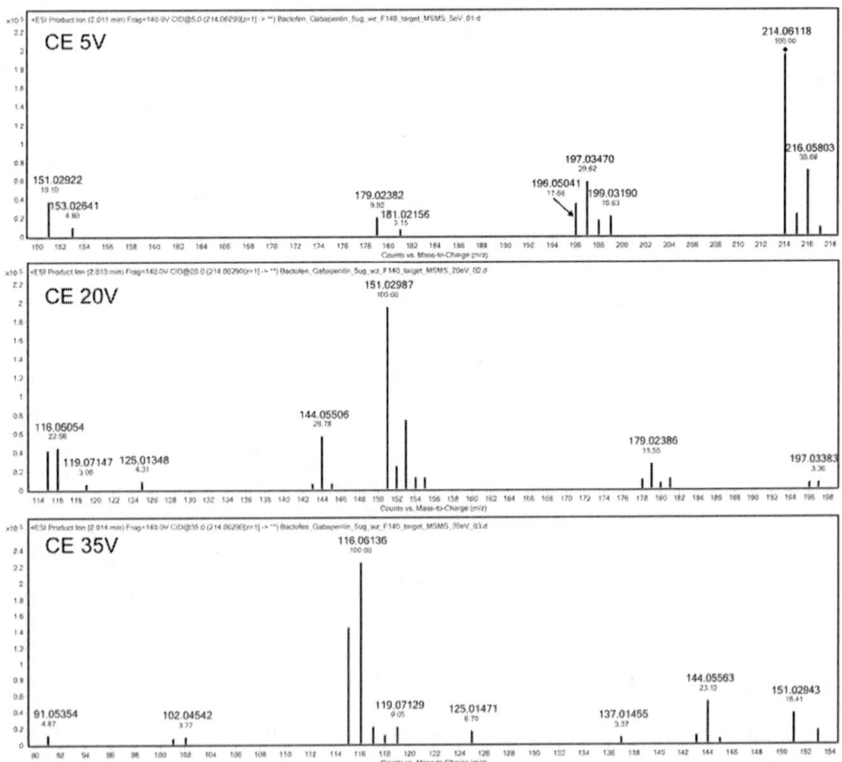

Figure 3: High-resolution product ion mass spectra of baclofen obtained at different CEs. The accurate mass of the precursor protonated baclofen was 214.0629

Based on the MS/MS spectra, the general formula of baclofen fragment ions were suggested (Table 3).

Table 3. Accurate mass data obtained by liquid chromatography–quadrupole time-of-flight-mass spectrometry and proposed composition of characteristic product ions of baclofen

No.	Proposed composition	Observed (*m/z*)	Calculated (*m/z*)	CE (V)	Difference (mDa)	Difference (ppm)	Mass loss	Loss formula
1	$C_{10}H_9ClO_2$	197.03383	197.03638	5	2.55	12.96	17.02655	NH_3
2	$C_{10}H_{10}ClNO$	196.05123	196.05237	5	1.14	5.81	18.01056	H_2O
3	$C_{10}H_7ClO$	179.02329	179.02582	20	7.86	14.12	35.03711	H_5NO
4	C_9H_7Cl	151.02968	151.0309	20	1.22	8.09	63.03203	CH_5NO_2
5	C_7H_5Cl	125.01471	125.01525	35	0.54	4.35	89.04768	$C_3H_7NO_2$
6	C_9H_7	116.06136	116.06205	35	0.69	5.96	98.00088	CH_5ClNO_2

CE collision energy

In its structure, baclofen contains one atom of chlorine, which causes the formation of a characteristic mass spectrum evidencing the presence of isotopic ions of this element $[M + H{:}^{35}Cl]^+$ and $[M + H{:}^{37}Cl]^+$ (Figure 3, top panel). The mass difference between chlorine isotopes ^{35}Cl and ^{37}Cl was 1.9970; therefore, ion fragments with ^{35}Cl should be accompanied by the ion containing ^{37}Cl, differing by the value mentioned above. Examinations of the MS/MS spectra revealed that several very intense fragments were formed as a result of baclofen fragmentation, depending on the CEs. General formulae were proposed for six of them (Table 3). The use of HRMS enabled to define the precise mass of product ions of interest and errors in their determination as well as to analyze their structure in detail. At a 5-V CE, the most intense fragment was at *m/z* 197.03638. The fragment occurred due to dissociation of the ammonium group from the baclofen molecule. Figure 3 also contains the 196.05237 ion, which results from dissociation of a water molecule. This ion was used by Miksa and Poppenga [33] as one of the confirmation ions for determinations of baclofen in bovine serum. In our study, the ion in question was less intense than *m/z* 197.03638, which is likely to be associated with a type of ionization different from that applied by Miksa and Poppenga, i.e. atmospheric pressure chemical ionization [33].

The ion at *m/z* 179.02582 ($C_{10}H_7{}^{35}ClO$) is formed due to dissociation of both the water molecule and ammonium group and not the loss of chlorine (as suggested by Miksa and Poppenga [33]), because the ion at *m/z* 181.02282 ($C_{10}H_7{}^{37}ClO$), characteristic of a chlorine isotope, was also visible in Figure 3 (top panel). The use of a medium collision energy (CE = 20 V) resulted in the most intense signal coming from the fragment C_9H_7Cl at *m/z* 151.0309. The signal occurs due to dissociation of the CO_2 group and ammonium group from the baclofen molecule. The next stages of product formation involve dissociation of the chlorine atom from the ion at *m/z* 151.0309 and cyclization of its propylene group, leading to formation of the C_9H_7 = 116.06205 fragment, which was the most intense ion at CE = 35 V (Figure 3, bottom panel and Figure 4).

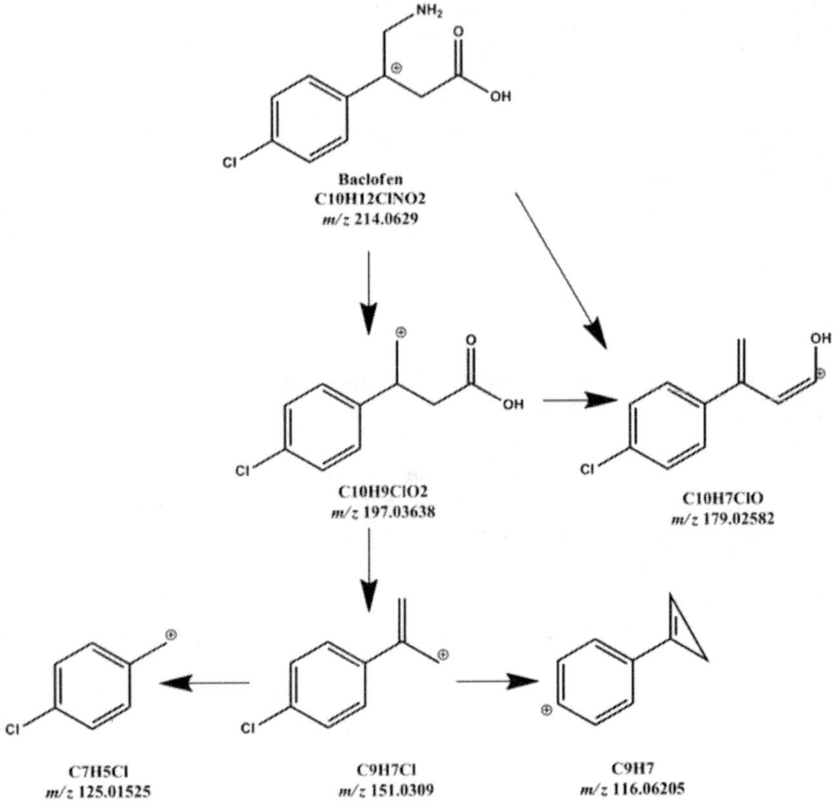

Figure 4. Probable product-ion formation pathways from the protonated baclofen precursor ion

In medico-legal toxicology, the cases of fatal intoxication with baclofen alone are rare, as compared to intoxications with other drugs affecting the CNS. The likely causes of baclofen overdose include adaptive and personality disorders as well as depression, leading to suicidal attempts. The literature reports also described cases of accidental consumption of baclofen by children [34, 35, 36]. Table 4 lists the cases of baclofen intoxication reported in literature. Only in one case of fatal baclofen intoxication were no other xenobiotics found, which could interact with baclofen. In the case described in the present study, toxicological analysis of the blood demonstrated a concentration of baclofen at 30.7 μg/mL, which markedly exceeds the lethal dose (6–9.6 μg/mL according to TIAFT data) [37]. In only one case described by Giovanni and d'Aloja [18] was the blood concentration of baclofen found to be higher than that in our case; moreover, they identified other xenobiotics in their material as well.

Table 4. Literature data regarding cases of baclofen intoxication

No.	Age/sex of victim	Sample/concentration	Method(s)	Other xenobiotic(s)	Outcome	References
1	47/female	Cerebrospinal fluid/0.00338 μg/mL	No data	Not detected	Survived	[38]
2	56/male	Blood/106 μg/mL Urine/774 μg/mL	GC–MS	Dipyrone and metabolites: 4-methyl-aminoantipyrine, oxybutynin	Death	[18]
3	51/female	Blood/2.7 μg/mL	No data	Clorazepate, hydroxyzine, paroxetine, phenobarbital, phenytoin, digoxin, warfarin	Survived	[39]
4	30/female	Urine/no data	No data	Ethanol 312 mg/dL	Survived	[40]
5	47/female	Blood/20 μg/mL	No data	Amitriptyline 1.8 μg/mL, alcohol 160 mg/dL	Death	[41]
6	30/male	Serum/17 μg/mL Urine/760 μg/mL	GC–MS HPLC	Not detected	Death	[19]
7	No data	Blood/0.032 to 2.53 μg/mL	LC–MS/MS	No data	No data	[22]
8	No data	Blood/0.16 to 0.35 μg/mL	LC–MS/MS	No data	No data	[23]
9	43/male	Blood/0.88 μg/mL	LC–MS	No data	Survived	[24]

GC–MS gas chromatography-mass spectrometry, *HPLC* high-performance liquid chromatography, *LC–MS/MS* liquid chromatography–tandem mass spectrometry, *LC–MS* liquid chromatography–single stage mass spectrometry

CONCLUSIONS

The application of HRMS enables reliable identification of baclofen in autopsy blood. The method designed is characterized by high-efficiency extraction. The study findings demonstrate that the specific, simple and quick procedure described by us for determination of baclofen in autopsy blood can be successfully used for routine toxicology testing in the cases of suspected baclofen intoxication. Unambiguous identification of baclofen is derived from the available product ion mass spectrum. In such a spectrum, we observed three product ions of baclofen (*m/z* 197.03638, 151.0309 and 116.06205) of high intensity. These product ions may be successfully employed as confirmative ions in QTOF-MS or triple-quadrupole-MS analysis. Using the obtained tandem mass spectra and defined accurate masses, we proposed the pathways of product ion formation from baclofen. However, the method has a limitation associated with the use of gabapentin as an IS instead of a stable isotope-labeled IS, which is why tested blood should be screened prior to analyzing to ensure that it is free from gabapentin. To our knowledge, this is the first report for analysis of baclofen by HRMS.

COMPLIANCE OF ETHICAL STANDARDS

Conflict of interest

The authors declare that they have no conflicts of interest.

Ethical standards

For this type of study, formal consent is not required. This article does not contain any studies with living human participants or animals performed by any of the authors. The analysis of toxic substances from the cadaver was officially requested by judicial authorities.

REFERENCES

1. Parnas I, Rashkovan G, Ong J (1999) Tonic activation of presynaptic $GABA_B$ receptors in the opener neuromuscular junction of crayfish. J Neurophysiol 81:1184–1191
2. Brogden RN, Speight TM, Avery GS (1974) Baclofen: a preliminary report of its pharmacological properties and therapeutic efficacy in spasticity. Drugs 8:1–14
3. Ellenhorn M (1997) Ellenhorn's medical toxicology. Diagnosis and treatment of human poisoning, 2nd edn. Williams & Wilkins, Baltimore, pp 937–943
4. Hugenholtz H, Nelson RF, Dehoux E, Bickerton R (1992) Intrathecal baclofen for intractable spinal spasticity—a double-blind crossover comparison with placebo in 6 patients. Can J Neurol Sci 19:188–915
5. Addolorato G, Leggio L, Abenavoli L, Agabio R, Caputo F, Capristo E, Colombo G, Gessa GL, Gasbarrini G (2006) Baclofen in the treatment of alcohol withdrawal syndrome: a comparative study vs diazepam. Am J Med 119:213–278
6. Houston AJ, Wong JC, Ebenezer IS (2012) Effects of the $GABA_B$ receptor agonist baclofen on primary drinking in rats. Eur J Pharmacol 674:327–331
7. Addolorato G, Leggio L (2010) Safety and efficacy of baclofen in the treatment of alcohol-dependent patients. Curr Pharm Des 16:2113–2117
8. Witczuk B, Khaunina RA, Kupryszewski G (1980) 3-(*p*-Chlorophenyl)-4-aminobutanoic acid–resolution into enantiomers and pharmacological activity. Pol J Pharmacol Pharm 32:187–196
9. Wuis EW, Dirks MJ, Termond EF, Vree TB, Van der Kleijn E (1989) Comparison of the pharmacokinetics of intravenously administered rac-baclofen and its (−)-(R)- and (+)-(S)-enantiomers in dogs. Int J Clin Pharmacol Res 9:239–246
10. Ferner R (1981) Atropine treatment for baclofen overdose. Postgrad Med J 57:580–581
11. Faigle JW, Keberle H (1972) The chemistry and kinetics of Lioresal. Postgrad Med J 48:9–13

12. Pommier P, Debaty G, Bartoli M, Viglino D, Carpentier C, Danel V, Maxime M (2014) Severity of deliberate acute baclofen poisoning: a nonconcurrent cohort study. Basic Clin Pharmacol Toxicol 114:360–364
13. Lee TH, Chen SS, Su SL, Yang SS (1992) Baclofen intoxication: report of four cases and review of the literature. Clin Neuropharmacol 15:56–62
14. Delhaas EM, Brouwers JR (1991) Intrathecal baclofen overdose: report of 7 events in 5 patients and review of the literature. Int J Clin Pharmacol Ther Toxicol 29:274–280
15. Dario A, Tomei G (2004) A benefit-risk assessment of baclofen in severe spinal spasticity. Drug Saf 27:799–818
16. Peng CT, Ger J, Yang CC (1998) Prolonged severe withdrawal symptoms after acute-on-chronic baclofen overdose. Clin Toxicol 36:359–363
17. Kochak G, Hunc F (1984) Improved gas-liquid chromatographic method for the determination of baclofen in plasma and urine. J Chromatogr 310:319–326
18. Giovanni ND, d'Aloja E (2001) Death due to baclofen and dipyrone ingestion. Forensic Sci Int 123:26–32
19. Fraser AD, MacNeil W, Isner AF (1991) Toxicological analysis of a fetal baclofen (Lioresal) ingestion. J Forensic Sci 36:1596–1602
20. Ban E, Park JS, Kim CK (2004) Semi-microbore HPLC for the determination of baclofen in human plasma using column switching. J Liq Chromatogr Relat Technol 27:3051–3064
21. Tosunoglu S, Ersoy L (1995) Determination of baclofen in human plasma and urine by high-performance liquid chromatography with fluorescence detection. Analyst 120:373–375
22. Dukova OA, Kotlovsky MY, Pokrovsky AA, Suvorova EV, Shivrina TG, Krasnov EA, Efremov AA (2015) Identification and quantitative determination of baclofen in human blood by HPLC with mass spectrometry detection. Biochemistry-Moscow Suppl Ser B Biomed Chem 9:137–142
23. Sørensen LK, Hasselstrøm JB (2014) Determination of therapeutic γ-aminobutyric acid analogs in forensic whole blood by hydrophilic interaction liquid chromatography-electrospray tandem mass spectrometry. J Anal Toxicol 38:177–183
24. Lanoux T, Lebrun D, Andreu P, Just B, Mateu P (2014) Baclofen poisoning. Toxicol Anal Clin 26:206–207
25. Di Rago M, Saar E, Rodda LN, Turfus S, Kotsos A, Gerostamoulos D, Drummer OH (2014) Fast targeted analysis of 132 acidic and neutral drugs and poisons in whole blood using LC–MS/MS. Forensic Sci Int 243:35–43
26. Chang SY, Zheng N-Y, Chen C-S (2004) Development and validation of a capillary electrophoresis method for the determination of baclofen in human plasma. Int J Appl Sci Eng 3:277–285
27. Shintani-Ishida K, Nakamura M, Tojo M, Idota N, Ikegaya H (2015) Identification and quantification of 4′-methoxy-α-pyrrolidinobutiophenone (4-MeOPBP) in human plasma and urine using LC–TOF-MS in an autopsy case. Forensic Toxicol 33:348–354
28. Hasegawa K, Wurita A, Minakata K, Gonmori K, Nozawa H, Yamagishi I, Suzuki O, Watanabe K (2014) Identification and quantitation of a new cathinone designer drug PV9 in an "aroma liquid" product, antemortem whole blood and urine specimens, and a postmortem whole blood specimen in a fatal poisoning case. Forensic Toxicol 32:243–250
29. Saito T, Namera A, Osawa M, Aoki H, Inokuchi S (2013) SPME–GC–MS analysis of α-pyrrolidinovaleorophenone in blood in a fatal poisoning case. Forensic Toxicol 31:328–332
30. CITAC/EURACHEM GUIDE (2002) Guide to quality in analytical chemistry. An aid to accreditation. https://www.eurachem.org/index.php/publications/guides/qa. Accessed Dec 2015

31. Chambers E, Wagrowski-Diehl DM, Lu Z, Mazzeo JR (2007) Systematic and comprehensive strategy for reducing matrix effects in LC/MS/MS analyses. J Chromatogr B 852:22–34
32. Kim TH, Shin S, Shin JC, Choi JH, Seo WS, Park G-Y, Kwon DR, Yoo SD, Lee A-R, Joo SH, Min BS, Yoo WY, Shin BS (2013) Liquid chromatography–tandem mass spectrometry determination of baclofen in various biological samples and application to a pharmacokinetic study. J Chromatogr B 938:43–50
33. Miksa IR, Poppenga RH (2003) Direct and rapid determination of baclofen (Lioresal) and carisoprodol (Soma) in bovine serum by liquid chromatography-mass spectrometry. J Anal Toxicol 27:275–283
34. Chapple D, Johnson D, Connors R (2001) Baclofen overdose in two siblings. Pediatr Emerg Care 17:110–112
35. Cooke DE, Glasstone MA (1994) Baclofen poisoning in children. Vet Hum Toxicol 36:448–450
36. Dasarwar N, Shanbag P, Kumbhare N (2009) Baclofen intoxication after accidental ingestion in a 3-year-old child. Indian J Pharmacol 41:89–90
37. TIAFT (2004) Reference blood level list of therapeutic and toxic substances, September 2004. https://www.yumpu.com/en/document/view/13423269/tiaft-reference-blood-level-list-of-therapeutic-and-toxic-gtfch. Accessed Dec 2015
38. Berger B, Vienenkoetter B, Korporal M, Rocco A, Meinck HM, Steiner T (2012) Accidental intoxication with 60 mg intrathecal baclofen: survived. Neurocrit Care 16:428–432
39. Sullivan R, Hodgman MJ, Kao L, Tormoehlen LM (2012) Baclofen overdose mimicking brain death. Clin Toxicol 50:141–144
40. Van Dierendonk DR, Dire DJ (1999) Baclofen and ethanol ingestion: a case report. J Emerg Med 17:989–993
41. Cooper DJ, Bergman J (2000) Massive baclofen overdose. Crit Care Resusc 2:195–197

CHAPTER 22

Forensic Comparison and Matching of Fingerprints: Using Quantitative Image Measures for Estimating Error Rates through Understanding and Predicting Difficulty

Philip J. Kellman[1], Jennifer L. Mnookin[2], Gennady Erlikhman[1] *, Patrick Garrigan[3], Tandra Ghose[4], Everett Mettler[1], David Charlton[5], Itiel E. Dror[6]

[1] *Department of Psychology, University of California Los Angeles, Los Angeles, California, United States of America,*

[2] *School of Law, University of California Los Angeles, Los Angeles, California, United States of America,*

[3] *Department of Psychology, St. Joseph's University, Philadelphia, Pennsylvania, United States of America,*

[4] *Department of Psychology, Technical University of Kaiserslautern, Kaiserslautern, Germany,*

[5] *Sussex Police, Sussex, United Kingdom,*

[6] *Institute of Cognitive Neuroscience, University College London, London, United Kingdom*

ABSTRACT

Latent fingerprint examination is a complex task that, despite advances in image processing, still fundamentally depends on the visual judgments of highly trained human examiners. Fingerprints collected from crime scenes typically contain less information than fingerprints collected under controlled conditions. Specifically, they are often noisy and distorted and may contain only a portion of the total fingerprint area. Expertise in fingerprint comparison, like other forms of perceptual expertise, such as face recognition or aircraft identification, depends on perceptual learning processes that lead to the discovery of features and relations that matter in comparing prints. Relatively little is known about the perceptual processes involved in making comparisons, and even less is known about what characteristics of fingerprint pairs make particular comparisons easy or difficult. We measured expert examiner performance and judgments of difficulty and confidence on a new fingerprint database. We developed a number of quantitative measures of image characteristics and used multiple regression techniques to discover objective predictors of error as well as perceived difficulty and confidence. A number of useful predictors emerged, and these included variables related to image quality metrics, such as intensity and contrast information, as well as measures of information quantity, such as the total fingerprint area. Also included were conFigureural features that fingerprint experts have noted, such as the presence and clarity of global features and fingerprint ridges. Within the constraints of the overall low error rates of experts, a regression model incorporating the derived predictors demonstrated reasonable success in predicting objective difficulty for print pairs, as shown both in goodness of fit measures to the original data set and in a cross validation test. The results indicate the plausibility of using objective image metrics to predict expert performance and subjective assessment of difficulty in fingerprint comparisons.

INTRODUCTION

There has been a longstanding belief in the scientific validity of fingerprint evidence, based on the apparent permanence and uniqueness of individual fingerprints, the experience-based claims of trained fingerprint examiners, and the longstanding courtroom acceptance of this forensic technique. Yet systematic scientific study of the accuracy of latent fingerprint identification is a very recent development, still very much in progress. In the past, fingerprint identification was sometimes even claimed to be "infallible" or to have a "zero error rate" so long as the method was appropriately applied by an experienced examiner [1], [2]. High-profile cases in which errors were discovered, along with the inherent implausibility of assertions of infallibility, led to doubts about such claims of accuracy, but only in the last few years have scientific efforts to assess the strengths and limitations of fingerprint identification gained traction. The 2009 National Academy of Sciences report on forensic science [3] emphasized and spotlighted both the limits of our knowledge and the need for basic research, and since that report. The available data suggest a low level of false positive errors by experts under experimental conditions and a substantially higher rate for false negatives [4], [5]. While these data suggest that well-trained, experienced examiners are highly accurate when making positive identifications, it is also clear that errors still occur. Understanding what characteristics of print pair comparisons make errors more or less likely is thus critical to assess both the power and limits of this important forensic technique.

Fingerprint examiners can specialize and become latent or tenprint examiners or both. A latent examiner focuses on comparing "chance" fingerprints left accidentally at crime scenes or elsewhere, to possible source prints. A tenprint examiner, by contrast, compares fingerprints purposefully collected in controlled circumstances (such as at a police station) with those on file in a database. In police stations, impressions from all ten fingers are often collected on a single sheet, which is why they are called tenprints. Tenprints are also referred to as "known prints" because the identity of the source of the impression is known. In this paper, we use the term known print to refer to such prints. Latent prints have to be processed in order to be made visible, and often contain only a portion of a finger or other friction ridge area. They are often smudged, distorted, and may contain artifacts or noise due to the surface upon which they were left, or as a result of processing. By contrast, known prints are collected in controlled situations where poor impressions can be retaken, so they are typically larger, clearer, and richer in information content than latent images. Latent prints tend to be highly variable in quality, while known prints generally capture fingerprint information with high fidelity. Known prints are often acquired by law enforcement agencies using ink or a scanner. A sample latent and known print are shown in Figureure 1.

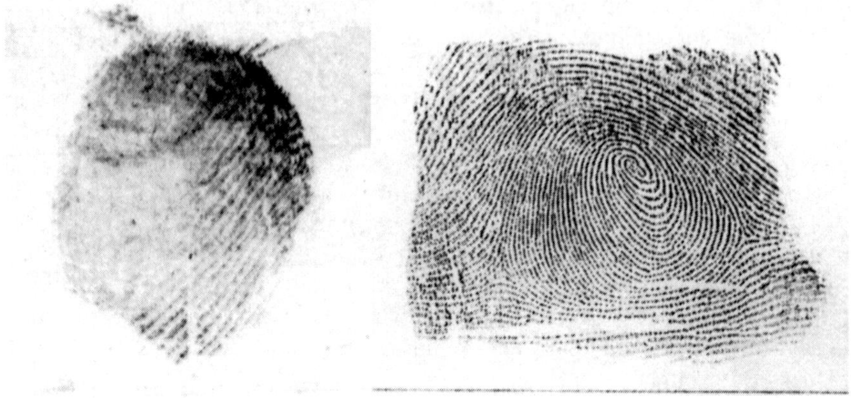

Figureure 1. Sample fingerprint images used in the study.

The image on the left is a latent print. Note large areas of the image that are smudged or missing. Contrast and ridge clarity vary greatly across the fingerprint area. These and other aspects of the image could make comparison difficult. The image on the right is a known print and is much clearer.

Until recently, there were virtually no scientific studies of how often fingerprint examiners made errors. However, recent studies have provided helpful information for this assessment [4]–[6]. Ulery, Hicklin, Buscaglia, and Roberts [4] had 169 latent print examiners compare an independent sampling of 100 fingerprint pairs (from a set of 744), each pair consisting of one latent print and one known print. Ulrey, et al. [4] found that 7.5% of matching pairs were labeled non-matches (false negatives), while only 0.1% of non-matching pairs were labeled matches (false positives). Similar results were found by Tangen, Thompson, and McCarthy [5]: 7.88% errors for matching pairs (false negatives) and 0.68% errors for non-matching pairs (false positives). These studies took place in experimental conditions quite different from actual casework. Error rates from these studies likely do not fully reflect real-world performance [7], but they do indicate high levels of performance by experts. Studies also indicate that experts perform far better than novices at fingerprint matching tasks [5], [6].

From a research point of view, the low false positive rates among fingerprint examiners make the discovery of determinants of such errors quite difficult. High accuracy leads to little variability in performance, undermining standard statistical analyses. However, the low number of these errors should not be taken as an indication that studying them has little practical importance. A false match can lead to a false conviction, and a false exclusion can lead investigators to focus their attention on erroneous leads or to fail to convict the actual perpetrator. Furthermore, the realities and pressures in real criminal casework may substantially increase error rates, including false positives. In addition, even if these experimental error rates were established to be similar to those in actual practice, these low error rates get multiplied by a very large number of fingerprint comparisons, so the absolute quantity of real-world errors would not be *de minimus*.

Ironically, the practical importance of understanding when and why fingerprint comparison errors occur is likely to *increase* as technology advances. It is common for a latent print to be submitted to an AFIS (automated fingerprint identification system) database, where automated routines return a number of most likely potential matches. Error rates (especially of the false-positive type) may increase as databases get larger (currently some databases include tens of millions of prints). The reason for this is that as a database grows, an AFIS searching that database is increasingly likely to find close non-matches, (prints that are highly similar to the latent, but are in fact from a different individual – what are often termed "look-alikes"). Obviously, searching larger databases also increases the chances of finding a true match, but such progress can also make the task of the human examiner more demanding and, potentially, error-prone [8].

From a visual information processing perspective, it is therefore interesting and important to determine what visual characteristics of fingerprints influence the ease and accuracy of comparisons. Ultimately, it may be possible to evaluate a fingerprint comparison in terms of the quantity and quality of visual information available [9] in order to predict likely error rates in comparisons. Better understanding of objective metrics could also help determine when a print pair contains or lacks sufficient information to make an identification or exclusion, that is, to determine when an "inconclusive" assessment is warranted. These considerations motivate the present study. Its primary goals are to: (1) measure expert examiner performance, and (2) to create a predictive framework by which one could assign an appropriate level of confidence in expert decisions, derived from an objective assessment of characteristics of the pair of images involved in a particular fingerprint comparison. These two goals are interconnected: examiner performance levels (error rates) are likely to depend on the complexity and difficulty of the comparison. Specifically, as comparisons become more difficult, errors are more likely to occur. A single overall 'error rate' for latent fingerprint comparison would be insufficiently granular, as it would fail to recognize that some comparisons are likely far easier than others, and thus far less prone to error. Hence, the characterization and prediction of error rates must be a function of the difficulty of the comparison. Notwithstanding this relationship, no previous research on fingerprint identification has attempted to generate objective models for the assessment of fingerprint difficulty.

Perceptual Aspects of Fingerprint Expertise

If asked to give reasons for a conclusion in a given comparison, fingerprint examiners would display significant explicit knowledge relating to certain image features, such as global conFigureurations, ridge patterns and minutiae, as these are often explicitly tagged in comparison procedures, and they are pointed out in training of examiners. It would be a mistake, however, to infer that the processes of pattern comparison and the determinants of difficulty are therefore fully available for conscious report or explicit description. As in many other complex tasks in which learning has led to generative pattern recognition (the ability to find relevant structure in new instances) and accurate classification, much of the relevant processing is likely to be at least partly implicit [10]–[12].

Like many other tasks in which humans, with practice and experience, attain high levels of expertise, feature extraction and pattern classification in fingerprint examination involves *perceptual learning* – experience-induced changes in the way perceivers pick up information [13], [14]. With extended practice, observers undergo task-specific changes in the information selected – coming to discover new features and relationships that facilitate classification in that domain. Evidence supporting this claim comes from increased perceptual learning when these features are exaggerated during training [15].

There are also profound changes in *fluency*: What initially requires effort, sustained attention, and high cognitive load comes to be done faster, with substantial parallel processing and reduced cognitive load [12]. In turn, becoming more automatic at extracting basic information frees up resources for observers to discover even more subtle or complex structural information, e.g., [16]. This iterative cycle of discovery and automaticity followed by higher-level discovery is believed to play a significant role in attaining the impressive levels of performance humans can attain in areas such as chess, chemistry, mathematics, and air traffic control, to name just a few domains [12], [17].

While several studies have explored the influence of bias and emotional context on fingerprint matching and classification [18]–[22], there has been relatively little work investigating perceptual aspects of expertise among examiners or perceptual learning processes that lead to expertise. One exception is a study by Busey and Vanderkolk [6], in which novice and expert fingerprint examiner performance was compared on a conFigureural processing task. Subjects were shown a small image patch from a fingerprint and, after a mask, attempted to match the image patch to a luminance- and orientation-adjusted version presented along with a distractor fingerprint image patch. Subjects did not have an opportunity to closely examine the fingerprint patches as they only appeared for one second. This presentation procedure required subjects to rely heavily on the broad patterns of ridge flow (conFigureural information) to perform the matching task. Experts exhibited nearly perfect performance, while novices had an accuracy of 0.8. Compared to the novices, experts may have utilized fingerprint information more efficiently, focused on entirely different information, and/or more effectively filtered out irrelevant information; the study did not provide information for distinguishing among these possibilities. However, the ease of information processing of task-relevant information is a hallmark of fluency effects in perceptual learning.

Similarly, Thompson et al. [23] found that the amount of visible area in a target print was positively correlated with classification accuracy among novices. Interestingly, this relationship also depended on the source of the print (e.g., index vs. pinky), although it was unclear why one finger should hold more information than another since presented areas were constant across prints. Marcon [24] had naïve observers rate "high quality" (known prints) and "low quality" latents for distinctiveness. Performance for categorizing pairs of prints as coming from the same source or a different source was higher for high-quality and high-distinctiveness images. Together, these studies show that performance suffers when fingerprint image quality is low, but reveal little about the specific nature of the information that correlates with low or high quality.

Fingerprint Features in the Standard Taxonomy

The first step in latent print examination is often manual preprocessing. For example, the region of the image that contains the fingerprint could be selected from the background and oriented upright. If a fingerprint is to be submitted to a database for automated comparison, key features need to be identified and labeled. Automated searches are then carried out by software that finds fingerprints on file with similar spatial relationships among the features labeled in the submitted fingerprint. This is the only part of the examination and comparison process that is automated. The software returns a list of potential matches, many of which can be quickly excluded. Some will be closer non-matches or a match, and these require further scrutiny by a human examiner.

Whether examiners are provided with potential matches via automated database searches or via investigative work, they often make their match decisions using the ACE-V approach: Analysis, Comparison, Evaluation, and Validation [25]. The examiner first looks at the latent print closely (analysis), then compares the two prints relative to each other, looking for both similarities and differences (comparison). They then evaluate those similarities and differences to arrive at a decision about whether the prints match or not. In the final step, a second examiner independently validates the conclusion. Mnookin [26] points out that there is no formalized process for any of these steps. There is no method or metric for specification of which features should be used for comparison, nor any general measure for what counts as sufficient information to make a decision. Examiners rely on their experience and training rather than formal methods or quantified rubrics at each step of the process.

Despite the lack of a formalized procedure, some attempts have been made to formally describe and classify the kinds of features that might be found in a fingerprint. Three types of features are commonly used to describe the information used for fingerprint comparison (for a complete discussion, see [27]). Level I features are global descriptors of ridge flow easily seen with the naked eye. These include patterns in the central region (the "core") of the fingerprint. Cores can be classified into a limited number of typical patterns such as left- and rightward loops, whorls, tented-arches and arches. Deltas are another Level I feature that are triangular patterns of ridge flow that often occur on the sides of loops and whorls. A leftward loop and a delta are indicated by the yellow and green boxes respectively in Figureure 2. Level I features are too common to be sufficient for identification, but they can be used for exclusion purposes as well as to guide inspection of the more detailed Level II and Level III features.

Level II features include *minutiae* such as ridge bifurcations and ridge endings. Level II features are found where fingerprint ridges and valleys split or end. *Minutiae* are highlighted in red circles in Figureure 2. The power of fingerprints for identification purposes is largely due to the high variability in the existence and relative positions of these features across fingers and individuals. Scarring, which occurs naturally with age and wear, can also add unique ridge patterns to a fingerprint. However, while scars can be used to compare the fingerprint found at a crime scene to that of a suspect in custody, they may not always exist in fingerprints on file that can be old and therefore predate the markings.

Level III features are the smallest fingerprint features used by some examiners for comparison. These include the positions of sweat pores and ridge thickness. Pores are indicated in light blue circles in Figureure 2. The visibility of Level III features depends on the quality of the prints and examiners do not uniformly make use of them for comparison purposes.

Figureure 2. Various image features commonly identified by expert examiners.

Red circles indicate *minutiae* (ridge bifurcations or endings); blue circles indicate pores (they appear as small white dots along a ridge); the yellow square indicates the delta; the green rectangle indicates the core, in this case a leftward loop.

Predictor Variables

What properties of the images in fingerprint pairs are most important and informative in comparing fingerprints? What visual qualities of individual prints or of print pairs make accurate matching performance more or less likely? Although we relied on regression methods to provide answers to this question, it was important to develop as inputs to the regression analyses a wide variety of possible image characteristics that could be relevant. To generate such factors, we were guided by vision science, intuition, insights from fingerprint examiners, and prior work on image processing of fingerprints [27], as well as the standard taxonomy of levels of pattern information in fingerprints (described above). Some variables intuitively seem likely to relate to the sheer quantity of available information; for example, having greater print area available for comparison might make comparisons more accurate. However, this might well be oversimplified; quality of information might matter as much or more than total print size. We created several image quality metrics that appear sensitive to smudging, missing regions, poor contrast, etc. These metrics were computed in an automated fashion on the fingerprint images themselves, and were designed to relate in a variety of ways to the presence or absence of visual information that examiners use, and could therefore function as independent variables that are predictive of examiner performance.

We hypothesized that difficulty would be a function both of the characteristics of the individual prints (the latent and the potential match) and also of the characteristics of the *pair*. Because known prints are obtained under relatively standardized conditions, they are subject to significantly less variability than latent prints obtained from crime scenes. Accordingly, we expected that more of the variability in visual information quality affecting fingerprint comparisons would be determined by characteristics of latent prints. An especially poor quality latent might be more difficult to assess than a higher quality one, all else being equal. However, we also believed that pair difficulty would be a function of interaction effects between the latent and the known, not simply a function of the information quality and quality of each independently. We therefore developed quantitative measures involving both individual prints and print *pairs*.

A general description and motivation for the image features we selected or developed is provided below. Except where noted, we assessed each predictor variable for both the latent print and the known print. For many variables, we also derived a variable that expressed an interaction or relationship of the values of a variable for the latent and known print combined (such as the ratio of latent print area to the known print area, or the Euclidean sum of contrast variability for the latent and known print combined). For details about the procedures used to derive the measures we used, please see the supplementary materials.

Total Area.

This variable was defined as the number of pixels in the fingerprint after the fingerprint was segmented from the background. Although machine vision algorithms exist that could have been used for determining the region of usable print image, those algorithms we examined were not as good as human segmentation, and different human observers in pilot work produced strong agreement. Accordingly, we segmented fingerprints from their surrounds by having human observers designate their boundaries (see supplementary materials for details). In general, we expected that larger areas, especially of latent prints, would provide more information for making comparisons.

Area Ratio.

To relate the relative area of a latent to a potentially matching known print, we divided the area of the latent fingerprint by the area of the known print. Typically the known print, obtained under controlled conditions, presents a more complete image. Thus, *Area Ratio* relates to the proportion of known print information potentially available in the latent print. However, for non-matching prints, the area of the latent may be larger than that of the known print because of differing finger sizes. Occasionally, even for a matching latent and known print, the latent could be larger than the known print due to smearing. The ratio was therefore not strictly in the range [0,1] and cannot be considered a true proportion.

Image Intensity.

We measured the mean and standard deviation of pixel intensity taking into account all of the pixels in each fingerprint image (with intensities scaled in the range of [0,255]). The mean intensity and standard deviation of intensity provide two related but different measures, sensitive to different image characteristics. Very dark images (low mean intensity) might indicate the presence of large smudges that produce large, dark areas. Low standard deviation in intensity would make ridges (transitions from light to dark) difficult to detect.

Block Intensity.

The image was divided into 50×50 pixel regions and the average pixel intensity was computed within each region. The mean of the block intensities is the same as the overall mean *Image Intensity*. The standard deviation of these regional averages (*standard deviation of block intensity*), however, can provide additional information about variability in image intensity across the image. Low variability is indicative of many similar areas across the image, but does not provide information about whether those regions have low or high contrast (i.e., an all black image and an image with 50% white and 50% black pixels, evenly distributed across the image would have low *Block Intensity* variability). When pixel intensities are not uniformly distributed across the image, variability of block intensity is high (i.e., some regions of the image are darker than others). For latent images, this may indicate the presence of a smudge or worse contact (lighter impression) in some regions of the image.

Deviation from Expected Average Intensity (DEAI).

Intensity, as coded above, may be a useful predictor variable, but both intuition and pilot work led us to believe that it might not capture some significant aspects of intensity variations. We therefore developed a separate intensity measure – deviation from expected average intensity. In an ideal fingerprint image, one might expect approximately half of the pixels to be white (valleys) and half to be black (ridges). The expected mean intensity would therefore be half of the range, or 127.5 (with the brightest pixel normalized to 255 and the darkest to 0). The absolute deviation of the observed average from the expected average was computed using the following formula: $DAEI = -|\text{mean pixel intensity} - 127.5|$

Using absolute value here ensures that deviations from the midpoint of the intensity range in either direction are scored as equivalent; the negative sign ensures that the measure increases as the mean pixel intensity approaches 127.5 (large deviations produce a large negative value of the measure). While ridges (black regions), on average, are thicker than valleys (white regions), making the average intensity slightly lower than 127.5, the difference was relatively small and was ignored.

Contrast.

Michelson contrast was computed for each segmented fingerprint. Michelson contrast is defined as: $Contrast = \frac{(L_{\max} - L_{\min})}{(L_{\max} + L_{\min})}$

This contrast measure produces a value between 0 (least contrast) and 1 (most) by dividing the difference of maximum and minimum intensity values by their sum. Michelson contrast is typically calculated from luminance values. In our images, we calculate Michelson contrast from pixel intensity values, which is appropriate given that fingerprint images may be displayed on a variety of monitors with different Gamma corrections.

Block Contrast.

The preceding measure obtained the Michelson contrast for an entire image. We also computed contrast for smaller image regions – block contrast – by segmenting the entire image into 50×50 pixel regions. *Block Contrast* is defined as the mean across the blocks. To illustrate the difference between overall contrast and block contrast, the Michelson contrast of an entire image containing all gray pixels except for one white and one black pixel, would be 1. *Block Contrast*, however, would be very low, since most regions of the image would have 0 contrast. If black and white pixels were distributed more evenly across the image such that they appeared in each block, then *Block Contrast* would be high. High values of the measure may indicate the presence of clear ridges and valleys in many areas of the fingerprint. A separate but related predictor was the *standard deviation of block contrast* across blocks. Small standard deviation values could indicate high information content throughout the image (*Block Contrast*close to 1 everywhere) or that the image was uniformly smudged (*Block Contrast* close to 0 everywhere).

Ridge Reliability.

Orientation-sensitive filters were used to detect edges in the fingerprint image. The relative responses of these filters were then used to identify "high reliability" regions where ridge orientation was uniquely specified (see supplementary materials for details). The proportion of high reliability regions was computed, resulting in an overall reliability score for each print. Ridge Reliability ranged between 0 and 1, with larger values indicating a greater proportion of print area with well-defined ridge orientation. An additional, relational predictor was computed by taking the Euclidean sum of the *Ridge Reliability* for the latent and known print (*Ridge Reliability Sum*). Large values of this measure indicate a high proportion of regions with well-defined ridge orientation in both the latent and known prints.

Visibility of Cores and Deltas.

Earlier we described global conFigureurations – *Cores* and *Deltas* – that provide Level I information to fingerprint examiners. The fact that ridge flow in fingerprints tends to follow a circular pattern dictates that there will be some global core (a whorl, loop, or arch) at or near the center of each print. Likewise the transition from a core, especially loops and whorls, to the circular ridge flow tends to give rise to deltas (see Figureure 2). As there will be only one core and at most a small number of deltas in any print, these serve as important reference points in making comparisons [27].

Unlike all of the other variables we used, which could take on a continuous range of values, *Cores* and *Deltas* are binary (either present or not).

Relations Among Basic Predictors

To remove effects on regression coefficients of differing scales of various predictors, we standardized all continuous metrics by subtracting the mean and dividing by the standard deviation. Standardization made some measures that were strictly non-negative (like *Standard Deviation of Intensity*) take on negative values. As is often recommended in using regression methods [28], we also examined the variables for collinearity and found that several predictors were highly correlated. For example, the mean and standard deviation *Intensity* measures were correlated (Pearson's r=−0.77 for latents and −0.44 for known prints). High correlation among predictors is an undesirable feature for regression models [28] because it makes it harder to assess the individual effect of those predictors. If two predictors had a correlation of greater than 0.5, we removed one of them. After removal, the variance inflation factor, a measure of collinearity, for all continuous metrics was less than 5, indicating that collinearity was sufficiently reduced [28]–[30].

In addition, we included two-way interactions between all predictors that applied to both a latent and known print. For example, in addition to the *Standard Deviation of Block Contrast* for the latent and known print, we included the interaction between the two terms. In addition to *Area Ratio* and *Ridge Reliability Sum*, these interactions are *relational* predictors that encode something about the relative quality of information in a latent and known print.

Overview of the Experiment

We developed a database of fingerprint images, both of the known prints and their corresponding latents, and computed a variety of metrics that we hypothesized would relate to image quality and information content. Our primary focus was accuracy, but we also measured response times and asked experts to provide subjective judgments of difficulty and confidence for each print pair. We tested expert fingerprint examiners in a task requiring a forced-choice judgment of whether two prints matched. As will be described below, the task approximated what examiners do in their real-life work in some ways but differed in others. For example, in the study reported here, images appeared on a computer monitor; examiners were limited in comparison time for each pair; and judgments were constrained to indicating that a pair of prints came from the same source or different sources, i.e., "inconclusive" was not a permitted response for difficult comparisons. These features of our design were chosen so that we could collect important data, including best-guess match determinations for difficult comparisons, and to permit us to obtain enough data to allow us to explore the set of image characteristics that might predict difficulty. We fit a regression model to measure how various image characteristics predict performance. To foreshadow some of the results, we found that a subset of image features such as measures of the reliability of ridge orientation information, the ratio of the visible area of the latent to the known print, and measures of contrast and intensity information were predictive of performance. The model accurately identified print pairs that had low accuracies, suggesting that it can be used as a valid tool for identifying potentially difficult comparisons and that in general, it may be feasible to use these methods to predict error rates for print pairs, as a function of comparison difficulty, with reasonable accuracy.

METHOD

Ethics Statement

This study was performed in accordance with the guidelines of the Declaration of Helsinki. All experts provided written, informed consent after the general purpose of study was explained and were fully aware of the purpose and procedure of the study. Participation was voluntary. The study was approved by the institutional review board of the University of California, Los Angeles.

Participants

Fifty-six fingerprint examiners (18 male, 35 female, three not reported) participated in the study. Forty participants self-reported as latent print examiners, three as known print examiners, ten as both, and three did not report. Years of experience were reported between the range of 1 and 25 years (Latent: Mean=9.54, SD=6.97; Ten-Print: Mean=10.45, SD=8.07). Twenty-seven participants reported being IAI certified. 32 reported that their labs were accredited.

Participants were either directly recruited at the 2011 IAI (International Association for Identification) Educational Conference or via a flyer sent out in advance of the conference. As incentive, all participants were entered into a raffle to win an iPad 2. All participants signed informed consent forms prior to participating. As indicated above, some limited demographic information was collected, but it was stored separately from individual participant IDs such that the two could not be linked.

Apparatus

All stimuli were displayed on laptop computers with 17-inch monitors at a resolution of 1024×768 pixels. Stimuli were presented using a program accessed online; data were stored on the website's server.

Stimuli

Fingerprints were collected from 103 individuals. Each individual first used a single finger to produce a clear, known print using ink as is often done in police stations. Then, using the same finger, they touched a number of surfaces in a variety of ways (with varying pressure, smudges, etc.), to create a range of latent fingerprint marks that reflect those found in a crime scene. Professional fingerprint examiners who participated in the study reported that these prints were similar to those that they encounter in their everyday casework). The latent fingerprints were lifted using powder and were scanned at 500 dpi using the FISH system. Image dimensions ranged from 826 pixels in height to 1845 pixels and from 745 pixels in width to 1825 pixels. The latent prints varied in clarity, contrast, and size. For each individual who contributed to the database, we collected a total of six prints – one known print and five matching latent prints. Across individuals we varied the fingers used. Each scanned fingerprint was oriented vertically and approximately centered. Some individuals contributed multiple sets of prints from different fingers.

To create the non-matching pair of prints, we did not want to randomly select a known and a latent, as such pairs would often be too obviously different. This would make the "non-match" decisions nearly uniformly easy, and would also, by default, indicate which were the "matching" pairs. Therefore, we obtained similar, but non-matching, known prints by submitting each latent print to an AFIS search process. An expert selected from the AFIS candidate list what he deemed to be the most similar non-matching print. That enabled us to produce non-matching pairs with a relatively high degree of similarity. The final database consisted of 1,133 fingerprint images – five latent prints from 103 fingers (515), 103 known prints that matched (103), and another 515 known prints, to provide a potential non-match for each of the latents. Since we used an AFIS database from a different country from where we collected the known prints, it was highly unlikely that an actual match would be presented by the AFIS database search as a candidate. Furthermore, the expert who selected the most similar print from the AFIS candidate list verified for each comparison that this was a similar print, but not an actual match.

Of the 1,133 fingerprint images, 200 latent and known print pairs were selected and used for the study; half were a match and half were a close non-match. Individual print metrics were computed for each image or image pair (see below) and prints were selected to (approximately) uniformly sample each feature space. Known prints were sampled without replacement, but multiple latent prints from the same finger were occasionally selected since each latent could be paired with a different known print image (the match or a close non-match). Print pairs were then grouped into batches of 20, each containing ten matches and ten non-matches. Latent prints from the same finger did not appear within the same batch.

Design

A group of experts made match/non-match judgments and provided confidence and difficulty ratings on a subset of 200 print pairs selected from a database of over a thousand fingerprint images. Two fingerprint images that were either from the same finger (match) or from two different fingers (non-match) were presented side-by-side. Images were presented on computer screens and were oriented upright. Examiners had a maximum of three minutes to evaluate each pair of images. Performance was recorded for each print-pair tested, and a model was fit predicting performance based on the set of image features computed for each image in the database.

Procedure

Participants were tested in a large room, seated at desks with individual laptop computers. Before data collection began, each participant was asked to sign a consent form, and then given written instructions detailing how the stimuli would be presented and the judgments they would be required to make. Participants were told that they would be asked to compare latent-known print pairs and determine whether they were matches or non-matches (without the option to choose "inconclusive" as a response). Participants were also told that they would be asked for confidence and difficulty ratings for each of their judgments. The instructions emphasized that this procedure was not intended to replicate real-world conditions and that participants should simply try to maximize accuracy. Participants were also instructed to refrain from using any fingerprint examiner tools not provided by the experimenter, such as a compass.

When the experimental program was initiated, participants were asked to report their age, gender, years of experience, specialization, IAI certification, lab accreditation, and lab affiliation. Reporting this information was optional.

Next, the experiment began. On each trial, two fingerprints were presented side-by-side. The latent print was always on the left. A button in the top-left corner of each image window allowed participants to zoom in on each image individually. Fingerprint image size was constrained within the bounds of each window, so that each print was always viewed through an aperture of 460 pixels by 530 pixels. The initial presentation of the images had them scaled to fit entirely in this window. A single level of zoom allowed participants to magnify the image. Participants could also translate each image independently within its window (both when the image was zoomed or unzoomed) either by dragging it with mouse or by using arrow buttons in the top-left corner of each image window. No other image manipulation features were available.

Participants made a match/non-match judgment by clicking a button at the bottom of the screen. Specifically, participants were asked: "Do these prints come from the same source or a different source?" Participants then made difficulty and confidence ratings by clicking on a Likert scale. The participants were asked: "How difficult is the comparison?" and "How confident are you in your decision?" On the Likert scales, "1" corresponded to least difficult/least confident and "6" corresponded to most difficult/most confident. Once all responses were recorded, an additional button appeared allowing the participant to advance to the next trial. Supplementary Figureure S1 shows a sample screenshot of the experiment.

Participants had three minutes to complete each trial. A message was given after two and a half minutes warning that the trial would end in 30 seconds. If the full three minutes elapsed without a decision, that trial was ended, and the participant moved on to the next trial. After presentation of a set of 20 print pairs, participants were given a short break and asked if they wanted to complete another set of 20 comparisons.

Each set of 20 print pairs contained ten match and ten non-match comparisons, though examiners were not provided with this information. The order in which print pairs were presented within a set was randomized across subjects. The sets were presented in a pseudo-random order so that approximately ten participants completed each set. Although the number of trials completed by individual participants varied based on their availability and willingness to do more comparisons, most participants completed two sets of prints (40 print pairs).

RESULTS

Data Preprocessing

If the participant made a match/non-match judgment, but time expired before they could make difficulty or confidence ratings, the data were retained. There were thirteen such trials. If only difficulty and confidence ratings were provided, but a comparison judgment was not made before time expired, the trial was excluded from the analyses. Twenty such trials were excluded from the total of 2,312 comparisons (fewer than 1%). For one subject, time expired on eight of the trials they completed. There was no consistency in which print pairs had time expire – for two of those pairs, time expired for two subjects, for the rest, time expired for only one subject.

Descriptive Statistics

Responses were aggregated across participants and prints. Overall accuracy (percent of correctly classified latent-known print pairs, averaged across subjects) was 91% (range: 8.3–100%, SD 17%). Overall accuracy was 86% for "match" trials (14% false negatives) and 97% for "non-match" trials (3% false positives). Of the 2,292 comparisons, there were 200 errors, resulting in an overall error rate of 9.6%. There was some variability in performance among experts (range: 79–100%, SD 5%).

Across all participants, 118 of the 200 print pairs produced 100% accuracy. Mean difficulty and confidence ratings for these pairs were 2.62 and 5.23 respectively, compared to ratings of 4.06 and 4.15 for prints that were misclassified by at least one participant. Of the118 pairs that produced no errors, 72 were non-matches and 46 were matches. The lowest accuracy, 8.3% (1/12), corresponded to false negatives for a "match" print-pair. Average accuracy for each print pair is shown in Figureure 3 sorted by increasing accuracy.

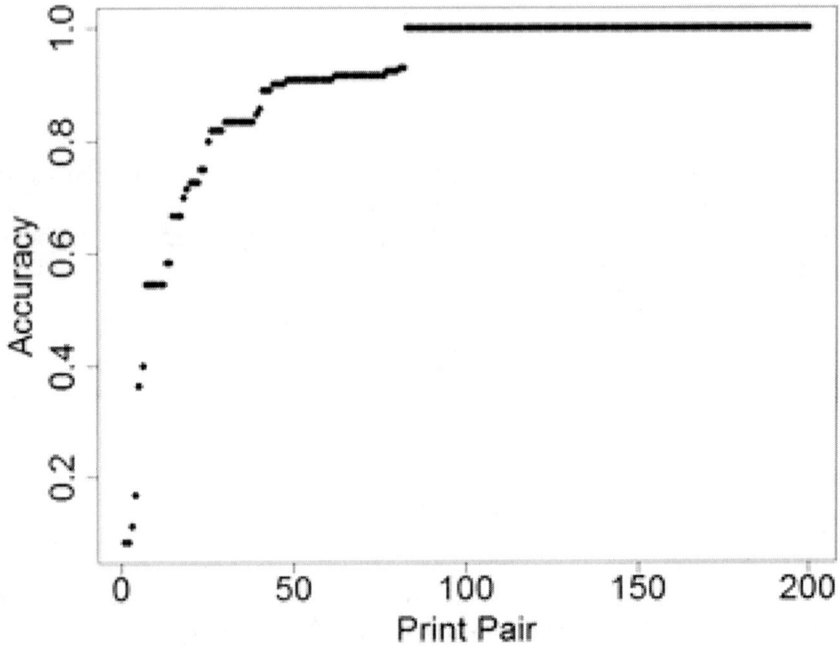

Figureure 3. Sorted average accuracy for each print pair.

Print pairs are numbered along the x-axis from 1–200 in order of increasing accuracy.

Correlations Among Dependent Measures

We measured the correlations among the three dependent measures. There was a strong negative correlation between average difficulty and confidence ratings ($r(198)=-0.91, p<0.001$) and weaker correlations between average accuracy and confidence ($r(198)=0.52, p<0.001$), and between average accuracy and difficulty ($r(198)=-0.50, p<0.001$). These results suggest that experts' confidence in their judgments is well matched to their perceived difficulty of the judgments, and further that both expert's perceived confidence and difficulty are predictive of performance. There was also a strong positive correlation between response time (RT) and difficulty ($r(198)=0.71, p<0.001$) and a negative correlation between response time and confidence ($r(198)=-0.59, p<0.001$). Accuracy was highest and RT lowest for prints that were rated least difficult. Accuracy decreased and RT increased as print difficulty ratings increased. Altogether, these results suggest that experts faced with difficult print comparisons tend to have lower confidence in their judgments, and take more time to ultimately make a match/non match decision. Excluding the 118 prints with

100% accuracy, the correlations between accuracy and confidence and between accuracy and difficulty were qualitatively weaker, but the difference did not reach significance. The full set of correlations is shown in Table 1.

Table 1. Correlations between dependent measures.

		Accuracy	Confidence	Difficulty
All Print Pairs	Confidence	0.52***		
	Difficulty	−0.50***	−0.91***	
	Response Time	−0.48***	−0.59***	0.71***
Print Pairs with Accuracy <100%	Confidence	0.36**		
	Difficulty	−0.32**	−0.89***	
	Response Time	−0.22*	−0.34**	0.45***

Note. *** $p<0.001$, ** $p<0.01$, *$p<0.5$.

Regression Analyses

We fit a crossed, logistic regression model in which print pair performance (1= accurate; 0= inaccurate) was crossed with expert and print identity. This is a type of mixed-effects model and is appropriate for analyzing these data for several reasons [31], [32]. First, not every subject evaluated every print pair. A mixed-effects approach enables the examination of both the predictor variables and the random effects due to inter-subject differences (i.e., differences between expert performance and differences between evaluations of the same print pair by multiple experts). Second, a mixed-effects approach allows one to model individual item differences by fitting data from individual trials instead of aggregating across all presentations of an item [33], [34]. Differing levels of expertise and experience, as well as differences in comparison strategy and decision thresholds, could give rise to variability in participant performance independent of the fingerprint features. Variability across items could occur if some comparisons were easier than others irrespective of differences in measured image features. Including these sources of variability in the model allows us to test whether print comparisons and experts differed from one another, instead of assuming they were all equivalent, and simply averaging across participants and items. Data were fit using the "arm" [35] and the "lme4" [36] R packages for R version 2.15.2.

For each of i print-pair comparisons (items) and j experts (subjects), we define $y_{i,j}$ as

$$y_{i,j} = \begin{cases} 1 & \text{if print pair } i \text{ is accurately classified} \\ & \text{(correct identification or rejection) by expert } j \\ & \quad \text{if print} - \text{pair } i \text{ is inaccurately classified} \\ 0 & \quad \text{(false alarm or miss) by expert } j \end{cases}$$

where $X_{i,j}$ is a vector describing the features measured on a print pair, β is a vector of coefficients (the fixed effects; one coefficient for each feature), expertID_j is the expert-specific random effect, which allows the intercepts to vary across experts, and printID_i is the item-specific random effect. expertID and printID were normally distributed.

The regression equation can be rewritten and expanded as:

$$\text{logit}(\Pr(y_{i,j}=1)) = \beta_0 + x_{1,i,j}\beta_1 + x_{2,i,j}\beta_2 + \ldots + x_{n,i,j}\beta_n + printID_i + expertID_j \quad (2)$$

where n is the number of predictors. In this form, it can be seen that printID and expertID can be grouped with β_0 as intercept terms. Because printID and expertID are vectors, the equation reflects that each combination of print and expert has its own intercept term. It is this combined term (β_0 + printID_i + expertID_j) that varies across experts and items. Multi-level modeling allows one to capture possible differences between individual subjects or test items without fitting a separate regression equation for each item (by applying a distribution over the terms that vary, in this case printID and expertID; see [37]).

Individual differences among experts may arise due to differences in experience, training, and other factors. These could manifest themselves as different baselines of performance, or intercept terms in the model. All else being equal, one expert might do better with the exact same print pair than another expert. This variability is captured by the expertID term in the model. It is also possible to model item-specific (in this case, print-pair-specific) effects; these are represented by printID. PrintID captures differences in print comparison difficulty inherent to individual print pairs and not related to the features used to predict print pair accuracy. In constructing a model, it is assumed that the error terms are uncorrelated; however, it is possible that print pair errors are correlated across participants. Inclusion of the item-specific term captures this potential non-independence (Baayen et al., 2008). A likelihood ratio test showed that the model with the predictors fit the data better than a null model with only the random effects terms ($\chi^2(17)=53.27$, $p<0.001$).

Comparing a model that included the random expert effect (expertID) to one that did not, we found that the Akaike Information Criterion (AIC) was slightly smaller for the model that included the effect, but the Bayes Information Criterion (BIC) was smaller for a model that did not. Both of these measures are information-theoretic metrics of goodness-of-fit that take into account overfitting of the data with excess parameters. Qualitatively, a more parsimonious model that fit the data almost as well would have a smaller AIC and BIC [38], [39]. The fact that the criteria move in opposite directions when the model includes expertID suggests that any differences between the models should be treated with caution [38]. A likelihood ratio test comparing the two models was significant ($\chi^2(1)=4.79, p<0.05$). ExpertID terms varied from between −0.52±0.69 to 0.44±0.77. All values of expertID were within two standard errors of zero. In terms of Equation 2, this means that β_0 + expertID was not reliably different from β_0. Based on these analyses, we felt justified in averaging across experts and ignoring between-expert differences in all subsequent modeling steps by removing the expertID term. This same analysis could not justify excluding the print-pair specific term, printID, and so it was retained in the model.

We simplified the model further by removing predictors (fixed effects) based on minimization of the AIC [40]. A likelihood ratio test revealed no statistically significant difference between a model that included all of the predictors and the reduced model ($\chi^2(11)=9.55, p>0.05$), indicating that the removal of predictors increased parsimony without significantly impacting predictive ability.

This analysis included all print pairs used in the study. This was done because the goal of the study was to create a model of difficulty for novel comparisons for which ground truth regarding whether or not a print pair shares a common source is unavailable. A separate analysis using only matching pairs showed highly similar results, including all of the predictors that proved to be reliable in the main analysis.

The Accuracy Model

The model obtained for accuracy was:

$$
\begin{aligned}
Accuracy = \mathrm{logit}^{-1}(3.385 + 0.798 * Delta\ (L) \\
+0.534 * Mean\ Block\ Contrast\ (K) - \\
0.471 * Area\ Ratio - 0.451 * SD\ Block\ Contrast\ (LxK) \\
+0.419 * Sum\ of\ Ridge\ Measures + \\
0.334 * DAEI\ (LxK) + printID
\end{aligned}
$$

Where L and K indicate whether the predictor applies to a latent or known print image respectively, and LxK indicates predictors that apply to print pairs. printID is the item-specific, random effect. The parameters of the fitted model are shown in Table 2. All predictors were significant (Wald's z, $ps<0.05$), except for Delta (L) and DEAI (LxK) which were marginally significant ($p=0.054$ and $p=0.053$ respectively). It should be noted that there is some disagreement on how to calculate p-values for the Wald statistic in unbalanced, mixed-effects data due to difficulty in determining the appropriate degrees of freedom and they should therefore be interpreted with caution [32], [41], [42].

Table 2. Predictors for accuracy model.

Fixed Effects	Coefficient Estimates	Standard Error	z
Intercept	3.385	0.197	17.167***
Delta (L)	0.798	0.415	1.923
Mean Block Contrast (K)	0.534	0.164	3.268**
Area Ratio	−0.471	0.156	−3.010**
SD Block Contrast (LxK)	−0.451	0.128	−3.530***
Ridge Sum	0.419	0.154	2.715**
DEAI (LxK)	0.334	0.173	1.938
Random Effects	Variance		
printID	2.154		

Note: *** $p<0.001$, ** $p<0.01$. p-values are reported here, but should be interpreted with caution. They were not used for model selection. Estimates are arranged by coefficient magnitude in descending order (see text). L – latent, K – known print, LxK – interaction.

To get a more intuitive notion of model performance, we used the predicted proportions from the logistic regression as estimates of average performance across experts. The resulting fit was very good ($R^2_{adj}=0.91$). We also computed the root mean squared error (RMSE) by taking the sum of the squared differences between predicted and observed values. Values closer to 0 indicated better performance. The error for the fitted model ($RMSE_{model}=0.06$) was lower than for a null model that only included the printID random effect ($RMSE_{null}=0.18$).

Validation of the Regression Model for Accuracy

The dataset was split into training and testing sets. The training set contained 180 (90%) of the print pairs (2063 individual observations), and the testing set contained the remaining 20 print-pairs (10%, 229 observations). The testing set print pairs were a representative sample of the overall dataset, containing 12 pairs with perfect accuracy and 8 pairs with less-than-perfect accuracy. This was important in order to ensure that the training set did not have too few pairs with low accuracies (there were only 24 pairs with average accuracies below 80%). We replicated the model selection procedure for data only from the training set. The same predictors were selected with comparable coefficients, except for Delta (L) which was replaced with Core (L). For both the full and training datasets, the coefficients for these two predictors, Delta (L) and Core (L), were not significantly different from zero and were within two standard deviations of zero. Nevertheless, they could not be excluded based on the selection procedure described above. The fit of the model to the training set was comparable to the fit of the model to the full set (R^2_{adj}=0.89, $RMSE_{train}$=0.07).

We used this regression model fitted to the training set to predict accuracy for the withheld testing set of 20 print pairs. The percentage of variance explained was worse for the testing set than for the training set, suggesting some amount of overfitting (R^2_{adj}=0.64). The error, however, was comparable between the training and testing sets ($RMSE_{test}$=0.07). The model's predictions are shown in Figureure 4.

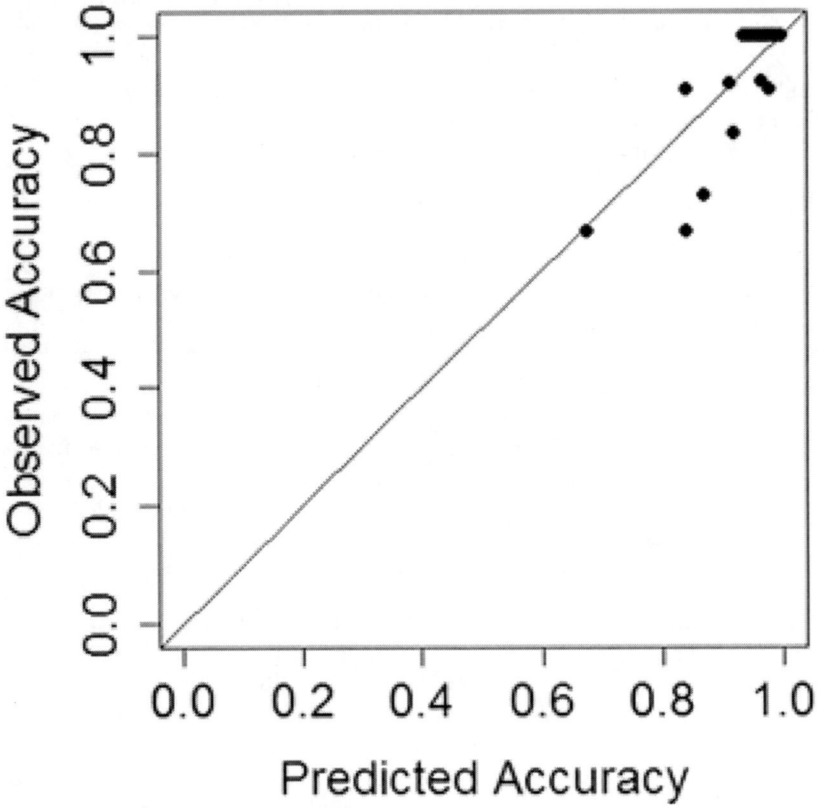

Figureure 4. Model performance.

Model predictions of average accuracy for 20 test print pairs plotted against observed average accuracy.

As a secondary assessment of model performance, we used the model to predict whether at least one expert made an error on a print pair. We divided the set of print pairs into two classes: those that had 100% accuracy (perfect pairs) and those that had less than 100% accuracy (non-perfect pairs). A naïve classification strategy not based on the model and that assumes no errors are ever made would have a classification accuracy of 107/180 or 59%. Using the model fitted to the training set, we parametrically varied a classification threshold such that print pairs with a predicted accuracy greater than or equal to that threshold were classified as perfect pairs and those below that predicted accuracy were classified as non-perfect pairs. A threshold setting of 94% resulted in the best classification performance of 164/180 or 91% correctly labeled pairs.

The classification procedure described above was repeated for predictions generated for each left out (testing) pair using the threshold optimized on the training set. 75% (15/20) of the pairs were correctly classified as either having perfect (9/15) or non-perfect accuracies (6/15). The classifier was slightly better at correctly identifying print pairs that had at least one error than those that were perfect: 3 perfect prints were misclassified as having an error and 2 non-perfect pairs were misclassified as perfect.

Difficulty Ratings

Difficulty ratings showed a reliable negative correlation with accuracy (see Descriptive Statistics, above), indicating that experts had reasonable metacognitive awareness (i.e., print pairs that were thought to be difficult tended to have lower accuracy across experts). Accuracy for prints with an average difficulty rating greater than 3 (on a scale of 1 to 6) was 84% compared to 91% for all prints. We compared the fitted model from the previous section to one that also included difficulty rating as a predictor. The resulting model had significantly better goodness of fit than the model described in the preceding section that did not include difficulty rating as a predictor ($\chi^2(1)=81.1$, $p<0.001$, $RMSE_{model+difficulty}=0.05$, $R^2_{adj}=0.95$).

We added difficulty rating as a predictor for the regression model applied to the training set described above. Predictive performance on the testing set was worse (decreased R^2) than when the difficulty rating was not included. However, classifier performance on the testing set was slightly improved, with 85% (17/20) of the pairs classified correctly. One perfect print was misclassified as non-perfect, and two non-perfect prints were misclassified as perfect. The discrepancy between the relatively worse regression fit and the improvement in classifier performance is due to two non-perfect print pairs that had a predicted accuracy that was much lower than their true accuracy. These were classified correctly as non-perfect, but contributed significantly to the error of the regression fit.

The inclusion of difficulty ratings in applications of this model must be made with caution. All other measures capture objective features of the fingerprint image, while difficulty ratings are subjective and therefore may vary across individuals and rely on the good faith of the raters. Therefore, while difficulty rating may be informative to include, in subsequent models we opt to exclusively deal with objective factors. We return to this point in the discussion.

Regression Analyses of Other Dependent Measures

Difficulty ratings, confidence ratings, and response times were reliably correlated with accuracy and so ought to also depend on print pair information content. If similar features are predictors for many measures, then they are likely capturing something important about the fingerprint images. Here, we fit models of the other dependent measures to the training dataset as a further validation step: the importance of particular image features as valid predictors of accuracy is bolstered if those same features are shared in models of other dependent measures.

Unlike accuracy, response time varied greatly across experts, with some experts taking much longer times on comparisons that other experts evaluated fairly quickly. There are several possible reasons for this variability. Less experienced examiners may take longer to come to the same conclusion than a seasoned examiner (a perceptual fluency that comes with expertise; see [12]). Some subjects may have completed the comparison quickly, but then took time to deliberate confidence and difficulty ratings since response time was recorded only once all answers were given, and not when the subject selected “match” or “non-match”. Also, the self-confidence of the examiners in their abilities may have affected response time. Only a small component of the variability in response time was likely to be due to differences in attention or interest since such differences would presumably have led to greater variability in accuracy, which was not observed.

We fit a linear, mixed-effects model to normalized response time data for the training set following the same model selection steps as for the accuracy model described above. Due the variability in response time across experts, the random effect of expertID was retained in the model. The results of the regression are shown in Table 3. Three features, Core (L), Mean Block Contrast (K), and SD Block Contrast (L) were found to be predictive of response time using the same model selection procedure that was used for the analysis of predictors of comparison accuracy. The latter two predictors were also selected in models of accuracy (SD Block Contrast as part of an interaction term). Visibility of cores instead of deltas was selected as a predictor of response time. Interestingly, core visibility also appears as predictor when the model is fit to a testing set. Visibility of a core might make comparison of latents and known prints much easier: if the cores do not match then no further examination is required, so a comparison can be made quickly. Absence of a core could also make it difficult to orient the latent and known prints, since, as noted earlier, these features could act as landmarks for orienting two prints during comparison.

Table 3. Predictors for response time model.

Fixed Effects	Coefficient Estimates	Standard Error	t
Core (L)	−0.234	0.074	−3.149**
Intercept	0.226	0.085	2.646**
Mean Block Contrast (K)	−0.177	0.038	−4.647**
SD Block Contrast (L)	−0.105	0.038	2.748**
Random Effects	Variance		
printID	0.216		
expertID	0.226		

Note: ** $p < 0.01$. Estimates are arranged by coefficient magnitude in descending order (see text). L – latent, K – known print, LxK – interaction.

Linear mixed-effects models were also fit separately for difficulty and confidence ratings. Like response time, there was a great deal of inter-subject variability for both measures. Variability in confidence and difficulty ratings may be due to differences in degree of expertise and self-confidence in the task. Variability in ratings may also be due to differences in interpretation of the rating task and therefore in response strategy. One expert, for example, responded with maximum confidence to all comparisons, saying to the experimenter that in real-world situations an expert would be 100% confident or rate a comparison as inconclusive.

Table 4 contains the coefficient estimates for the model of difficulty rating. As in the model of accuracy, Ridge Sum, Area Ratio, and Core (L) were selected as predictors. Similar to response time, difficulty was also negatively correlated with accuracy, so the regression coefficients have opposite sign to those in the accuracy model. In addition, visibility of Cores in the known print and the interaction of the Core terms were also selected. Delta (L) appears in this model as well as in the model of accuracy.

A similar model was fit for confidence ratings. The results are shown in Table 5. Identical predictors with comparable magnitudes were selected as for the difficulty rating model. The coefficients have opposite sign since high difficulty ratings correspond to low confidence ratings. Because difficulty and confidence are so strongly correlated (−0.91), it is not surprising that the exact same predictors are selected for in both models.

Table 4. Predictors for difficulty rating model.

Fixed Effects	Coefficient Estimates	Standard Error	t
Intercept	2.748	0.301	9.121***
Core (L x K)	−2.104	0.722	−2.913**
Core (L)	1.719	0.705	2.437**
Core (K)	0.935	0.324	2.883**
Delta (L)	−0.778	0.191	−4.082***
Ridge Sum	−0.207	0.079	−2.631**
Area Ratio	0.202	0.078	2.571**
Random Effects	Variance		
printID	1.076		
expertID	0.301		

Note: *** $p<0.001$, ** $p<0.01$. Estimates are arranged by coefficient magnitude in descending order (see text). L – latent, K – known print, LxK – interaction.

Table 5. Predictors for confidence rating model.

Fixed Effects	Coefficient Estimates	Standard Error	t
Intercept	5.248	0.247	21.255***
Core (L x K)	2.034	0.564	3.604***
Core (L)	−1.644	0.551	−2.983**
Core (K)	−0.920	0.253	−3.631***
Delta (L)	0.581	0.149	3.899***
Area Ratio	−0.162	0.062	−2.647**
Ridge Sum	0.155	0.062	2.517**
Random Effects	Variance		
printID	0.616		
expertID	0.488		

Note: *** $p<0.001$, ** $p<0.01$. Estimates are arranged by coefficient magnitude in descending order (see text). L – latent, K – known print, LxK – interaction.

DISCUSSION

We evaluated expert performance on a fingerprint matching task. Experts were highly accurate, committing relatively few errors despite limited access to resources and restricted viewing time. Using a number of potential predictors derived from image processing analyses, we were able to identify, using regression analyses, several image characteristics predictive of expert performance. Six features in particular were found to be important predictors of accuracy: Ridge Sum, Area Ratio, visibility of Deltas in the latent print, Mean Block Contrast of the known print, interaction between SD Block Contrast for latents and known prints, and the interaction between DEAI (deviation from expected average intensity) for the latents and known prints. Taken together, these features explain 64% of the variance in performance accuracy on a novel set of print pairs that were withheld from those used to train the model. A classifier derived from the full data set identified the pairs on which at least one expert made a mistake with 91% accuracy, and a similar model derived from 90% of the data classified novel pairs with 75% accuracy.

Many of the same image characteristics were also predictors of subjective difficulty ratings, confidence ratings, and response times. We also found that difficulty ratings, a subjective measure, were moderately correlated with accuracy and could improve the performance of the classifier on novel print pairs.

There are several interesting observations that can be made about the set of features that were found to be predictive of accuracy (Table 2). First, four of the six features were relational, in the sense that they were calculated based on information contained in both prints that make up a comparison pair. This is a desirable feature of the model since a particular print could arise in two separate comparisons (e.g., a latent print compared to a matching and a non-matching print). In real world scenarios, a single latent print may be compared to many known prints. In cases where one of the prints in the study was of very poor quality, then such relational features might not matter. For example, if Mean Block Contrast (K) is low (i.e., for a very washed out or very dark print), then a comparison would be difficult irrespective of some relational features such as Area Ratio. Conversely, if two prints do not share Level 1 pattern type, they will not make for a difficult comparison regardless of the quality and quantity of information in each. In general, however, error rates and difficulty seem likely to be primarily characteristics of print comparisons, rather than individual prints, as difficulty for actual non-match comparisons will be most acute when the prints share significant similarities, and difficulty for actual matches will be most acute when latent quality or quantity is limited or misleading. Results in our regression models support this idea.

Second, the features within the model correspond to many types of information content. Mean Block Contrast (K), SD Block Contrast (L x K), and DEAI (L x K) capture properties of the image itself (i.e., dark or light, uniform or not). Area Ratio and Delta (L) reflect large-scale or conFigureural (Level I) characteristics of prints, and Ridge Sum relates to visibility of fine detail in the image such as Level II features (see Introduction). These outcomes fit broadly with the idea that fingerprint examiners access different kinds of information in making comparisons and that basic image characteristics determine the detectability of relevant features and patterns.

Third, the signs of the coefficients provide appealing interpretations. That high contrast and clarity of ridges are predictors of accuracy should not be surprising. The DEAI measure increases as the average pixel intensity approaches 127.5, the mean expected pixel intensity for an image that contains 50% white and 50% black pixels. We assumed that this proportion would correspond to greater clarity, since a mostly light or dark image could be difficult to analyze. The positive coefficient found for this measure in the accuracy model indicates that as the proportion of white to black pixels approaches 0.5 in the latent and known print, accuracy increases.

Visibility of deltas in the latent image also had a positive effect on accuracy perhaps because they provided orienting information, making it easier to match and compare feature locations on the latent and known print. Accuracy decreased as SD Block Contrast (L x K) and Area Ratio increased. When SD Block Contrast is high in both the latent and known print, accuracy is low. In general, high variability in Block Contrast picks up variable image quality across image regions (e.g., due to gaps or smudging in portions of a print). In smudged regions, pixels would be uniformly dark, while in clear regions pixel intensity would be more variable, leading to higher contrast measures in those areas. If an image were more uniform in pixel intensities, it would have lower variability in contrast across regions and therefore lower SD Block Contrast measures. Area Ratio had a large, negative coefficient. This at first seems counterintuitive; higher area ratios tend to correspond to larger areas of latent prints. One possible explanation might be that in the process of creating a database of print pairs with varying size latents that were not too difficult, small latents could perhaps have tended to be of higher quality while larger latents tended to be of lower quality. However, the evidence suggests that this explanation is not probable, because in fact, print area ratio and ridge reliability (perhaps the variable most directly corresponding to subjective image quality) were not correlated (r=0.017, p=0.415). Instead, we believe that latent prints with larger areas may be associated with more difficult comparisons because it is more difficult to identify distinctive regions of the image when the print is large. Since non-matching known prints were chosen by submitting the latents to an AFIS system, and selecting a non-match deemed to be the most similar of those available, the non-matches likely shared many features. If experts were only shown a small latent region, it might have been easier to compare that region to the corresponding region on the known print and quickly exclude mismatched pairs as compared to a larger latent image with more or larger, accidentally matching regions.

Difficulty ratings were used in two ways to add to the modeling results. We used difficulty rating itself as a predictor of accuracy. Difficulty ratings improved the fit of a model trained on all of the print pairs, but did not improve the predictive power of a model on a testing set of withheld prints. Classification performance, however, was improved. While ratings are not objective, there was nevertheless a moderate correlation between difficulty ratings and accuracy, suggesting that experts were reasonably aware of which comparisons were difficult. However, outside the experimental setting, it may be impractical to expect to be able to get a group of experts to provide ratings.

Difficulty ratings, confidence ratings, and response times were also evaluated as separate dependent measures. Because these measures correlated moderately with accuracy, we expected that similar features should be selected for when the same features were used to predict other dependent measures. Four of the six features that appeared in the accuracy model also appeared in the other models. A fifth feature, SD Block Contrast (K), which was included as part of an interaction term in the accuracy model appeared in the model of response time. Some features, such as visibility of cores, appeared in the other models but not in the model of accuracy. Cores and deltas are global features. Their presence or absence can be used as a quick measure of assessing difficulty. However, global features on their own are not sufficient to make a comparison. Accuracy, therefore, depends to a greater extent on image quality, relational information, and ridge information.

These results suggest that physical characteristics derived through image processing methods may be valuable in predicting expert difficulty and error rates for print pairs. Given that the present work is the first effort we know of to systematically predict errors from physical characteristics of print pairs, the predictive results are highly encouraging. Validation across larger data sets would be desirable for practical use of a predictive model such as the one derived here, further developments along these lines, along with continuing progress in characterizing the physical quality of prints (e.g., [9]) will likely prove to have practical value in quantifying the likely evidentiary value of expert assessments of fingerprint matches.

While these results on modeling print-pair difficulty are encouraging, there are also many differences between the paradigm used in the present study and the actual process of fingerprint comparison. In forensic settings, experts typically have unlimited evaluation time and access to image processing tools that were not available in the present study. In addition, examiners typically are not in a 'forced-choice' situation, and may decide that a real-world comparison is inconclusive rather than reaching a conclusion about match or non-match.

Despite these limitations, there are several important dimensions to these results. The experiment shows that even under constraints, experts were highly accurate. More than half of the print pairs had perfect accuracy, even in circumstances where the examiners' time was limited, their access to processing tools constrained, and in which they were not permitted to select the option of "inconclusive". Relatively few studies have examined expert performance in fingerprint matching tasks, and this study adds to that body of research. Given the constraints imposed on examiners in this study, we would suspect that error rates in forensic laboratory settings could well be lower than those that we observed. For example, it is possible that had examiners had the option of choosing "inconclusive," they might have elected to make that choice for some of the prints for which we saw the highest error rates. However, it is also possible that they might have selected "inconclusive" for prints for which they performed, individually or in aggregate, with a high degree of accuracy. Our data do not permit us to assess either of these possibilities; but this example illustrates why taking these data as offering an "error rate" would be both misleading and inappropriate.

Experiments in ecologically valid settings are difficult to conduct. Compared to this experiment, fingerprint examiners in actual practice encounter multiple factors that may improve accuracy (such as more time to conduct the comparisons, more access to tools, verification checks, etc.), as well as factors that can reduce accuracy (such as biasing influences from extraneous contextual case information, or pressures from investigators [19], [43]). Given the significant differences between our experimental conditions and ecologically-valid fingerprint identification, it is important to state clearly that the point of this experiment is *not* to measure error rates, and it would be a mistake to take these data as direct evidence of a specific error rate for the field [7]. Rather, we are interested in identifying the features that correlate with difficulty, in order to understand both what features of print pairs affect difficulty, and to begin to understand how error rate might *vary* with print difficulty.

This present study is therefore an important step in "unpacking" error rates, an endeavor that has great importance to forensic science and the legal system. The mere fact that some fingerprint comparisons are highly accurate whereas others are prone to error has a wide range of implications. First, it demonstrates that error rates are indeed a function of comparison difficulty (as well as other factors), and it is therefore very limited (and can even be misleading) to talk about an overall "error rate" for the field as a whole. In this study, more than half the prints were evaluated with perfect accuracy by examiners, while one print was misevaluated by 91 percent of those examiners evaluating it, and numerous others were misevaluated by several examiners. This distribution of errors strongly indicates that error rates do vary depending on the visual content of the specific comparisons. This experiment therefore provides strong evidence that print comparisons do vary in difficulty and that these variations also affect the likelihood of error.

Second, this study lays down a foundation for finding objective print characteristics that can quantify the difficulty of the comparison. The model we offer provides both evidence for what specific visual criteria seem to affect difficulty, as well as a model for combining these criteria to best predict accuracy. While further study regarding this model and its effectiveness in circumstances more ecologically valid than what was undertaken here is warranted before making definitive claims, this model illustrates the significant potential for creating objective measures of difficulty for print pairs.

Third, a more sophisticated understanding of the relationship between error rate and difficulty should be important for the courts in weighing fingerprint evidence (and the need for better information about the strengths and limitations of latent fingerprint comparison and other forensic techniques that has been highlighted by the NAS inquiry into forensic science [3]). Courts are instructed, when assessing expert evidence, to focus on the "task at hand", and this research helps to show that fingerprint examination may vary in difficulty in ways that may be relevant to its evaluation as legal evidence [44], [45]. More nuanced assessments of fingerprint task difficulty might, for example, affect how a judge understands the admissibility of a specific fingerprint comparison, or what degree of certainty the expert will be allowed to express about a conclusion, or it might impact the weight or probative value given to a specific match conclusion by the fact-finder [46].

Fourth, the implication of these findings go beyond the courtroom; they also provide vital insights that can considerably enhance the work of forensic laboratories. For example, similar to medical triage, forensic laboratories may benefit from considering how the need for different procedures and checks can be made to fit the difficulty of the comparison (e.g., "the need for verification, and what sort of verification, may be highly dependent on the difficulty of the decision and the type and likelihood for a potential error. In cases with greater cognitive difficulty, when errors are more likely, more stringent verification procedures are needed; whereas more simple and straightforward prints may not require the same level and type of verification." [47]).

Fifth, the understanding of what makes some comparisons more difficult than others has implications for the selection and training of fingerprint examiners. During examiner training, benchmarks and skill sets can be set as criteria to ensure candidates have the cognitive abilities needed to perform the tasks. Better understanding of which print comparisons are easier or more difficult can improve both training materials and the assessments of trainees. Training needs to have clear cognitive goals, and must use appropriate materials to develop and evaluate trainees' perceptual learning [12]. Understanding which comparisons are more error-prone can help improve both the design of training materials, to ensure a range of difficulty, and can also help evaluators assess performance in a more nuanced way.

ACKNOWLEDGMENTS

Portions of this research were presented at the annual meeting of the *Vision Sciences Society (VSS)* and *European Conference on Visual Perception (ECVP)* in 2013. We thank Clara Sao, Jenny Chun, and Rachel Older for help in the data collection. And we gratefully acknowledge all of the participants who volunteered to take part in the construction of the database and the experimental task.

AUTHOR CONTRIBUTIONS

Conceived and designed the experiments: IED JLM PJK GE PG EM DC TG. Performed the experiments: GE IED DC. Analyzed the data: GE PG TG. Contributed reagents/materials/analysis tools: IED GE PG TG EM DC. Wrote the paper: IED JLM PJK GE PG TG.

REFERENCES

1. Cole SA (2005) Is fingerprint identification valid? Rhetorics of reliability in fingerprint proponents' discourse. J Law Policy 28: 109–135.
2. Mnookin JL (2008b) The validity of latent fingerprint identification: Confessions of a fingerprinting moderate. Law Prob Risk 7: 127–141.
3. National Research Council, National Academy of Sciences (2009) Strengthening Forensic Science in the United States: A Path Forward. Washington, DC: National Academy Press.
4. Ulrey BT, Hicklin RA, Buscaglia J, Roberts MA (2011) Accuracy and reliability of forensic latent fingerprint decision. Proc Natl Acad Sci U S A 108: 7733–7738.
5. Tangen JM, Thompson MB, McCarthy DJ (2011) Identifying fingerprint expertise. Psychol Sci 22(8): 995–997.
6. Busey TA, Vanderkolk JR (2005) Behavioral and electrophysiological evidence for conFigureural processing in fingerprint experts. Vision Res 45: 431–448.
7. Koehler JJ (2008) Fingerprint error rates and proficiency tests: What they are and why they matter. Hastings Law J 59: 1077–1100.
8. Dror IE, Mnookin JL (2010) The use of technology in human expert domains: challenges and risks arising from the use of automated fingerprint identification systems in forensic science. Law Probab Risk 9: 47–67.
9. Pulsifer DP, Muhlberger SA, Williams SF, Shaler RC, Lakhtakia A (2013) An objective fingerprint quality-grading system, Forensic Sci Int. 231: 204–207.
10. Chase WG, Simon HA (1973) Perception in Chess. Cogn Psychol 4: 55–81.
11. 11.Schneider W, Shiffrin RM (1977) Controlled and automatic human information processing: I. Detection, search, and attention. Psychol Rev 84: 1–66.
12. Kellman PJ, Garrigan P (2009) Perceptual learning and human expertise. Phys Life Rev 6: 53–84.
13. Gibson EJ (1969) Principles of perceptual learning and development. New York: Prentice Hall.

14. Kellman PJ (2002) Perceptual learning. In: Pashler H, Gallistel CR, editors. Stevens' handbook of experimental psychology, vol. 3 (3rd edition). New York: John Wiley & Sons. pp. 259–299.
15. Dror IE, Stevenage SV, Ashworth A (2008) Helping the cognitive system learn: Exaggerating distinctiveness and uniqueness. Appl Cogn Psychol 22: 573–584.
16. Bryan WL, Harter N (1899) Studies on the telegraphic language: The acquisition of a hierarchy of habits. Psychol Rev 6: 345–375.
17. Kellman PJ, Massey CM (2013) Perceptual Learning, cognition, and expertise. In: Ross BH, editor. The Psychology of Learning and Motivation, Vol. 58 . Amsterdam: Elsevier Inc. pp. 117–165.
18. Dror IE, Cole SA (2010) The vision in "blind" justice: Expert perception, judgment, and visual cognition in forensic pattern recognition. Psychon Bull Rev 17: 161–167.
19. Dror IE, Rosenthal R (2008) Meta-analytically quantifying the reliability and bias ability of forensic experts. J Forensic Sci 53: 900–903.
20. Hall LJ, Player E (2008) Will the introduction of an emotional context affect fingerprint analysis and decision-making? Forensic Sci Int 181: 36–39.
21. Dror IE, Charlton D (2006) Why experts make errors. J Forensic Ident 56: 600–616.
22. Dror IE, Charlton D, Péron AE (2006) Contextual information renders experts vulnerable to making erroneous identifications. Forensic Sci Int 156: 74–78.
23. Thompson MV, Tangen JM, Treloar R, Ivison K (2010) Humans matching fingerprints: Sequence and size. Proc Hum Fact Ergon Soc Annu Meet 54: 478–481.
24. Marcon JL (2009) The distinctiveness effect in fingerprint identification: How the role of distinctiveness, information loss, and informational bias influence fingerprint identification (Doctoral dissertation). Available from ETD Collection for University of Texas, El Paso (Paper AAI3358893).
25. Ashbaugh DR (1999) Quantitative-qualitative friction ridge analysis: An introduction to basic and advanced ridgeology. Florida: CRC Press.
26. Mnookin JL (2008a) Of black boxes, instruments, and experts: Testing the validity of forensic evidence. Episteme (Edinb) 5: 343–358.
27. Maltoni D, Maio D, Jain AK, Prabhakar S (2009) Handbook of fingerprint recognition. London: Springer.
28. Neter J, Kutner MH, Wasserman W, Nachtsheim C (1996) Applied linear statistical models. Fourth Edition. USA: McGraw-Hill/Irwin.
29. Chatterjee S, Price B (1991) Regression analysis by example. Second Edition. New York: John Wiley & Sons.
30. Booth GD, Niccolucci MJ, Schuster EG (1994) Identifying proxy sets in multiple linear regression: an aid to better coefficient interpretation. Research paper INT-470. United States Department of Agriculture, Forest Service, Ogden USA.
31. Breslow NE, Clayton DG (1993) Approximate inference in generalized linear mixed models. J Am Stat Assoc 88: 9–25.
32. Baayen RH, Davidson DJ, Bates DM (2008) Mixed-effects modeling with crossed random effects for subjects and items. J Mem Lang 59: 390–412.
33. Dixon P (2008) Models of accuracy in repeated-measures designs. J Mem Lang 59: 447–456.
34. Jaeger TF (2008) Categorical data analysis: Away from ANOVAs (transformation or not) and towards logit mixed models. J Mem Lang 59: 434–446.
35. Gelman A, Su Y-S (2013) arm: Data analysis using regression and multilevel/hierarchical models. R package version 1.6-04. http://CRAN.R-project.org/package=arm
36. Bates D, Maechler M, Bolker B (2012) Lme4: Linear mixed-effects models using S4 classes. R package version 0.999999-0. http://CRAN.R-project.org/package=lme4
37. Gelman A, Hill J (2007) Data analysis using regression and multilevel/hierarchical models. New York: Cambridge University Press.
38. Akaike H (1973) Information theory as an extension of the maximum likelihood principle. In Petrov BN, Csaki F, editors. Second International Symposium on Information Theory. Budapest, Hungary: Akadémiai Kiadó. pp. 267–281.

39. Burnham KP, Anderson DR (2002) Model selection and multimodel inference: A practical information-theoretic approach. Second edition. New York: Springer-Verlag.
40. Zuur AF, Ieno EN, Walker NJ, Saveliev AA, Smith GM (2009) Mixed effects models and extensions in ecology with R. New York: Springer Science+Business Media, LLC.
41. Agresti A (1996) An introduction to categorical data analysis. New Jersey: John Wiley & Sons Inc.
42. Agresti A (2002) Categorical data analysis. New Jersey: John Wiley & Sons Inc.
43. Kassin SM, Dror IE, Kukucka J (2013) The forensic confirmation bias: Problems, perspectives, and proposed solutions. J Appl Res Mem Cogn 2: 42–52.
44. Daubert v. Merrell Dow Pharmaceuticals 509 US 579 (1993)
45. Kumho Tire Co. v. Carmichael 526 US 137 (1999)
46. Faigman DL, Bllumenthal JA, Cheng EK, Mnookin JL, Murphy EE, et al. (2012) Modern Scientific Evidence: The Law and Science of Expert Testimony: Fingerprints, Vol. 5 (Ch. 33)
47. Dror IE (2009) How can Francis Bacon help forensic Science? The four idols of human biases. Jurimetrics 50: 93–110.

CHAPTER 23

A New Multiplex Assay of 17 Autosomal STRs and Amelogenin for Forensic Application

Suhua Zhang[1] , Huaizhou Tian[2] , Jun Wu[2] , Shumin Zhao[1] , Chengtao Li[1]

[1] *Shanghai Key Laboratory of Forensic Medicine, Institute of Forensic Sciences, Ministry of Justice, Shanghai, People's Republic of China,*

[2] *PEOPLESPOTINC Research & Development, Beijing, People's Republic of China*

ABSTRACT

This paper describes a newly devised autosomal short tandem repeat (STR) multiplex polymerase chain reaction (PCR) systems for 17 autosomal loci (D1S1656, D2S441, D3S1358, D3S3045, D6S477, D7S3048, D8S1132, D10S1435, D10S1248, D11S2368, D13S325, D14S608, D15S659, D17S1290, D18S535, D19S253 and D22-GATA198B05) and Amelogenin. Primers for the loci were designed and optimized so that all of the amplicons were distributed from 50 base pairs (bp) to less than 500 bp within a five-dye chemistry design with the fifth dye reserved for the sizing standard. Strategies were developed to overcome challenges that encountered in creating the final assay. The limits of the multiplex were tested, resulting in the successful amplification of genomic DNA range from 0.25–4 ng with 30 PCR cycles. A total of 681 individuals from the Chinese Han population were studied and forensic genetic data were present. No significant deviations from Hardy–Weinberg equilibrium were observed. A total of 180 alleles were detected for the 17 autosomal STRs. The cumulative mean exclusion chance in duos ($CMEC_D$) was 0.999967, and cumulative mean exclusion chance in trios ($CMEC_T$) was 0.99999995. We conclude that the present 17plex autosomal STRs assay provides an additional powerful tool for forensic applications

INTRODUCTION

The study of genetic variation, using DNA polymorphisms distributed throughout the genome, has allowed better understanding of the history and diversity of human populations as well as providing a system for forensic testing [1], [2]. At present, STR typing can be considered a standard approach and the method of choice in the forensic field, allowing a high discrimination power adequate for addressing most problems of human identification and paternity testing [3]. Analysis of STR loci is an essential component of forensic genetics. 13 core STR loci of Combined DNA Index System (CODIS) which used for a DNA database of felon were widely used in commercial kits [4]. On the other hand, many highly polymorphic STR loci exist, which are unlinked to the core STR loci (non-CODIS loci). Analyses of such polymorphic non-CODIS loci in addition to the core CODIS loci may increase the power of exclusion in complex kinship testing [4], [5].

The present study sought to develop a new multiplex PCR system for simultaneous typing of highly polymorphic non-CODIS loci. We selected 16 polymorphic non-CODIS STR loci (D1S1656, D2S441, D3S3045, D6S477, D7S3048, D8S1132, D10S1435, D10S1248, D11S2368, D13S325, D14S608, D15S659, D17S1290, D18S535, D19S253 and D22-GATA198B05) and a CODIS STR locus (D3S1358) for total 17 STR loci. This paper described the new development of the 17plex autosomal STRs assay and surveyed their efficacy for forensic application.

MATERIALS AND METHODS

Ethics Statement

Human blood samples were collected with the approval of the Ethics Committee of the Institute of Forensic Sciences, Ministry of Justice, P.R. China. Samples were obtained from unrelated volunteers after receiving written informed consents. Animal blood samples were collected with the approval of the Animal Use Committee of the Institute of Forensic Sciences, Ministry of Justice, P. R. China.

(a) Candidate STRs Selection

The initial candidate pool of autosomal STRs for this study was based on the STR DNA Internet DataBase created by John M. Butler et al (http://www.cstl.nist.gov/strbase/) and National Center for Biotechnology Information (NCBI) (http://www.ncbi.nlm.nih.gov/).

STRs were then selected according to the following criteria: (I) non-coding, non-CODIS, autosomal STRs; (II) average heterozygosity ≥0.30 in European, African and Asian population groups; (III) located on different chromosomes, if not, at most two STRs were selected from the same chromosome and make sure on different arms; and (IV) physical distance between the selected STRs and CODIS STRs above 10 Mb.

After applying the described criteria, 16 ideal candidate STRs were obtained. Flanking sequences of these STRs, as well as reported sequence variants within these region, were obtained using the University of California Santa Cruz Genome Browser (Human February 2009 Assembly; GRCh37/hg19) at http://genome.ucsc.edu/.

Locus D3S1358 which belonged to CODIS STRs was added to our final assay, for comparing the accuracy of the final multiplex and usually used commercial kit (e.g. AmpFℓSTR® Identifiler® Kit). The sex-typing locus Amelogenin was also included in the final assay.

(b) Primer and Multiplex Assay Design

PCR primer designing was performed using software of primer premier v5.0 and Oligo v6.0, applying the following main criteria: primer length of 18–30 bp; amplicon length of 50–500 bp; optimum Tm from 56°C to 64°C; optimum CG content range from 45% to 55%. Obtained primer pairs were then checked for non-specific hybridizations in other genome regions using NCBI Basic Local Alignment Search Tool (BLAST) at http://blast.ncbi.nlm.nih.gov/. AutoDimer software was also applied to screen candidate primer sequences for inter-primers compatibility.

Forward primer of each STR was labelled at the 5′ end with either 6 FAM™, HEX™, TEMRA™ and ROX™ (Applied Biosystems, Foster City, CA). All primers were synthesized and labelled by Sangon Biotech. Co. Ltd (Shanghai, China).

All selected STRs were then organized by expected amplicon length and assigned into four different dye-labelling fluorochromes in order to achieve an evenly balanced genotyping assay for a single PCR and electrophoretic separation.

(c) PCR Amplification and STRs Detection

All selected STRs were initially amplified in singleplex, in order to evaluate primer performance and expected allele sizes. After optimization, the amplification of all STRs was performed in a single multiplex PCR including 1×PCR buffer, 200 μM dNTP, 1.5 mM $MgCl_2$, 1 U GoldTaq DNA polymerase, 5 μL primer mix and 0.25–4 ng of genomic DNA in a 25 μL final reaction volume. Thermocycling condition for the multiplex PCR assay was initial incubation at 95°C for 11 min; 30 cycles at 94°C for 30 s, 60°C for 60 s and 70°C for 60 s; with a final extension at 60°C for 30 min.

PCR products were subsequently prepared for capillary electrophoresis by adding 1 μL of each amplified product to 9 μL of a 17:1 mixture of Hi-Di formamide and Orange-500 size standard, respectively. Separation and detection were performed with AB 3130*xl* Genetic Analyzer (Applied Biosystems, Foster City, CA) using filter set E5 and POP4 polymer (Applied Biosystems, Foster City, CA). Samples were injected electrokinetically for 5 sec at 3 kV. The STR alleles were then separated at 15 kV at a run temperature of 60°C. Genotyping data of studied samples were then collected with GeneMapper v3.2.1 software (Applied Biosystems, Foster City, CA). Positive control DNA of 9947A human cell line samples (Applied Biosystems, Foster City, CA) was used to test the overall performance of the multiplex genotyping assay.

(d) Sensitivity Testing

To determine the minimum quantity of DNA required to achieve reliable results (peak highs at all loci exceeded 100 relative fluorescence units (RFU)) with the multiplex assay, we used aliquots of the cell line 9947A DNA (50, 100, 250, 500 pg, 1, 2, 4 and 5 ng). Every concentration tested for three replicates.

(e) Specificity Testing

To determine the specificity of this multiplex assay, several kinds of living creatures' bloods (Human, Pig, Chick, Duck, Dog and Cat) were tested.

(f) Population Samples for STR Polymorphism Analysis

Peripheral bloods were collected from a total of 681 unrelated healthy individuals (343 males and 338 females) of Chinese Han population, the largest nation of China. All samples of the volunteers were genotyped with commercial kit of AmpFℓSTR® Identifiler® PCR Amplification Kit before.

Genomic DNA was extracted using the Chelex-100 protocol as described by Walsh et al [6]. The quantity of recovered DNA was determined by spectrophotometric method.

(g) Statistical Analysis

Hardy-Weinberg equilibrium and allele frequencies of the 17 autosomal STR were calculated by GeneAlEx 6.5 software (Genetic analysis in Excel) [7]. Statistical parameters to evaluate the forensic efficiency, such as power of discrimination (DP), heterozygosity (HET), mean exclusion chance in duos (MEC_D), mean exclusion chance in trios (MEC_T),polymorphism information content (PIC), and Genediversity (GD) were assessed by PowerMarker V3.25 and Cervus V2.0 software [1], [8].

RESULTS AND DISCUSSION

(a) Primer and Multiplex PCR Optimization

The primer sequences, T_m values and concentrations of the 17 autosomal STR loci and Amelogenin locus in the final multiplex assay were listed in Table 1. Primers for each STR locus were initially tested in singleplex to evaluate the performance. The criteria for primer "failure" in this study is defined as those that produce profiles that exhibit incomplete adenylation, the presence of PCR artifacts, nonspecific products, low signal, or no PCR product at all [5]. With those "failure" ones, singleplex PCR optimization were adopted, otherwise, re-designing the PCR primers. Once the successful primers at locus were determined, those were combined together and tested in multiplex. At first, the tested multiplex assay indicated a high degree of baseline noises and many profiles displayed the features of failure. This was due to multiplex primer incompatibility. Based on the data of primer concentrations and peak heights, optimizations of the multiplex assay were performed.

The final assay design resulted in a good working base to set up the multiplex reaction. In the initial multiplex tests some primer pairs revealed weaker performance than others, as would be expected since the optimum annealing temperature was not exactly the same for all primers. To minimize these effects and aiming for a more balanced result, we tested different primer mix concentrations and annealing temperatures. After the multiplex development and optimization, 17 autosomal STRs plus Amelogenin were successfully amplified in a single PCR reaction, following the final optimum conditions. The genotyping profiles of positive control DNA of 9947A and ladder were showed in Figureure 1 and Figureure 2, respectively.

Table 1. Primer information of STR loci included in the new studied kit.

STR locus	Dye Label	Primer Sequence:(5′-3′)	Primer Concentration(μM)	Tm	fragment size range
D6S477	6-FAM™	F: GGCTGATGAGGTGAAATATTTGC	0.14	60.7	100–165
		R: GATATCTCAAACAACCTCAACAACA	0.14	58.2	
D22-GATA198B05	6-FAM™	F: AGCCAGCATTCTTAAAACTCTAAGG	0.2	60.5	170–240
		R: ATCAGCCCTGTGACAGAAGTAAATA	0.2	59.5	
D15S659	6-FAM™	F: TGATTTGCCATGATAGATGGCT	0.16	59.8	245–330
		R: GTTTGGGACTTCAAAGACAAAAGA	0.16	59.8	
D8S1132	6-FAM™	F: ATCACATCCTTGTTTCCTCATTTTATT	0.5	59.4	335–450
		R: AGATTGCGTCACTGCACTCC	0.5	58.2	
Amelogenin	HEX™	F: CCCTGGGCTCTGTAAAGAATAG	0.2	58	102–108
		R: ATCAGAGCTTAAACTGGGAAGCTG	0.2	61.5	
D3S1358	HEX™	F: TATGATTCCCCCACTGCAGT	0.1	57.5	115–160
		R: ATGAAATCAACAGAGGCTTGC	0.1	57.1	
D3S3045	HEX™	F: TAGATGGCTTCCTAGGCTCAATTAG	0.24	61	186–255
		R: ATAAATATTTCTGTTTCTCCTGGGG	0.24	59.9	
D17S1290	HEX™	F: TTCAGTTAGCCAAGATAATGCCA	0.26	59.7	265–380
		R: GTAAGGCTGAGTTCCTCTGGGT	0.26	59.3	
D14S608	HEX™	F: TACTACCTCTTCAGTGAGCTTTCGT	0.5	59.2	388–450
		R: CTGGCAACAATGATTCTATTTTCTC	0.5	59.7	
D2S441	TEMRA™	F: GAACTGTGGCTCATCTATGAAAACT	0.15	59.1	80–122
		R: GCTAAGTGGCTGTGGTGTTATGATA	0.15	60.7	
D18S535	TEMRA™	F: CACACCCATAACTTTTTTCCTCTAG	0.25	59.1	130–170
		R: GCAGGCTAAGAGTTACCCATAATTA	0.25	59.1	
D13S325	TEMRA™	F: TATCTCCAGAATCCTTCCAGCTAAT	0.15	60.3	175–265
		R: CTGGAATAACTATTCATGCTAACCA	0.15	58.1	
D10S1435	TEMRA™	F: TGCATTGAGTTTTATTCTGTTTATC	0.5	56.2	275–350
		R: AGTACCATGTGAATCTTCATCTTCC	0.5	58.7	
D11S2368	TEMRA™	F: GCATTGGCTGTCATAATATTCACTC	0.5	60.2	355–420
		R: TATACATCTCCTCCAAGAGCTTTCC	0.5	60.6	
D1S1656	ROX™	F: GTGTTGCTCAAGGGTCAACTGTATG	0.2	62.8	115–170
		R: AGAGAAATAGAATCACTAGGGAACC	0.2	57.1	
D7S3048	ROX™	F: GCATTGCCACTTCCCCCTG	0.5	63.1	175–233
		R: CCCCCGCAGTCAAAAATCT	0.5	59.8	
D10S1248	ROX™	F: GAGCATTAGCCCCAGGACCA	0.5	62.2	234–280
		R: TGCAGTGCTTGGCAAAGAGC	0.5	62.2	
D19S253	ROX™	F: CTCCATGAACAGACAGTTTGTCTTC	0.5	59.9	355–420
		R: CTTTTCAGCTTCCATAAACATGAGT	0.5	59.4	

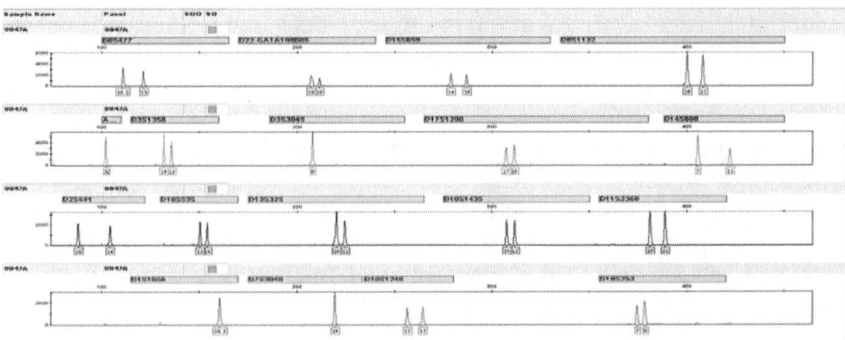

Figureure 1. The genotyping profile of positive control DNA of 9947A with the 17plex assay.

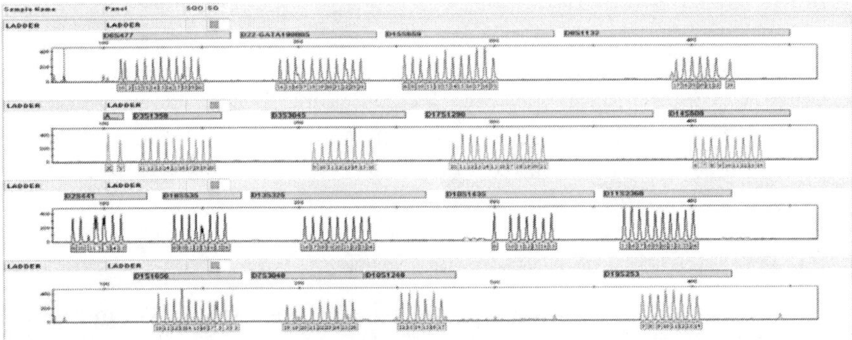

Figureure 2. The genotyping profile of ladder with the 17plex assay.

(b) Multiplex PCR Performance

In general, during the genotyping of 681 unrelated population samples, the multiplex reaction revealed to be robust and able to successfully amplify all STRs in samples with 0.25–4 ng DNA in a 25 μL final reaction volume. The genotyping results of locus D3S1358 by our new multiplex assay and the commercial kit of AmpFℓSTR® Identifiler® Kit were identical with each other.

Several DNA concentrations ranging from 50 pg to 5 ng were tested with the multiplex, using dilution series of positive control sample 9947A. The best results, revealing an improved peak balance among STRs, were observed when using 0.5–1 ng of DNA in a 25 μL PCR final volume, although good results for all STRs were obtained for concentrations from 0.25 to 4 ng. Nevertheless, when using highly concentrated DNA samples (>4 ng), we observed strong signals and over-scaled peaks; in these cases, the presence of pull-ups and abnormally shaped peaks due to ineffective fluorescence correction can make interpretation of the genotyping results difficult. Furthermore, the use of higher amounts of DNA in the reaction occasionally caused the inhibition of amplification of longer size amplicons. In samples with DNA amounts as low as 0.1–0.25 ng, it was still possible to amplify all markers; nonetheless, the results were not always consistent, sometimes presenting unusual allele imbalances or allele dropouts observed from the known consensus genotypes of the control DNAs used. The marked heterozygote imbalance or allele dropouts with small quantities of target DNA have been previously reported and are a consequence of stochastic variation in PCR amplification when the number of template molecules is very low [9]. We highlight the fact that, in low copy number applications of this multiplex, the results must be interpreted with caution, and duplicate analyses are required to minimize errors.

As a practical application for forensic use, the multiplex assay was employed in several kinds of living creatures' bloods (Human, Pig, Chick, Duck, Dog, Cat), aiming to confirm the specificity. Only human blood can have positive results. There are no detected signals for other creatures at any of these 18 STRs included in the kit, which proved that this kit is only suitable for human testing.

(c) LD Information of 17 Autosomal STRs

Detail information of 17 autosomal STRs (16 non-CODIS STR and a CODIS STR of D3S1358) in the multiplex was listed in Table S1. According to the STRs selection criteria, 14 STRs (D1S1656, D2S441, D3S3045, D6S477, D7S3048, D8S1132, D11S2368, D13S325, D14S608, D15S659, D17S1290, D18S535, D19S253 and D22GATA198B05) located on discrete chromosomes were selected. Although D10S1435 and D10S1248 are found on the same chromosomes (chromosome 10), they are located on different arms (10p15.3 and 10q26.5, respectively) and separated by a considerable distance. In addition, the physical distances between the 16 selected non-CODIS STRs and 13 CODIS STRs are more than 10 Mb.

Genetic linkage disequilibrium (LD) in the human genome would not for consideration if the physical distances are more than 10 Mb [10], [11], [12]. Thus, LD between the current 16 non-CODIS STR and the CODIS core loci may be excluded from consideration. Analysis of extended STR loci using both the present multiplex system and commercial kits which include the CODIS core loci have been demonstrated to be remarkably effective in cases involving difficult kinship testing.

(d) Forensic Efficiency in Han Populations

The polymorphism analysis of 17 autosomal STRs was studied in the Han population of China. All studied STRs followed Hardy–Weinberg equilibrium. In Table 2 we present allele frequencies derived from the population and the forensic efficiency values for each STR. A total of 180 alleles were detected for the 17 autosomal STR loci while locus D1S1656 had the greatest number of variants, with 15 alleles. All loci show high forensic efficiency with DP values above 0.8. The mean HET value of the STR set was 0.7934, and all loci showed HET values higher than 0.7 except locus D2S411. The highest PIC was 0.8571, at locus D7S3048, while the lowest PIC was 0.6747, at D3S1358. And GD values range from 0.7219 (D3S1358) to 0.8709 (D7S3048).

Table 2. Frequencies and forensic parameters of 17 autosomal STRs.

Allele	D1S-1656	D2S-441	D3S-1358	D3S-3045	D6S-477	D7S-3048	D8S-1132	D10-S1435	D10-S1248	D11-S2368	D13-S325	D14-S608	D15-S659	D17S-1290	D18S-535	D19S-253	D22GATA-198B05
6												0.0646					
7												0.1931				0.1696	
8								0.0308	0.0007			0.0250	0.0051		0.0308	0.0345	
9		0.0007		0.3774								0.1322	0.0015		0.1858	0.0022	
9.3		0.0022															
10	0.0073	0.2504		0.0228	0.0103			0.0433	0.0007			0.2394	0.0059	0.0411	0.0419	0.0206	
10.3		0.0132															
11	0.0609	0.3414	0.0007	0.0308	0.0051			0.1557	0.0066			0.1909	0.1571	0.0477	0.0184	0.1586	
11.2					0.0015												
11.3		0.0455															
12	0.0617	0.1711	0.0007	0.1358	0.0690			0.3789	0.0756			0.1131	0.2173	0.0044	0.0808	0.3311	
12.2															0.0110		
12.3		0.0066															
13	0.1050	0.0316	0.0073	0.2070	0.2078			0.2386	0.3796			0.0367	0.1043	0.0103	0.1997	0.2070	
13.2															0.0007		
13.3		0.0029															
14	0.0786	0.1138	0.0573	0.1681	0.1931			0.1322	0.2093			0.0051	0.0360	0.0198	0.2915	0.0624	0.0066
14.3	0.0044																
15	0.2812	0.0184	0.3392	0.0529	0.3025			0.0176	0.2137	0.0022			0.1762	0.2115	0.1322	0.0140	0.0184
15.3	0.0132																
16	0.2107	0.0022	0.3561	0.0051	0.1615	0.0022	0.0140	0.0029	0.0960	0.0330	0.0044		0.1667	0.3040	0.0073		0.0325
16.3	0.0125																
17	0.0727		0.1696		0.0374	0.0044	0.1028		0.0169	0.1329	0.0044		0.1035	0.1689			0.1557
17.3	0.0470																
18	0.0132		0.0646		0.0073	0.0932	0.2070			0.1182	0.0389		0.0242	0.1167			0.0675
18.3	0.0286																
19			0.0029		0.0044	0.0712	0.2137			0.1791	0.2548		0.0022	0.0565			0.0778
19.3	0.0029																
20			0.0015			0.1872	0.1549			0.1960	0.2651			0.0117			0.0991
21						0.1204	0.1424			0.2093	0.2129			0.0073			0.2930
22						0.0903	0.1021			0.0756	0.1468						0.1659
23						0.1571	0.0485			0.0433	0.0448						0.0184
24						0.1564	0.0147			0.0103	0.0213						0.0051
25						0.0881					0.0051						
26						0.0257											
27						0.0037					0.0015						
DP	0.9591	0.9172	0.8754	0.9116	0.9288	0.9696	0.9561	0.9044	0.9009	0.9571	0.9267	0.9504	0.9572	0.9415	0.9407	0.9253	0.9529
HET	0.8238	0.6990	0.7327	0.7592	0.7518	0.8840	0.8399	0.7239	0.7504	0.8385	0.8135	0.8399	0.8385	0.7974	0.7827	0.7827	0.8311
MEC_D	0.5325	0.3937	0.3097	0.3770	0.4227	0.5829	0.5174	0.3615	0.3533	0.5220	0.4181	0.4947	0.5222	0.4641	0.4607	0.4128	0.5040
MEC_T	0.6971	0.5722	0.4825	0.5574	0.6005	0.7387	0.6856	0.5411	0.5326	0.6893	0.5954	0.6659	0.6896	0.6385	0.6356	0.5915	0.6742
PIC	0.8267	0.7428	0.6747	0.7311	0.7661	0.8571	0.8241	0.7193	0.7139	0.8263	0.7631	0.8110	0.8267	0.7894	0.7890	0.7582	0.8136
GD	0.8429	0.7749	0.7219	0.7637	0.7956	0.8709	0.8435	0.7547	0.7510	0.8453	0.7939	0.8326	0.8457	0.8129	0.8134	0.7879	0.8327

Since all of the 17 autosomal STR loci were independent from each other, the combined forensic efficiency parameters were calculated based on allele frequencies while cumulative mean exclusion chance in duos ($CMEC_D$) was 0.999967, and cumulative mean exclusion chance in trios ($CMEC_T$) was 0.99999995. These results suggest that the 17 autosomal STR loci included in the multiplex assay are highly polymorphic and informative in the Han population of China.

Concluding Remarks

In this study, we displayed a simple and sensitive 17plex STRs new assay for forensic application, which allows the simultaneous genotyping of 17 autosomal STRs plus Amelogenin presenting high polymorphic in Chinese Han population. The development of this 17plex assay provides support for the use of additional loci for complex kinship testing involving deficient paternity cases and for use in missing person/mass disaster cases.

AUTHOR CONTRIBUTIONS

Conceived and designed the experiments: CTL HZT. Performed the experiments: SHZ JW. Analyzed the data: SMZ SHZ. Contributed reagents/materials/analysis tools: CTL HZT. Wrote the paper: SHZ.

REFERENCES

1. Butler JM (2006) Genetics and genomics of core STR loci used in human identity testing. J Forensic Sci 51: 253–265.
2. Wenk RE, Chiafari FA, Gorlin J, Polesky HF (2003) Better tools are needed for parentage and kinship studies. Transfusion 43: 979–981.
3. Chakraborty R, Stivers DN, Su B, Zhong Y, Budowle B (1999) The utility of short tandem repeat loci beyond human identification: implications for development of new DNA typing systems. Electrophoresis 20: 1682–1696.
4. Opel KL, Chung DT, Drábek J, Butler JM, McCord BR (2007) Developmental validation of reduced-size STR Miniplex primer sets. J Forensic Sci 52: 1263–1271.
5. Hill CR, Butler JM, Vallone PM (2009) A 26plex autosomal STR assay to aid human identity testing. J Forensic Sci 54: 1008–15.
6. Walsh PS, Petzger DA, Higuchi R (1991) Chelex-100 as medium for simple extraction of DNA for PCR-based typing from forensic material. Biotechniques 10: 506–513.
7. Peakall R, Smouse PE (2012) GenAlEx 6.5: genetic analysis in Excel. Population genetic software for teaching and research-an update. Bioinformatics 28: 2537–2539.
8. Liu K, Muse SV (2005) PowerMarker: an integrated analysis environment for genetic marker analysis. Bioinformatics 21: 2128–2129.
9. Jobling MA, Gill P (2004) Encoded evidence: DNA in forensic analysis. Nat Rev Genet 5: 739–751.
10. Reich DE, Cargill M, Bolk S, Ireland J, Sabeti PC, et al. (2001) Linkage disequilibrium in the human genome. Nature 411: 199–204.
11. Service SK, Ophoff RA, Freimer NB (2001) The genome-wide distribution of background linkage disequilibrium in a population isolate. Hum Mol Genet 10: 545–551.
12. Weiss KM, Clark AG (2002) Linkage disequilibrium and the mapping of complex human traits. Trends Genet 18: 19–24.

Index